11th International Conference on Intelligent Technologies in Logistics and Mechatronics Systems (ITELMS'2016)

AA005097

Panevezys, Lithuania
28 – 29 April 2016

Editors:

Laszlo Koczy
Dalia Susnienė
Daiva Žostautienė

ISBN: 978-1-7138-4112-8

Printed from e-media with permission by:

Curran Associates, Inc.
57 Morehouse Lane
Red Hook, NY 12571

CURRAN ASSOCIATES INC. proceedings.com

Some format issues inherent in the e-media version may also appear in this print version.

Copyright© (2016) by MEDIMOND s.r.l.
All rights reserved.

Printed with permission by Curran Associates, Inc. (2022)

For permission requests, please contact Kaunas University of Technologies at the address below.

Kaunas University of Technology
K. Donelaičio St. 73
44249 Kaunas, Lithuania

Phone: +370 (37) 300 000
Fax: +370 (37) 324 144
Email: ktu@ktu.it

https://itms.ktu.edu

Additional copies of this publication are available from:

Curran Associates, Inc.
57 Morehouse Lane
Red Hook, NY 12571 USA
Phone: 845-758-0400
Fax: 845-758-2633
Email: curran@proceedings.com
Web: www.proceedings.com

Contents

© Medimond

11th International Conference on Intelligent Technologies in Logistics and Mechatronics Systems

ITELMS'2016

April 28-29, 2016
Panevėžys, Lithuania

Editors

Laszlo Koczy
Dalia Susnienė
Daiva Žostautienė

MEDIMOND
PUBLISHING COMPANY

International Scientific Committee

- Assoc. Prof. **Torma ANDRÁS** (Széchenyi István University, HUNGARY)
- Prof. **Kiril ANGUELOV** (Technical University of Sofia, BULGARIA)
- Prof. **Paulo Tavares de CASTRO** (Porto University, PORTUGAL)
- Prof. **Jan FURCH** (Brno University of Defence, CZECH REPUBLIC)
- Assoc. Prof. **Anda GHEORGHIU** (National University of Political and Administrative Studies & Hyperion University, ROMANIA)
- Dr. **Ausma GOLUBEVA** (University of Latvia, LATVIA)
- Dr. **Simona GRAMA** (Groupe ESC Troyes, FRANCE)
- Dr. **Volodymyr HUTSAYLYUK** (Military University of Technology, POLAND)
- Prof. **Isak KARABEGOVIC** (BOSNIA AND HERZEGOVINA)
- Assoc. Prof. **Marios KASINOPOULOS** (University of Central Lancashire, CYPRUS)
- Dr. **Hilal Yildirir KESER** (Uludag University, TURKEY)
- Prof. **Laszlo KOCZY** (**Chair**, Széchenyi István University and Budapest University of Technology and Economics, HUNGARY)
- Prof. **Piotr KULCZYCKI** (AGH University of Science and Technology, POLAND)
- Assoc.Prof. **Nida KVEDARAITĖ** (Kaunas University of Technology, LITHUANIA).
- Prof. **Anatoly LEVCHENKOV** (Technical University, LATVIA)
- Prof. **Zygmunt MIERCZYK** (Military University of Technology, POLAND)
- Prof. **Vytautas OSTAŠEVIČIUS** (Kaunas University of Technology, LITHUANIA)
- Prof. **Leonīds RIBICKIS** (Riga Technical University, LATVIA)
- Prof. **Marek SITARZ** (Silesian Technical University, TRANSMEC, POLAND)
- Prof. **Aleksandr SLADKOWSKY** (Silesian Technical University, POLAND)
- Prof. **Lucjan SNIEZEK** (Military University of Technology, POLAND)
- Prof. **Daniel SPRECIC** (BOSNIA AND HERZEGOVINA)
- Assoc. Prof. **Brigita STANIKŪNIENĖ** (Kaunas University of Technology, LITHUANIA)
- Assoc. Prof. **Dalia SUSNIENĖ** (Kaunas University of Technology, LITHUANIA)
- Dr. **Arūnas TAUTKUS** (Kaunas University of Technology, LITHUANIA)
- Prof. **Panayiotis TEODOSOIOU** (Cyprus University of Technology, CYPRUS)
- Dr. **Dainius VAIČIULIS** (Kaunas University of Technology, LITHUANIA)
- Dr. **Dmitri VINNIKOV** (Tallinn University of Technology, ESTONIA)
- Prof. **Darius VIRŽONIS** (Kaunas University of Technology, LITHUANIA)
- Dr. **Marzena WOJCEK-AUGUSTYNIAK** (University of Natural Sciences and Humanities, POLAND)
- Prof. **Daiva ŽOSTAUTIENĖ** (Kaunas University of Technology, LITHUANIA)

Foreword

The eleventh international scientific conference Intelligent Technologies in Logistics and Mechatronics Systems is organized in the Forum on Innovative Technologies and Management for Sustainability that was an umbrella event for two conferences, i.e., Intelligent Technologies in Logistics and Mechatronics Systems (ITELEMS'11) and Changes in Social and Business Environment. The main idea of this joint event was to provide a worldwide forum and built bridges among international participants working in engineering and social fields and to find joint topics for future research as to better suit industry, business and society needs.

It is evident that a transition to a sustainable society poses a number of serious technological and socio-economic challenges. Sustainability involves more than just the technical issue of eco-efficiency. Progressive competition, increasing customer demand, growing production and consumption scale, as well as environment degradation cause a lot of social, environmental, and economic problems. These different issues of sustainability require joint efforts of scientists, industry, business, and political leaders and our choice will depend on our values and vision of the future.

Conferences topics included:

- Intelligent Logistics Systems;
- Multi Criteria Decision Making;
- Composites in Infrastructures;
- Intelligent Mechatronics Systems;
- Sensors and sensing phenomena;
- Mechanisms of Transport;
- Transport Systems;
- Modern Building Technologies;
- Military technologies;
- Production Engineering.

The conference is the result of a collective effort and we take the opportunity through this written welcome to express our recognition of the effort and work put in by all those people who have made it possible to organise ITELEMS'11: we pay tribute to the Scientific Committee who have assured the quality of the accepted papers, to the members of the organising committee for their keen motivation and to all the people who have directly or indirectly influenced the smooth progress towards the conference. Finally, we would like to express much gratitude to all of the authors for contributing their papers.

Prof. Laszlo Koczy
Assoc. prof. Dalia Susnienė
Prof. Daiva Žostautienė

Forum on Innovative Technologies and Management for Sustainability (ITMS'2016)

ktu 1922 panevėžys faculty of technologies and business

The 11th International Conference Intelligent Technologies in Logistics and Mechatronics Systems (ITELMS'2016), 28-29 April 2016, Panevėžys, Lithuania

Concept of RES-based microgrid for apartment building

Adomavičius V.[a], Valickas J.[b]*, Petrauskas G.[a], Pušinaitis L.[a]

[a]*Kaunas University of Technology, Studentų 48–205, LT–51367 Kaunas, Lithuania*
[b]*Kaunas University of Technology, S. Daukanto 12–113, LT– 35212 Panevėžys, Lithuania*

Abstract

Concept and case study of energetically self-sufficient eco-friendly apartment building is presented in this paper. Topicality of the eco-friendly buildings, structure of a microgrid for energy supply to the building's loads, reasoning of local renewable energy sources choice for power and heat energy production in the microgrid are analysed. Findings of the study are concluded and possible measures to be taken in order to encourage and accelerate implementation of eco-friendly buildings running on the RES-based power and heat supply systems are discussed.

© 2016 V. Adomavičius, J. Valickas, G. Petrauskas, L. Pušinaitis.
Peer-review under responsibility of the Kaunas University of Technology, Panevėžys Faculty of Technologies and Business

Keywords: Energetically self-sufficient buildings; energy efficiency; renewable energy; microgrid; heat and power.

1. Introduction

Presently pollution of atmosphere by the greenhouse gases (GHG) is one of the main problems and reasons of destroying ecosystem, climate and infrastructure in the Earth planet. The main sources of GHG emissions are heat and power production plants, industry, agriculture, transport, forestry and buildings. Dangerous negative impact of the GHG emission to the climate change is disclosed by the climatologists in thousands of scientific publications, e.g. [1, 2]. Further unlimited GHG emissions will result in more often natural disasters of increased force.

Direct negative impact of air pollution to the health of people is disclosed by medical scientists. Air pollution is already the fourth-highest risk factor for death globally. According to the American Association for the Advancement of Science, it led to more than 5.5 million premature deaths in 2013: in China - 1.6, in India - 1.4, in the US – 0.3, in the rest of world – 2.2 millions of people [3].

It is considered that the main remedy for solving the climate change problem is massive sustainable development in all sectors of economy by the decisive, urgent and deep reduction of the GHG emissions into Earth atmosphere. Similar recommendations can be found in many scientific papers, documents and reports, including the report of international team of scientists The International Panel for Climate Change at the United Nations [4]. In general, it is considered that a phenomenon of the global climate change is already proved by scientist climatologists and many perceptive people can confirm it on basis of their own experience even without knowing discoveries of the climatologists. The main problem now is how to accelerate the transition from fossil fuels to the renewable energy. This problem is analysed and discussed in many publications as well, e.g. [5, 6]. It is clear that unlimited pollution of environment and the climate change can be stopped or at least substantially mitigated only in case of global mobilisation of efforts to eliminate fossil fuels from all sectors of economy. It is very important to do it urgently – as soon as possible.

Usage of energy in the buildings makes up about 40 % of the total energy consumed in Lithuania and about

* Corresponding author. Tel.: +370 656 24494
E-mail address: jonas.valickas@ktu.lt

41 % in the EU. Therefore buildings have a high potential for reduction of the GHG. However, a great inertia exists in this sector and, unfortunately, the pace of phasing out fossil fuels from the buildings is very slow. Concept of the apartment building fed from RES-based microgrid and case study is presented in this paper in order to show the promising possibilities of renewable energy systems in buildings sector and to encourage start-up of their implementation in the nearest future.

Nomenclature

Eai	total power demanded by all electric appliances of the building per month *i*
*E*d*i*	total demands of power by all loads per month *i*
*E*ev*i*	total power demanded by all electric vehicles per month *i*
*E*hw*i*	total power demanded for producing of hot water per month *i*
Ei	total amount of power produced by wind turbines and photovoltaic power plant per month *i*
*E*sh*i*	total power demanded for space heating in the building per month *i*
Esi	amount of power produced by photovoltaic power plant per month *i*
*E*w*i*	total amount of power produced by all wind turbines per month *i*
*H*mo	average perennial monthly sum of global irradiation per m^2 on the plane of optimal inclination (35 deg.)
*H*mv	average perennial monthly sum of global irradiation per m^2 on the vertical plane
*T*ap	average perennial monthly temperature of air in surroundings of the building
*T*al	average monthly temperature of air in loggias of the building

2. Choice of renewable energy sources for energy production in building's microgrid

In general case, energy in buildings is necessary for space heating (cooling), water heating (domestic hot water – DHW) and feeding of domestic electric appliances where electric vehicles can be included too. Many primary energy sources are available in nowadays for heat and power production in buildings, but taking into consideration the climate change and the United Nations Framework Convention on Climate Change (UNFCCC) carried in Paris in December 2015, the best option would be to use non-polluting energy sources for this purpose. First of all wind and solar energy sources can be exploited in many cases for electricity production in the buildings or in a close vicinity of the buildings. These energy sources are not only almost universally available, smokeless, but also self-delivering and absolutely free from payment: users of these energy sources will never have any troubles regarding the orders, supply of energy resources and increase of their prices, because it always will be for free independently from the amounts of energy consumption.

Electricity is universal type of energy, which can be used not only for feeding various electric appliances but also for space and water heating and for charging batteries of electric vehicles. Space heating by means of electric heaters is quite acceptable and affordable, if building has high energetic class due to the relevant thermal insulation. However, better energy efficiency of space and water heating can be achieved when free renewable energy sources suitable for heat energy production are used.

Non-polluting and free renewable energy sources for heat production also exist in a surrounding area of buildings. It is renewable heat sources of low temperature – the air, ground and water bodies (if available). Heat pumps of various types can be used for heat energy production by exploiting energy of the mentioned above heat energy sources of low temperature.

Resources of renewable energy suitable for power and heat production practically are inexhaustible and they newer will come to the end. Their global technical potentials in exajoules per year are presented in Table 1 [7].

Table 1. Global technical potential of the main renewable energy sources.

Renewable energy sources	Unit	Global technical potential, EJ/y Min	**Max**	Total global energy demands, EJ 2008
		Electricity production		
Solar energy		60	1200	
Geothermal energy		118	1 109	
Wind energy	EJ/y	85	580	61
Ocean energy		7	331	
Hydro energy		50	52	
		Heat energy production		
Geothermal energy	EJ/y	10	312	164
		Primary energy supply		
Direct solar energy	EJ/y	1 575	49 837	492
Biomass energy		50	500	
		1EJ = 278 TWh		

Hydro energy, if it is available in a close vicinity of buildings, is also could be acceptable renewable energy source, however, it is not so universally available as solar, wind energy and renewable heat energy sources of low temperature and, besides, possibilities to use hydro energy are restricted because of the environmental restrictions. Biomass, liquid biofuel and biogas energy sources are renewable as well, but they are not self-delivering and free from payment. They can be used in some cases too, however they are less attractive. Operation and maintenance expenditures of the renewable energy systems running on biofuels in general case are higher in comparison with solar and wind energy systems and heat pumps.

3. Case study

In general, apartment buildings may be very different in many aspects: as per number of apartments, number of floors, type of roofs, styles of architecture and so on. Energetically self-sufficient and efficient buildings have to be designed according to the principles of solar architecture. For example, it would be desirable to have a roof of building with optimal slope to the horizontal plane, if solar modules of PV power plant are intended to put on it. Besides, the best orientation of the roof slope would be southward. Our experimental apartment building with a microgrid for feeding domestic electric loads is depicted in Fig. 1.

Fig. 1. General view of apartment building running on universally available local renewable energy sources.

The building has usual electric loads of domestic appliances in all apartments, ten electric vehicles and two heat pumps "air to water" for domestic hot water (DHW) and space heating. Electricity generating equipment comprises small-scale wind turbines (WT) and rooftop photovoltaic power plant (PVPP).

3.1. Characteristics of apartment building

Necessary data for determination of energy demands required for the building itself and the main parameters of power and heat generation equipment designed to install in the building's microgrid are presented in Table 2. Capacity and number of necessary wind turbines for covering energy demands of the building together with PVPP can be chosen more or less freely, however, capacity of PV array is restricted by the roof area.

Table 2. Characteristics of apartment building and power and heat generation equipment to be installed.

Characteristics of apartment building		Power and heat generation installations	
Number of apartments	10	Wind power plants	Horizontal axes rotor
Number of inhabitants	40	– capacity of wind turbine	9.5 kW
Number of electric vehicles	10	– number of wind turbines	3
Area of one apartment	73 m^2	– hub height	24 m
Total heating area	730 m^2	Photovoltaic power plant	Crystalline Silicon
Temperature of heating space air	18–22℃	– number of PV modules	252
Temperature of DHW	50℃	– capacity of one module	300 Wp
Consumption of DHW per inhabitant	2 m^3	– capacity of PV power plant	75.6 kWp
Consumption of hot water per month	80 m^3	Space heating and DHW	Two HP „air to water"
Number of glassed-in loggia's	8	– range of ambient temperature	From –15 C to +25℃
Area of one glassed-in wall of loggia	14 m^2	– range of power consumption	~ 5-6 kW
Total area of all loggias glassed-in walls	112 m^2	– range of heat generation capacity	~ 9–28 kW
Roof area	520 m^2	Locality: Taurage region, Lithuania	55°18'46" North
Building's energetic class	A+		22°01'52" East

Solar energy resources are distributed rather evenly in territory of Lithuania, but wind energy resources strongly depend on the region of country. Therefore geographic coordinates of the microgrids locality are important for evaluation of available amounts of produced power.

3.2. Local renewable energy resources

Local renewable energy resources are described in Table 3. Solar energy resources are evaluated by the average perennial monthly sum of global irradiation per m^2 on the plane of roof slope, the average perennial monthly sum of global irradiation per m^2 on the vertical plane (windows of loggias). Renewable heat energy resources of low temperature existing in outdoor air are evaluated by the average perennial monthly temperature of air in the locality of building and wind energy resources – by the wind speed, roughness of the locality and Weibull parameters.

Table 3. Resources of renewable energy available in a close vicinity of building.

Parameters	Conditions	Months 01	02	03	04	05	06	07	08	09	10	11	12	Per year
		Average monthly sum of global solar irradiation, kWh/m^2												
H_{mo}	Inclination 35°	23.9	43.1	112	152	176	168	167	149	113	66.4	27.2	17.2	1120
H_{mv}	Inclination 90°	25.6	42.1	98.1	108	103	90.8	94.2	97.1	91.2	62.5	28.5	18.7	860
		Average perennial outdoor temperature of air, ℃												
T_{ap}	In shadow	–4.0	–3.7	–0.4	5.1	11.3	14.9	16.3	15.9	11.9	7.6	2.5	–1.5	6.3
T_{al}	In loggias	To be determined experimentally in future during the implementation of pilot project												
		Average annual parameters of wind speed												
Wind speed	Hub height 25 m	Mean annual wind speed 5.7 m/s at height 25 m, roughness 0.1 m. Weibull parameters: a = 6.4 m/s; K=2.07												

Some simple improvements can be implemented in order to enhance efficiency of heat production when heat pump "air to water" is used. As it is known, the heat pump of this type operates more efficiently when temperature of air in the environment is higher. Therefore it is expedient to preheat the air before it is percolated through the outdoor module of the heat pump. For example, it could be glassed-in loggias of the building or greenhouses added on to the building and used as solar collectors for air heating. Simultaneously they could be used as hothouses too. The glassed-in loggia collects heat energy of solar radiation and does not allow its dissipation. Preheating of air is reducing consumption of electricity by the heat pump in sunny and half-sunny days.

3.3. Structure of the microgrid

Power circuits of the apartment building microgrid, which are depicted in Fig. 2, consists of power generation system (small wind turbines SWT1–SWT3 and PV array), power conversion system (rectifiers R1–R3, converters DC/DC1–DC/DD4, battery chargers Ch1–Ch4 and mutual grid-tied inverter), control system CS including switches S, Sc, Si, Sb, S1–S4, power storage battery B and capacitor C. Scheme of DC/DC converters is disclosed under the PV array. Similar innovative power conversion systems with mutual inverter are described in references [8,9].

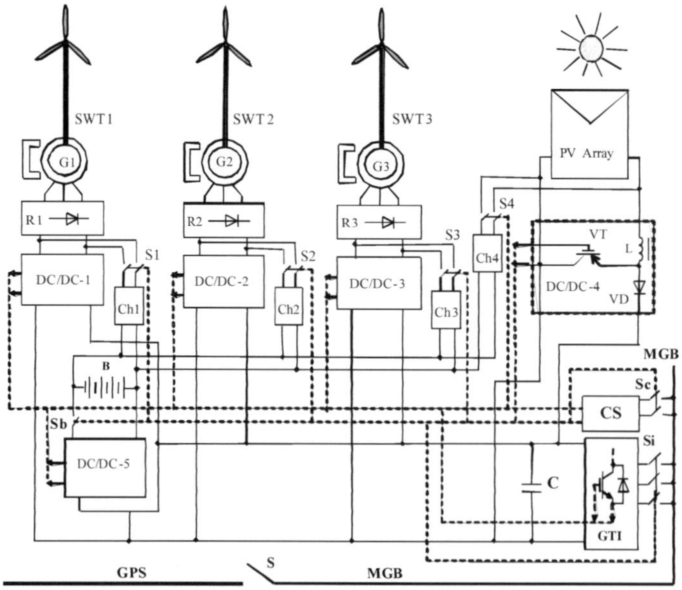

Fig. 2. Electrical scheme of the apartment building's microgrid power circuits with mutual inverter.

Usage of mutual inverter allows simplifying of scheme and reducing of investments. Microgrid's loads are not shown in Fig. 2 because it is well known matter. Microgrid's bus MGB can be connected to the grid of power system GPS. Thus, microgrid can function both in grid-connected and stand-alone modes of operation.

3.4. Power production in the microgrid and demands

Probable power production in the microgrid (Fig. 2) of the apartment building (Fig. 1) was computed in order to determine the limits of locally available power and to compare it with reasonable average power demands. Results of computations are presented in Table 4. Analysis of the received data show that power production greatly exceeds total power demands of all apartments almost in every month and total overproduction of power per year makes up around 53 MWh or about 59 % of total power demands. Surplus of electricity can be sold and supplied into GPS or it can be used for charging of batteries of additional electric vehicles. Connection to the GPS enhances flexibility and reliability of power supply to electricity users in the apartment building. If the microgrid is connected to the GPS, then capacity of the battery for power storage can be significantly reduced.

Table 4. Comparison of the possible power production in the building's microgrid and total power demands.

MONTH	POWER PRODUCTION, kWh			POWER DEMAND, kWh					
	Ewi	*Esi*	*Ei*	*Edi*	*Eai*	*Eshi*	*Ehwi*	*Eevi*	*Ei-Edi*
1	7 068	1 889	8 957	8 500	4 000	1 500	800	2 200	+457
2	6 025	3 175	9 200	8 300	4 000	1 300	800	2 200	+900
3	6 847	7 560	14 408	7 600	4 000	600	800	2 200	+6 808
4	5 130	9 224	14 354	7 350	4 000	350	800	2 200	+7 004
5	4 020	10 312	14 332	7 000	4 000	0	800	2 200	+7 332
6	3 677	9 737	13 413	7 000	4 000	0	800	2 200	+6 413
7	3 446	9 555	13 001	7 000	4 000	0	800	2 200	+6 001
8	4 285	8 831	13 116	7 000	4 000	0	800	2 200	+6 116
9	5 216	6 985	12 201	7 000	4 000	0	800	2 200	+5 201
10	6 185	4 446	10 630	7 350	4 000	350	800	2 200	+3 280
11	6 224	1 826	8 050	7 650	4 000	650	800	2 200	+400
12	9 630	1 261	10 891	8 000	4 000	1 000	800	2 200	+2 891
PER YEAR	67 752	74 800	142 552	89 750	48 000	5 750	9 600	26 400	+52 802

4. Conclusions and discussion

The main objective of this paper was to disclose possibilities and benefits of microgrids installation in apartment buildings and to encourage their implementation. Analysis of the case study results confirmed this possibility even in locality with moderate wind and solar energy resources. If it would be necessary, amounts of the produced power could be simply increased by installing additional small wind turbines or increasing their capacity. Besides, efficiency of solar cells is increasing step by step and amount of power produced from the same area of PV array is increasing year-by-year. Installation of microgrids of this type in apartment buildings and other buildings [10] has many advantages: bulk of produced intermittent power is not supplied into the grid of power system but consumed locally and it allows avoiding of losses due to the power supply over long lines; the grid of power system is not burdened by the intermittent power; imported dirty fossil fuels are not consumed; emissions of greenhouse gases are avoided; energy independence and reliability of power supply increases and etc.

Promising results and findings of the case study described in this paper allow taking the next steps: designing a pilot project of energetically self-sufficient apartment building, its implementation and experimental research of the RES-based microgrid operation in natural conditions. Researches of the microgrid of apartment building in natural conditions will allow receiving more exact results both in technical and economic aspects and will give reliable answers to many questions regarding mass implementation of energetically self-sufficient buildings.

References

[1] Benestad RE. A mental picture of the greenhouse effect. *Theoretical and Applied Climatology;* 2016. Available online: http://dx.doi.org/10.1007/s00704- 016-1732-y.

[2] Thorne PW, Vose RS. Reanalyses Suitable for Characterizing Long-Term Trends, *Bull. Amer. Meteor. Soc.* 2010; 91: 353–361 Available online: http://dx.doi.org/10.1175/2009BAMS2858.1

[3] Yuhas A. Scientists: air pollution led to more than 5.5 million premature deaths in 2013. *The Guardian*, February 12; 2016.

[4] IPCC WGIII Fifth Assessment Report. Summary for Policy Makers. United Nations; 2014. p.31 Available online: file:///C:/DOCUME~1/TEST/LOCALS~1/Temp/ipcc_wg3_ar5_summary-for-policymakers_approved.pdf
[5] Tjell JCh. Is Climate Change Essentially Unavoidable? *Environmental Research, Engineering and Management* 2014; 3(69): 3–4.
[6] Hvelplund F. Innovative Democracy, Political Economy, and the Transition to Renewable Energy. A full-Scale Experiment in Denmark 1976–2013. *Environmental Research, Engineering and Management;* No. 4(66), 2013. p. 5–21
[7] Adomavicius V, Kaminickas M. Analysis of PV power future development possibilities. *Energetika* 2014; 60 (4):233–248.
[8] Kepalas V, Ramonas Č, Adomavičius V. Connection of the RES-based power plants into the electric grid. *Electronics and Electrical Engineering* 2007; 8(80): 67-72.
[9] Ramonas Č, Adomavičius V, Kepalas V. Research of the power conversion processes in the system of power supply from a number of wind turbines over the one grid-tird inverter. Proceedings of International Conference *Electrical and Control Technologies; 2009.* KTU, Kaunas; 2009. p. 368-373
[10] Adomavicius V, Kharchenko V, Valickas J, Gusarov V. RES-based microgrids for environmentally friendly energy supply in agriculture. *5th International Conference Trends in Agriculture Engineering.* Prague, Czech Republic; September 3–6, 2013. p. 51–55

Forum on Innovative Technologies and Management for Sustainability (ITMS'2016)

ktu 1922 panevėžys faculty of technologies and business

The 11th International Conference Intelligent Technologies in Logistics and Mechatronics Systems (ITELMS'2016), 28-29 April 2016, Panevėžys, Lithuania

A Model for Simulation of Binary Customer Behaviour in a Bursty Business Process

Ahrens A.[a]*, Purvinis O.[b], Zaščerinska J.[c], Andreeva N.[d]

[a]*Hochschule Wismar, University of Applied Sciences, Technology, Business and Design, Philipp-Müller-Straße 14, 23966 Wismar*
[b]*Kaunas University of Technology, Panevėžys faculty of Technologies and Business, Nemuno str. 33 - 219, LT- 37164 Panevėžys, Lithuania*
[c]*Centre for Education and Innovation Research, Dammes iela 33-102, Riga LV-1069, Latvia*
[d]*Immanuel Kant Baltic Federal University, Nevskogo street 14, Kaliningrad, 236016, Russian Federation*

Abstract

Earning profits in business requires analysis of binary customer behaviour. The aim of the research is to carry out mathematical analysis of gap processes underpinning elaboration of a simulation model of binary customer behaviour in a bursty business process. The meaning of such key concepts as *bursty business process*, *binary customer behaviour* and *gap* is studied. The novel contribution of the paper is the created model of binary customer behavior in a bursty business process based on gap processes. Directions of further research are proposed.

© 2016 A. Ahrens, O. Purvinis, J. Zaščerinska, N. Andreeva.
Peer-review under responsibility of the Kaunas University of Technology, Panevėžys Faculty of Technologies and Business.

Keywords: bursty business process; gap-processes; binary customer behaviour; model; simulation model.

1. Introduction

Business environment has shaped and been shaped by various social and political forces within the couple of past decades. Table 1 provides a historical perspective on business development during the past decades.

Table 1. Business practice in different historical periods of the past decades.

Phase	Historical period	Business approach
1.	1980s	De-Industrialization (a process of social and economic change caused by the removal or reduction of industrial capacity or activity in a country or region, especially heavy industry or manufacturing industry)
2.	1980s	Deregulation (the process of removing or reducing state regulations)
3.	1990s	Global Business (international trade or a company doing business across the world)
4.	2000s	Hybrid business (enterprise which makes use of traditional methods of distribution and Internet)

Changing business environment promotes the advancement of business. As business is traditionally oriented to earning profits via the process of buying and/or selling of goods and/or services, a number of profitable business models such as successful business model, sustainable business model, winning business model, dream business model and etc. as shown in Figure 1 are demonstrated by the business community.

* Corresponding author. Tel.: +49 3841 7537 330.
E-mail address: andreas.ahrens@hs-wismar.de

© 2016 A. Ahrens, O.Purvinis, J. Zaščerinska, N. Andreeva.
Peer-review under responsibility of the Kaunas University of Technology, Panevėžys Faculty of Technologies and Business

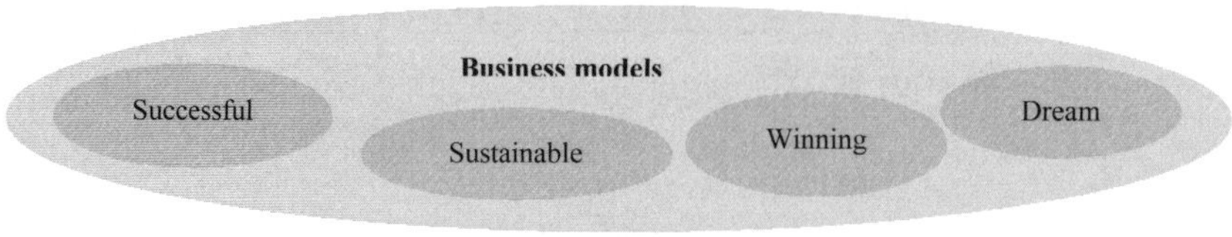

Fig. 1. Types of profitable business models.

For the purposes of the present contribution, the terms "the process of buying and/or selling of goods and/or services" and "the business process" are used synonymously. It should be noted that the business process is the fundamental element of a business environment as indicated in Figure 2.

Fig. 2. The relationship between business environment and business process.

The business process is permanently optimized in order to increase the profit. Optimization of the business process implies choices about quantity of goods to be delivered, number of the staff to be employed, etc. Additionally, such a result of the business process as purchase and/or sale of a good or service indicates the output of this process.

It should be noted that by model, a pattern of individual's or individuals' interpretation of a phenomenon is meant [3]. Models can be presented in a variety of forms such as verbal, graphic, computer, etc. A model can be characterized or featured as demonstrated in Figure 3 [3].

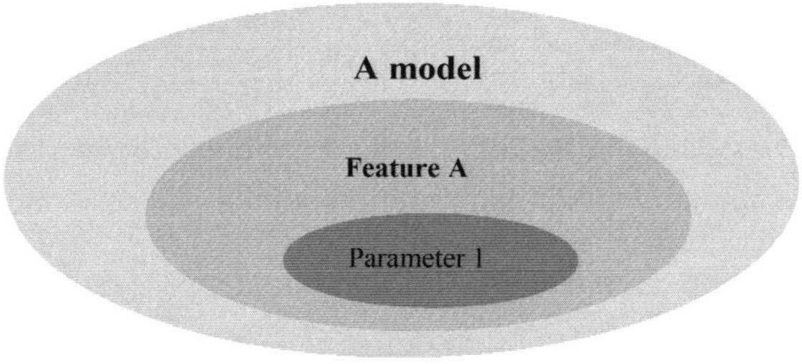

Fig. 3. Model elements.

The business models such as successful business model, sustainable business model, winning business model, dream business model and etc. are usually featured by such characteristics as competitiveness, efficiency, innovation, sustainability, profitability, scalability, flexibility, mobility, etc. as illustrated in Figure 4. It should be noted that business models focus on organizing a business process. For the evaluation of the business processes as a part of a particular business model, realistic simulation models are an important prerequisite. Such simulation models as business planning and economic forecasting [4] have been developed. Analysis of the simulation models for business planning and economic forecasting [5] revealed that

- on the one hand, these simulation models for business planning and economic forecasting focused on nonlinear and simultaneous relationships as well as dynamic processes [5], and,
- on the other hand, these simulation models for business planning and economic forecasting used such economic indicators as the gross national product as business planning is closely related with the economic system, [5].

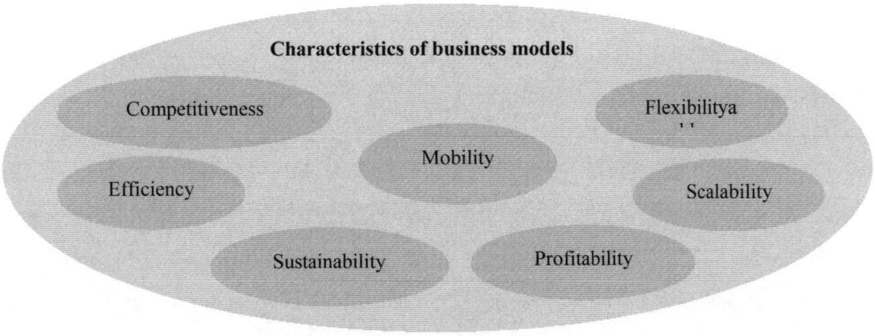

Fig. 4. Characteristics of business models.

Analysis of the simulation models for business planning and economic forecasting revealed that these simulation models focused on the process dynamism

- on the one hand, do not take into consideration the bursty nature of phenomenon, and
- on the other hand, use indicators that show phenomenon's developmental dynamism rather than criteria that serve to structure, assess and evaluate [9].

For the creation of a new simulation model, such an everyday business situation is considered as potential customers have to solve an issue formulated already in 1603 by William Shakespeare in his play *Hamlet* such as "To be, or not to be" [8]. Regarding a modern interpretation of potential customers' contemporary problems, Shakespeare's words may sound as "to buy, or not to buy". It should be noted that "to buy, or not to buy" is considered as binary customer behavior depicted in Figure 5.

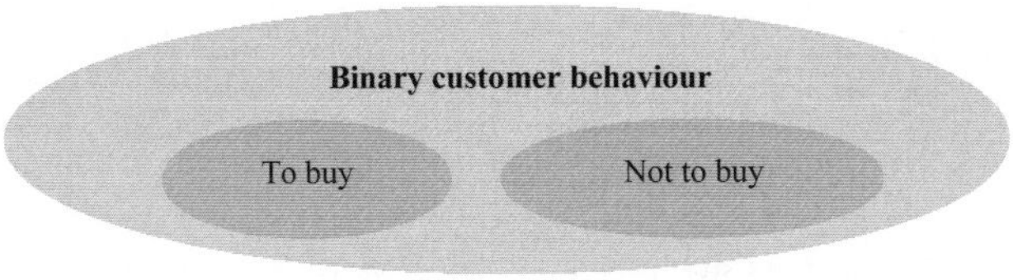

Fig. 5. Elements of customers' binary option.

Binary customer behavior influences the organization of the business process. Figure 6 shows a typical scenario in which a purchase as the output of the process of buying and/or selling of goods and/or services and/or the business process is highlighted (represented by "x") within a sequence of people (represented by "-") who visited a shop.

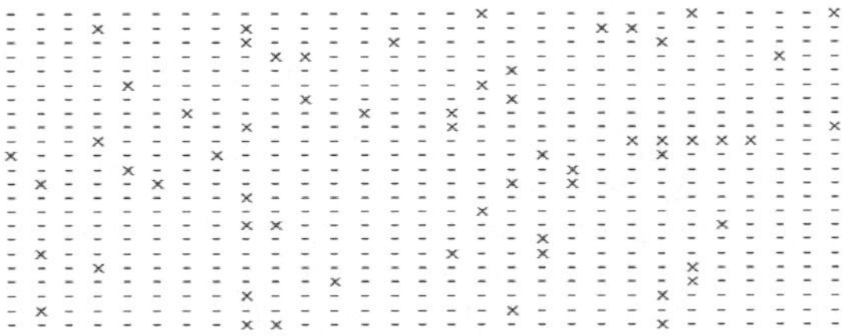

Fig. 6. Highlighting purchases (represented by "x") within a sequence of shop visitors (represented by "-").

However, the buyers can be more independently distributed over e.g. a day or they can appear really concentrated as highlighted in Figure 7.

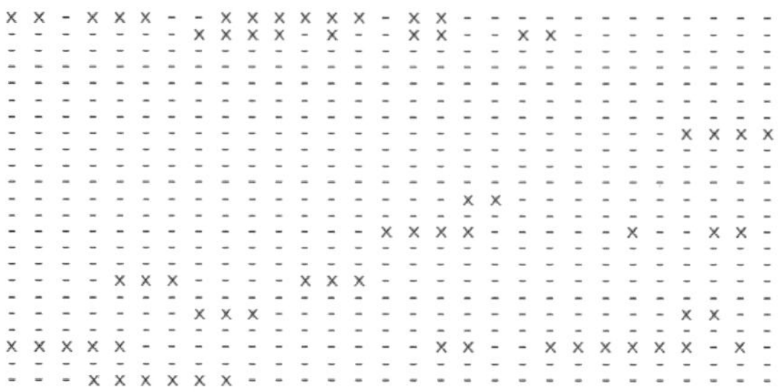

Fig. 7. Highlighting the bursty nature of purchases (represented by "x") within a sequence of shop visitors (represented by "-").

In general, the buyers probability can serve as a clear indicator of how often people decide to buy e.g. a product. However, the buyers probability does not deliver any information about how concentrated the purchases are.

The research question is as follows: How can binary customer behavior in a bursty business process be realistically modeled via gap processes? The aim of the research is to carry out mathematical analysis of gap processes underpinning elaboration of a simulation model of binary customer behaviour in a bursty business process.

The meaning of such key concepts as *model*, *simulation model*, *bursty business process*, and *gap* is studied. Moreover, the analysis demonstrates how the key concepts are related to the idea of *binary customer behaviour* and shows a potential model for development, indicating how the steps of the process are related following a logical chain: conceptual framework → simulation model development → conclusions.

The present contribution employs interdisciplinary research as interdisciplinary research assists in synthesizing, connecting and blending ideas, data and information, methods, tools, concepts, and/or theories from two or more disciplines in order "to make whole" [7]. The process of interdisciplinary research is organized as revealed in Figure 8 [2]:

- In Phase 1 of the interdisciplinary research, an issue is separately explored by two or more scientific disciplines.
- In Phase 2, the same issue is examined by the synergetic point of view of these two or more scientific disciplines.
- In Phase 3, results of the analysis are interpreted.

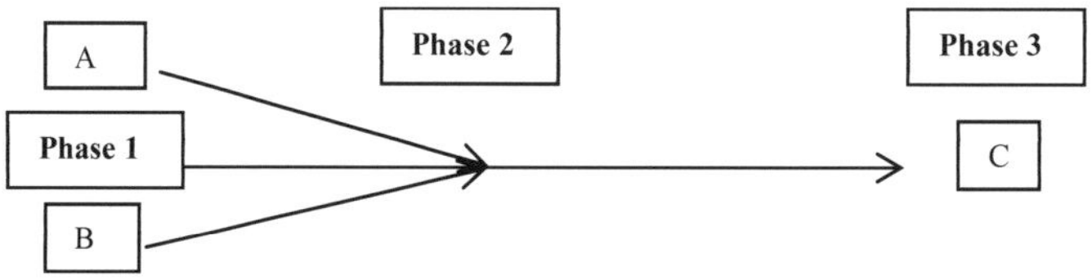

Fig 8. The interdisciplinary process by Repko [7].

For the creation of a model for simulation of binary customer behavior in a bursty business process based on gap processes, the synergy between business and telecommunications is promoted as the phenomenon of customers in the business process as well as bit-errors in data transmission appear to be of a similar nature, namely, the bursty nature. It should be noted that by bursty nature of phenomenon, intervals of high-activity alternating with long low-activity periods is meant [6]. Such models that describe the bursty nature of bit-errors in data transmission have been successfully implemented in telecommunications for optimizing data communication protocols and will be adopted in this work to the bursty business process. It should be noted that the present research is not limited to only two scientific disciplines, namely, business and telecommunication, but is based on a number of scientific disciplines such as education, entrepreneurship, pedagogy, psychology, sociology, etc.

In this work a realistic simulation model for binary customer behavior within a bursty business process is presented.

2. Gap processes for modeling binary customer behaviour

2.1. Conceptual framework

The present part of the contribution delivers a conceptual framework on a model for simulation of binary customer behaviour in a bursty business process. It should be noted that simulation model is identified as patterns of the management of phenomenon change in real-world situations. A simulation model should map the characteristic of the real world process with the required precision. The real world process in this work refers to the business process that includes any process such as selling or buying where binary decisions are made [3].

A model being created for simulation of binary customer behavior in a bursty business process is featured by burstiness. This model characteristic, namely burstiness, is described by parameters or criteria. By a parameter, definable, measurable, and constant or variable characteristic, dimension, property, or value, selected from a set of data (or population) to understanding a situation (or in solving a problem) is understood [4]. Criteria serve to structure, assess and evaluate [9]. Criteria is identified by analysis of [9]

- definition of the research object,
- structure of the research object and
- factors.

For determining a criterion, analysis of the bursty nature of the buyers in the business process is considered. As buyers' burstiness shapes the structure of the business process, burstiness is determined to be the criterion. The criterion burstiness describes the model for simulation of binary customer behaviour in a bursty business process as pointed in Table 2.

Table 2. Criterion that describes the model for simulation of binary customer behaviour in a bursty business process.

Model	Fundamental element	Characteristics	Criterion
Business	The business process: the process of buying and/or selling of goods and/or services	Burstiness	Burstiness

Further on, description of phenomenon's developmental dynamism requires indicators [9] to be defined. The search for indicators based on taking the bursty nature of the buyers into account reveals that the buyers' ratio is not any longer sufficient to describe the characteristic of the buying process [3]. The buyers ratio (in the following referred to the buyers' probability p_B) is defined as the number of buyers divided by the number of the people entering e.g. a shop. Hence, using a model with only one parameter is difficult to project the real buyers' characteristic onto the parameters of the model. This leads undoubtedly to an inaccuracy between the model setup and the real world characteristics. Hence, an additional parameter has to be introduced to describe the concentration of purchases in the buying process.

Within the binary decision paradigm, the business process such as selling or buying is a success if it finishes with a deal such as a sale or a purchase. Gap in the present contribution means the buying process which ends without a purchase. By a gap process, the buyer's concentration as well as the buyer's probability can be taken into account in a realistic way. Therein, the gaps between two buyers are assumed to be statistically independent from each other. Consequently, the model parameter or, in other words, criterion, namely burstiness, is described by such indicators as the buyers' probability and the buyers' concentration as delivered in Table 3.

Table 3. Criterion and indicators that describe the model for simulation of binary customer behaviour in a bursty business process.

Model	Fundamental element	Characteristics	Criterion	Indicators
Business	The business process: the process of buying and/or selling of goods and/or services	Burstiness	Burstiness	buyers' probability
				buyers' concentration

Figure 9 highlights the theoretical basis for the new simulation model, where the business process is described by gap-processes and illustrates the process between two buyers. i.e. how often visitors buy a product and how concentrated they appear. For the optimization of the business process related parameters e.g. the expected time between two buyers, a simulation model such as the proposed one can be helpful.

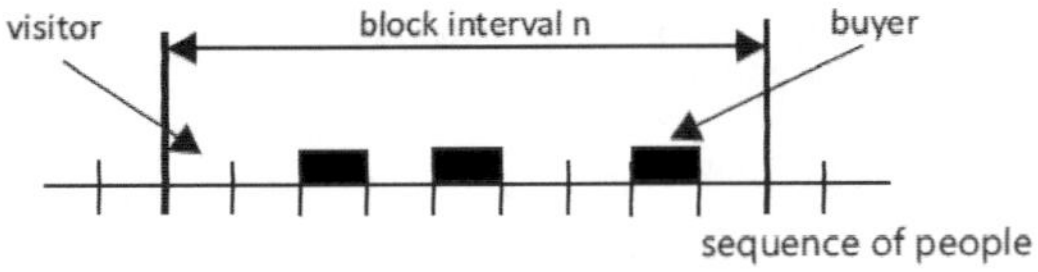

Fig. 9. Buyer's gap for describing binary customer behavior.

2.2. Model development for simulation of binary customer behavior in a bursty business process

Analyzing the buyers' characteristic, we can define a block interval n (identified as the probability $p_B(n)$)

where at least one buyer appears. The parameter *n* refers e.g. to the total number of people entering a shop in a given time e.g. a day. Choosing the parameter $n = 1$ the probability $p_B(n)$ equals the buyers probability p_e.

Now we can assume that the probability $p_B(n)$ can be described as a function of the buyers' probability p_e and the block interval length n. Here the following approximation is used [10, 1]

$$p_B(n) = \begin{cases} p_e \cdot n^{\alpha} & 1 \leq n \leq n_0 \\ 1 & n > n_0 \end{cases} . \quad (1)$$

The value α denotes the linear dependence between $\log_{10} p_B(n)$ and $\log_{10} n$ and is a measure for the buyers' concentration (also referred to the concentration of buying). The value of n_0 indicates the maximum block length to which the linear-dependence can be maintained (see Figure 10).

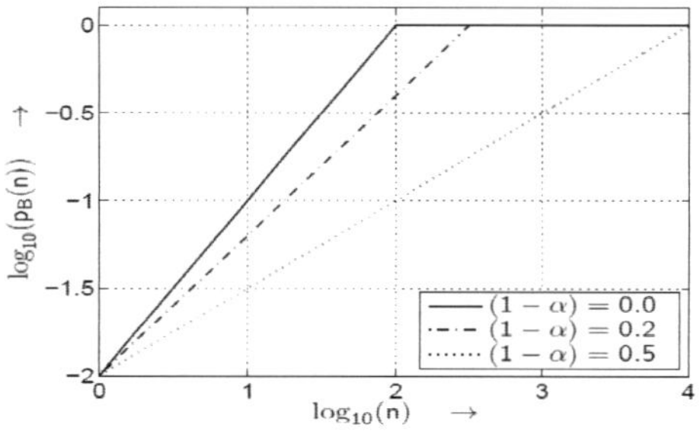

Fig. 10. *Relationship between the probability $p_B(n)$ and the block interval n for different parameters of the $(1-\alpha)$ at a buyer's probability of $p_e = 10^{-2}$.*

The analysis of concentration parameters $(1-\alpha)$ (referred to the concentration of buying) has shown that parameters in the range of 0.0 until 0.5 describe realistic scenarios. Thereby, a parameter $(1-\alpha) = 0$ describes the situation where the potential buyers appear independently distributed from each other. With increasing parameter $(1-\alpha)$ the buyers appear more and more concentrated and the probability $p_B(n)$ decreased for a given n. With the assumption that the distances (gaps k) between neighboring buyers are statistically independent from each other, the buyers' characteristic, namely the occurrence of bursty buyers, is defined by the buyers' gap-distribution function $u(k) = P(X \geq k)$, which describes the probability of a gap larger than *k*. The setup

$$p_B(n) = \begin{cases} p_e \cdot \sum_{k=0}^{n-1} u(k) & 1 \leq n \leq n_0 \\ 1 & n > n_0 \end{cases} \quad (2)$$

is used to develop the buyer's gap distribution function $u(k)$ for the buyers' gaps step by step.

Comparing (1) and (2), one gets:

$$\sum_{k=0}^{n-1} u(k) = n^{\alpha} \quad 1 \leq n \leq n_0 \quad (3)$$

and for the searched error-gap distribution $u(k)$ we yield:

$$\begin{array}{lcll} n = 1 & : & u(0) & = 1^{\alpha} \\ n = 2 & : & u(0) + u(1) & = 2^{\alpha} \\ n = 3 & : & u(0) + u(1) + u(2) & = 3^{\alpha} \\ \ldots & : & \ldots & = \ldots \\ n \leq n_0 & : & u(0) + u(1) + \cdots + u(n-1) & = n^{\alpha} \end{array}$$

The buyer's-gap distribution function $u(k)$ can be defined as follows:

$$u(k)=\begin{cases}(k+1)^{\alpha}-k^{\alpha} & 0\le k<n_0\\ 0 & k\ge n_0\end{cases} \tag{4}$$

Re-writing of $u(k)$ leads to the buyers-gap density function $v(k)=P(X=k)$, which describes the probability of a gap X equal to k :

$$u(k)=v(k)+v(k+1)+v(k+2)+\cdots$$

$$u(k+1)=v(k+1)+v(k+2)+\cdots$$

and by calculating the difference between $u(k)$ and $u(k+1)$ the buyers-gap density function $v(k)=P(X=k)$ can be obtained

$$v(k)=u(k)-u(k+1)\,. \tag{5}$$

Assuming that the buyers are independently distributed, i.e. $(1-\alpha)=0$, and using equation (4) and (5) one gets the following result for the buyers-gap density function $v(k)$:

$$v(k)=\begin{cases}1 & k=(n_0-1)\\ 0 & k\ne(n_0-1)\end{cases}. \tag{6}$$

With this result, the disadvantage of the model setup becomes evident. The model setup defined in (1) leads to a deterministic buyers-gap process. In situations, where the buyers appear concentrated, i.e. $(1-\alpha)>0$, one can also find an enlarged value at $v(n_0-1)$. This error leads to engraving inaccuracies in the simulation process. The reason is the discontinuity at $n=n_0$ in equation (1). A modification of this model setup is necessary. The following solution can be assumed: The linear increases of $\log_{10}p_B(n)$ can only be accepted for small parameters of n . The value of $\log_{10}p_B(n)$ has to change steadily into the value $\log_{10}p_B(n)=0$ for larger n. To the minimization of the model inaccuracy at $v(n_0-1)$ equation (4) has to be multiplied by the value $e^{-\beta\cdot k}$ [7]. For the buyers-gap distribution function $u(k)$ the following expression arises:

$$u(k)=\left((k+1)^{\alpha}-k^{\alpha}\right)\cdot e^{-\beta\cdot k}\quad 0\le k\le\infty$$

with

$$\lim_{k\to\infty}e^{-\beta\cdot k}=0\qquad \beta>0$$

and

$$\beta\approx p_e^{\frac{1}{\alpha}}$$

Figure 11 illustrates the buyers-gap distribution function $u(k)$ for different parameters $(1-\alpha)$ assuming a buyer's probability of $p_e=10^{-2}$.

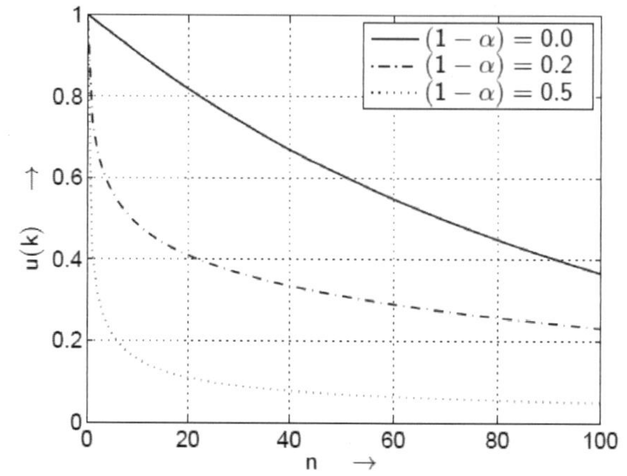

Fig. 11. *Buyers-gap distribution function* $u(k)$ *for different parameters of the* $(1-\alpha)$ *at a buyer's probability of* $p_e = 10^{-2}$.

The resultant buyers-gap density function $v(k)$ is depicted in Figure 12.

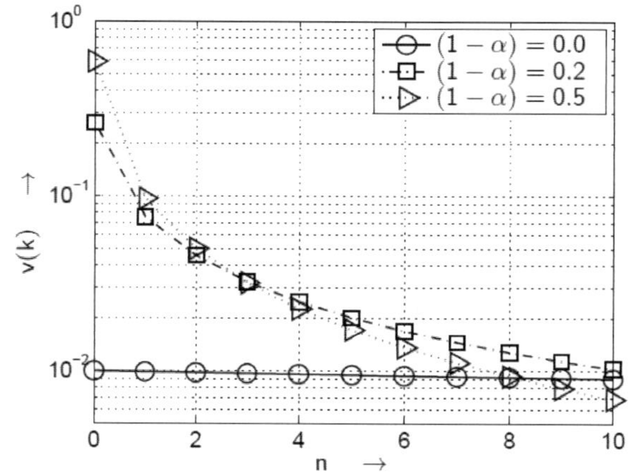

Fig. 12. *Buyers-gap density function* $v(k)$ *for different parameters of the* $(1-\alpha)$ *at a buyer's probability of* $p_e = 10^{-2}$.

Finally, the proposed system setup is highlighted in Figure 13.

Now, the buyers characteristic can be modelled by two parameters (the buyer's probability p_e and the buyer's concentration value $(1-\alpha)$).

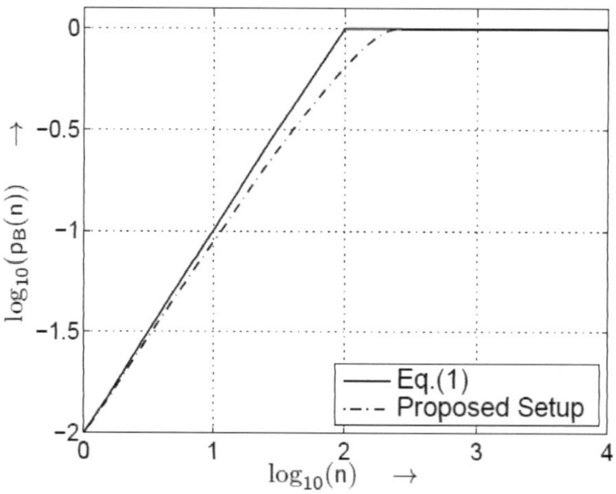

Fig. 13. *Approximated relationship between the probability* $p_B(n)$ *and the block interval n for different parameters of the* $(1-\alpha)$.

With the assumption that the distances between neighboring buyers are statistically independent from each other the model characteristic is described completely by the buyer's distribution function $u(k)$. For the creation of the gap processes a uniformly distributed random number Y is identical to the function $u(k)$ and the corresponding value of the buyer's gap is determined. For this, the following equation

$$Y \equiv u(k) \tag{6}$$

has to be solved numerically.

3. Conclusions

The theoretical findings on the inter-relationship between *model, model elements, the business process, the process of buying and/or selling of goods and/or services, the bursty nature of buyers* and *gap processes* allow determining such

- a criterion for simulation of binary customer behavior in the business process as *burstiness* as well as
- its indicators such as *buyers' probability* and *buyers' concentration.*

The theoretical analysis of the term *simulation model* assists in the development of a simulation model of binary customer behaviour in a bursty business process.

The conceptual framework on binary customer behaviour in a bursty business process elaborated within the present research serves as the basis for the development of the theoretical framework on the advancement of the elaborated model for simulation of binary customer behaviour in a bursty business process.

The present research has limitations. The inter-connections between *model, model elements, simulation model, binary customer behaviour, the bursty nature of buyers, the business process, the process of buying and/or selling of goods and/or services* and *gap processes* have been set. Another limitation is the theoretical analysis carried out only. Therein, the results of the study cannot be representative for the whole area. Nevertheless, the results of the research, namely the identified criterion such as *burstiness* and indicators such as *buyers' probability and buyers' concentration*, may be used as a basis of the promotion of the theoretical framework on a model for simulation of binary customer behaviour in a bursty business process. If the simulation results had been available for analysis, different results could have been attained. There is a possibility to continue the study.

Further research tends to facilitate the advancement of the theoretical framework on the development of a model for simulation of binary customer behaviour in a bursty business process. The search for relevant methods, tools and techniques for evaluation of binary customer behaviour in a bursty business process as well as the simulation model is proposed. Future research tends to analyse the implementation of the simulation model characterized by two indicators such as buyers' probability and buyers' concentration. A comparative research of simulation models for the analysis of binary customer behaviour in a bursty business process could be carried out, too.

References

[1] Ahrens A. A new digital channel model suitable for the simulation and evaluation of channel error effects. In: *Colloquium on Speech Coding Algorithms for Radio Channels*, IEE Electronics and Communications, London (UK); 17. April 2000. p. 2.1-2.15

[2] Ahrens A, Bassus O, Zaščerinska J. *Interdisciplinary Research on Current Issues in Higher Education.* Verlag: Mensch & Buch; 2014.

[3] Ahrens A, Purvinis O, Zaščerinska J, Andreeva N. Gap Processes for Modelling Binary Customer Behavior. *Proceedings of the 8th International Conference on Engineering and Business Education 8-9 October 2015*, Østfold University College, Fredrikstad, Norway. Published by University of Wismar, Germany; 2015. p. 8-13

[4] *Business Dictionary.* Parameter. BusinessDictionary.com; 2015.

[5] Erdman D. Simulation Models for Business Planning and Economic Forecasting. *SUGI Proceedings, Econometrics and Time Series Papers*; 1993.

[6] Karsai M, Kaski K, Barabási A, & Kertész J. Universal features of correlated bursty behavior. *Scientific Reports* 2, Article number:397 2012. doi:10.1038/srep00397.

[7] Repko A. *Interdisciplinary Research Process and Theory.* Second Edition. SAGE; 2012.

[8] Shakespeare W. *The Tragicall Historie of Hamlet Prince of Denmarke. As it hath beene diuerse times acted by his Highnesse seruants in the Cittie of London: as also in the two Vniuersities of Cambridge and Oxford, and else-where* [The "First Quarto"], pp. 35 ff. Nicholas Ling & J. Trundell (London), 1603. Reprinted as *The First Edition of the Tragedy of Hamlet: London, 1603.* The Shakespeare Press; 1825.

[9] Špona A, Čehlova Z. *Pētniecība pedagoģijā (Research in Pedagogy).* Rīga: Raka, (in Latvian); 2004.

[10] Wilhelm C. *Datenübertragung (in German).* Berlin: Militärverlag; 1976.

Forum on Innovative Technologies and Management for Sustainability (ITMS'2016)

ktu 1922 panevėžys faculty of technologies and business

The 11th International Conference Intelligent Technologies in Logistics and Mechatronics Systems (ITELMS'2016), 28-29 April 2016, Panevėžys, Lithuania

Capacitive micromachined ultrasound transducer for greenhouse gas detection

Barauskas D.[a]*, Pelenis D.[a], Viržonis D.[a], Sapeliauskas E.[a]

[a]*Kaunas University of Technology, Daukanto st. 12, Panevėžys, Lithuania.*

Abstract

This paper presents a creation of reliable and cross-selective platform of greenhouse gas sensing based on capacitive micromachined ultrasound transducers (CMUT) technology which main focus is in environmental monitoring and safety. We emphasize the need of cross-selectivity between different gases such as: CO_2, CH_4 and H_2S, since mixture of these gases is very common in various different gas emissions, and their impact to the greenhouse effect remains underexplored. CMUT sensors containing five elements with resonance frequencies ranging from 15 to 35 MHz were designed and fabricated by surface micromachining technology which is fully CMOS compatible. Two parallel measurement channels, one reference and one sensing were arranged to measure the mass loading of the CMUT structure and eliminate the non-informative changes of the working environment such as: temperature, pressure and humidity. CO_2 sensing experiments show that present sensing setup, when CMUT structure is functionalized by thin polyethyleneimine film, has the 4 Hz/ppm sensitivity.

© 2016 D. Barauskas, D. Pelenis, D. Viržonis, E. Sapeliauskas.
Peer-review under responsibility of the Kaunas University of Technology, Panevėžys Faculty of Technologies and Business

Keywords: CMUT; polyethyleneimine; gas sensor.

1. Introduction

Gas mixtures of so called "sour gases", namely carbon dioxide (CO_2), methane (CH_4) and hydrogen sulfide (H_2S) are typically emitted from many anthropogenic and natural sources and have crucial impact to the human health and global warming effect. The magnitude of the impact of each of these gases to global warming is still underexplored mainly because existing practical gas sensing approaches have low cross-sensitivity between CO_2, CH_4 and H_2S. Optical spectroscopy methods not always provide significant and distinct absorption, emission or scattering in a convenient region of the optical spectrum, and spectral overlap problems complicate direct quantification of the spectra. Also, the need for multiple wavelength tunable lasers may raise the cost of the sensor [1]. As alternative to spectroscopy, several versions of capacitive micromachined ultrasound transducers (CMUT) were demonstrated to be promising gas, chemical and biochemical sensing platforms [2-5]. Therefore in our work we address creation of reliable and cross-sensitive platform of greenhouse gas sensing for environmental monitoring and safety purposes based on CMUT technology. This approach was selected because of ability to sense the very low mass loading of the transducer structure. In this paper we report the fabrication and characterization of polyethyleneimine (PEI) functionalized CMUT for CO_2 gas sensing.

* Dovydas Barauskas. Tel.: +370 64585125
E-mail address: Dovydas.Barauskas@ktu.edu

© 2016 D. Barauskas, D. Pelenis, D. Viržonis, E. Sapeliauskas.
Peer-review under responsibility of the Kaunas University of Technology, Panevėžys Faculty of Technologies and Business

2. Device design, fabrication and assembly

Each CMUT device was designed to contain five different arrays of capacitive cells as shown in Fig. 1. The cross-section of a single CMUT cell is shown in Fig. 2. These cells, electrically connected in parallel, create arrays with different properties. This design lets us develop a sensor for cross-sensitivity between CO_2, CH_4 and H_2S on the same chip by functionalizing the surfaces of the fabricated arrays with different absorbers, specific for each gas.

Each of the CMUT arrays have different resonance frequency, starting from 15 MHz to 35 MHz. This was achieved by designing and fabricating cells of different diameter. We used fully CMOS compatible surface micromachining process, involving low-stress PECVD silicon nitride sacrificial release [6-10]. First, lithography on the highly doped silicon substrate was used to create cavities on oxidized silicon wafers by etching them with buffered oxide etchant (BOE). Thin sacrificial chromium film was filled in the cavities by physical vapor deposition process. Then low-stress PECVD silicon nitride layer of critical thickness was deposited. Titanium/aluminum top electrodes were formed by physical vapor deposition and next lithography step. After covering these electrodes by protective silicon nitride and opening of the etching vias wafers were submerged into highly selective CR-7 etchant for sacrificial release of the CMUT structure. Critical point drying was carried out to release the free standing membranes and last PECVD Si_xN_y layer was deposited to cover the etching holes and isolate the membranes from environment [6, 7].

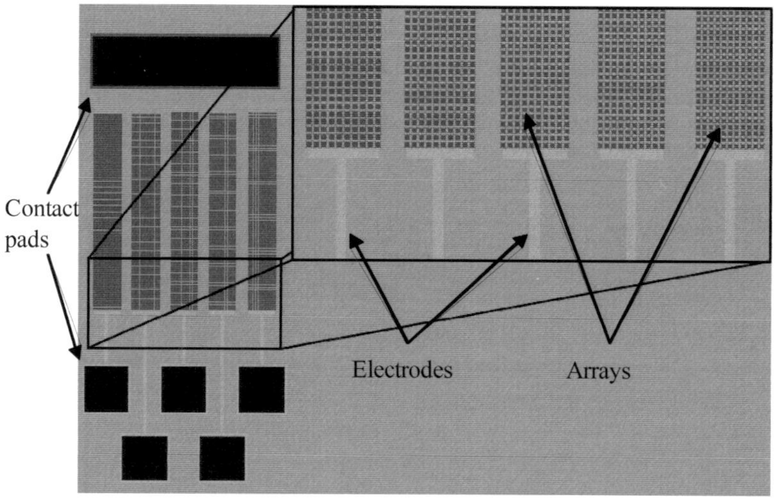

Fig. 1. CMUT structures of the gas sensor.

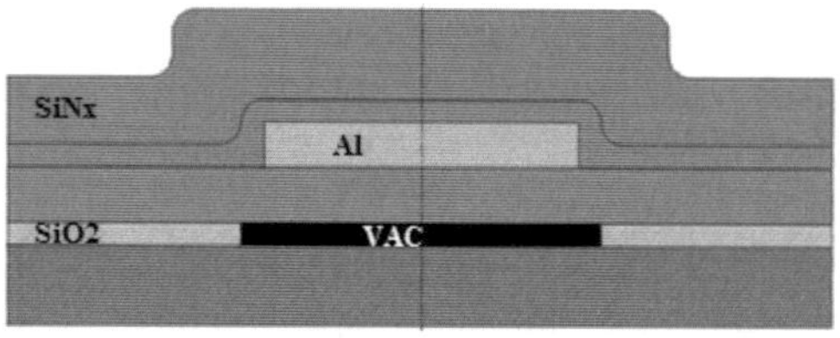

Fig. 2. Cross section of a single membrane: Si_xN_y – silicon nitride layer; SiO_2 – silicon dioxide layer; Al- aluminum; VAC – vacuum; black lines separate different deposition stages.

Fig. 3. Sensor chip wire bonded to PCB with the socket. Dimensions of entire assembly 10x20 mm; dimensions of the CMUT chip is 3x5 mm.

Fabricated sensors then were assembled to specifically designed printed circuit boards for evaluation and testing purposes. The contact pads of the top electrodes of the device were wire bonded with 30 μm gold wires and protected with epoxy resin. For connecting the device to electronics a special socket was made. Fully assembled chip is shown in fig. 3.

3. Experimental setup

3.1. Measurement equipment

The frequency spectra of the real part of the impedance ("resistance") of the fabricated devices were measured by the network analyzer Agilent 4395A equipped with the impedance measurement kit. Direct current bias was applied using Agilent N5752A power supply. To reduce the noise of the resonance frequency signal and amplify the output data, we developed a Colpitts type oscillator circuit for use with the CMUT element as electromechanical resonator [3]. Two of these circuits were integrated into the testing equipment and their output were connected to two independent frequency meters. These frequency meters were interfaced with STM32F400 microcontroller for real-time data acquisition. This arrangement enabled the transfer of the sensing data from two independent sensors to a regular personal computer via Universal Serial Bus (USB) with appropriate sampling frequency.

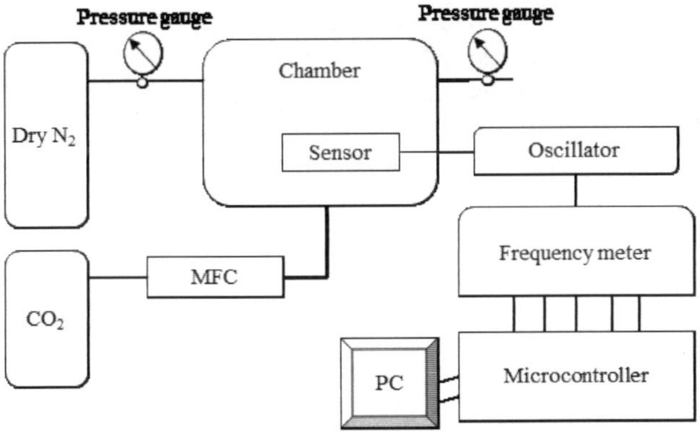

Fig. 4. Principle diagram of testing and measurement equipment.

Fig. 5. Photograph of the testing chamber showing oscillators, precision mass flow controller, pressure gauges and connections.

3.2. Testing chamber

For initial testing with CO2 gases a chamber and gas control hardware was assembled (Fig. 4 shows schematic diagram of the assembly). This arrangement assured constant dry nitrogen (N2) flow and precise control of CO2 concentration inside of the chamber by precision mass flow controller (MFC GFC17). Overall view of the testing chamber and external connections are shown in fig. 5.

4. Functionalization

For CO_2 sensing we functionalized the surface of the CMUT element with polyethyleneimine (PEI), which is known to be excellent for CO_2 absorption [11]. Prior to the modification of the device the surface of the CMUT was cleaned in isopropanol and dried using nitrogen stream. Then PEI was diluted in ethanol to decrease its viscosity and create adequately thin layer on top of the CMUT structure. The solution of ethanol and PEI was spin coated at 2000 rpm for 30 seconds. The thin layer on top of the device created a shift of the frequency spectra of the resistance and reduced its amplitude at the resonance frequency (fig. 6.). To compensate for this, oscillator was adjusted to get the output signal with adequately lower noise. Finally, two fully assembled CMUT devices, one with modified surface using PEI solution and another with no modification, were placed in a testing chamber. Dry nitrogen gas was filled in the chamber and flow of dry nitrogen was kept constant while various concentrations of CO_2 gases were generated with precision mass flow controller.

Experimental steps used during the experiments:

- Two CMUTs, one functionalized with PEI, are placed inside the chamber for parralel real-time measurements, using non-functionalized CMUT as a reference;
- Chamber is filled with dry N_2 gas and its flow is maintained constant during all the measurement process;
- Waiting for some time while the readings of functionalized CMUT stabilize during desorption of CO_2 absorbed from the ambient air;

- Flow of CO_2 gas is introduced into the chamber, absorption starts causing change of the resonance frequency;
- Resonance frequency is recorded in the real time by 5 samples per second rate;
- Flow of CO_2 is stopped, desorption starts;
- Resonant frequency of the sensor reach initial values.

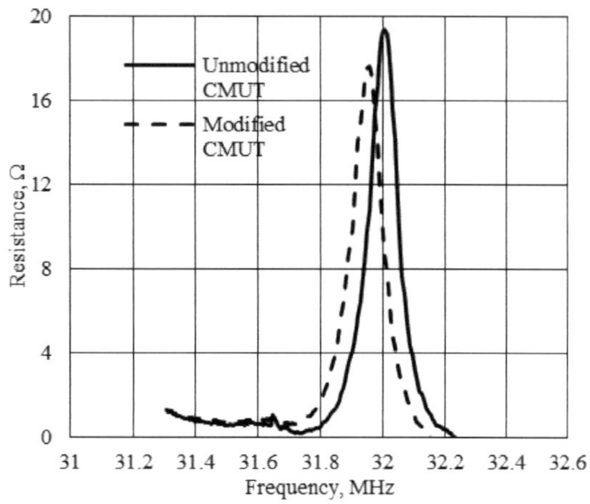

Fig. 6. Resonance frequency before and after modification with PEI.

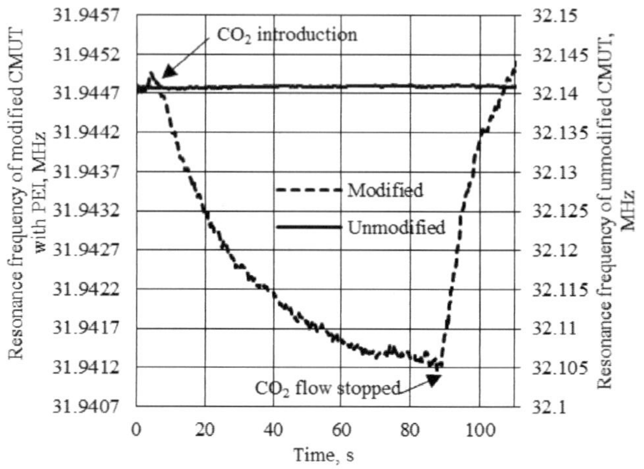

Fig. 7. Resonance frequencies of reference and sensing CMUT.

5. Results

Significant frequency shifts after introduction of CO_2 to the chamber can be observed only for the sensor, which surface was functionalized with PEI. Record of reference (unmodified) and sensing channels is shown in fig. 7. This confirms the viability of selected approach and method of detecting CO_2 gases. By using two of these devices at the same time and calculating the difference of resonant frequency shift between modified and unmodified devices we can remove the noise data coming from temperature changes, changes in pressure inside the chamber and humidity. Sensor readings during different CO_2 concentrations in the chamber are shown in fig 8. The concentration of CO_2 inside a chamber was calculated as a ratio between the CO_2 and N_2 flows measured at the chamber input.

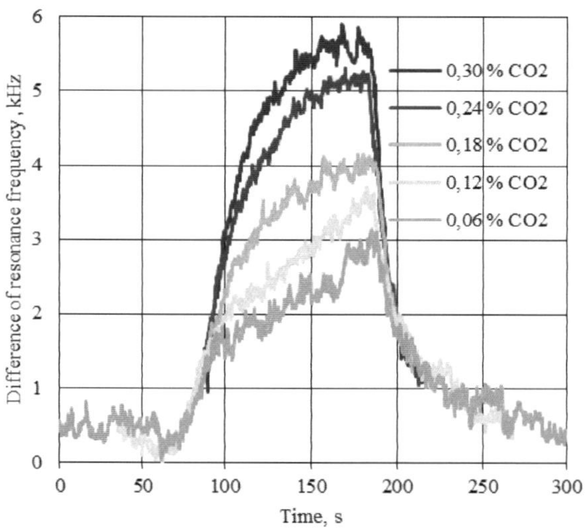

Fig. 8. Difference in resonant frequency of reference and sensing CMUTs during CO_2 gas introduction into the chamber with different flow rates [12].

Conclusions

Results from the experiments show excellent sensitivity of selected CO_2 sensing platform, reaching 4 Hz/ppm (2.5 kHz/600 ppm). This demonstrates that CO_2 gases can be detected using mass loading when sensor is functionalized by CO_2 specific absorber. As expected, absorption of the gas molecules reduces device frequency, thus it can be easily explained by the linear model as an impact of increased moving mass. In addition, by using two of the devices with the same electromechanical properties, one of which is used as a reference, we were able to suppress the noise data coming from changes in temperature, pressure and humidity.

Acknowledgements

We would like to acknowledge Panevėžys Mechatronics Center for providing workspace, testing equipment, measurement equipment, materials and chemicals needed for the experimental work.

References

[1] Khan A, Schaefer D, Tao L, Miller DJ, Sun K, Zondlo MA, Harrison WA, Roscoe B, Lary DJ. Low Power Greenhouse Gas Sensors for Unmanned Aerial Vehicles. *Remote Sensing*; 2012; 4 (12):. 1355-1368.

[2] Howarth RW, Santoro R, Ingraffea A Methane and the greenhouse-gas footprint of natural gas from shale formations. *Climatic Change*; 2011; 106 (4): 679-690.

[3] Barauskas D, Vanagas G, Virzonis D, Makaraviciute A, Ramanaviciene A, Ieee. *Capacitive micromachined ultrasound transducers (CMUT) for resonant gravimetric immunosensing*. New York : IEEE; 2014.

[4] Barauskas D, Vanagas G, Virzonis D, Ramanaviciene A, Makaraviciute A, Ramanavicius A. Application of CMUT as immunosensor, 2013. In: *Ieee International Ultrasonics Symposium (Ius)*; 2013. p. 1708-1711

[5] Kumar KS, Saha S, Pradhan PC, Sarkar SK. Capacitive Micromachined Ultrasonic Transducer Based Gas Sensor Modeling and Simulation. *Applied Mechanics and Materials;* 2011. 110-116, p. 5146-5149

[6] Mikolajūnas M, Kaliasas R, Andrulevičius M, Grigaliūnas V, Baltrušaitis J, Viržonis D. A study of stacked PECVD silicon nitride films used for surface micromachined membranes. *Thin Solid Films;*2008; 516 (23):8788-8792.

[7] Mikolajūnas M, Baltrušaitis J, Kopustinskas V, Vanagas G, Grigaliūnas V, Viržonis D. Plasma etching of virtually stress-free stacked silicon nitride films. *Thin Solid Films*; 2009; 517 (19): 5769-5772

[8] Xuecheng J, Ladabaum I, Khuri-Yakub BT. The microfabrication of capacitive ultrasonic transducers. *Journal of Microelectromechanical Systems;* 1998; 7 (3): 295-302.

[9] Jin XC, Ladabaum I, Khuri-Yakub BT. The microfabrication of capacitive ultrasonic transducers. *International Conference on Solid-SState Sensors and Actuators Chicago, June 16.1997;* 1997; 1: 437-440.

[10] Ladabaum I., Jin X., Soh H.T., Atalar A., Khuri-Yakub B.T. Surface micromachined capacitive ultrasonic transducers. *IEEE Trans Ultrason Ferroelectr Freq Control*; 1998; 45 (3): 678-690

[11] Zhu T, Yang S, Choi DK, Row KH. Adsorption of carbon dioxide using polyethyleneimine modified silica gel. *Korean Journal of Chemical Engineering*; 2010; 27 (6): 1910-1915.

[12] Barauskas D, Pelenis D, Sergalis G, Vanagas G, Mikolajunas M, Virzonis D, Baltrusaitis J. CMUT for high sensitivity greenhouse gas sensing; 2015. p. 1-4

Forum on Innovative Technologies and Management for Sustainability (ITMS'2016)

panevėžys faculty of technologies and business

The 11th International Conference Intelligent Technologies in Logistics and Mechatronics Systems (ITELMS'2016), 28-29 April 2016, Panevėžys, Lithuania

Concept of a high mobility platform with a hydrostatic drive system and active slip control

Bartnicki A.[a], Konopka S.[a], Muszyński T.[a], Spadło K.[a]*

[a]Military University of Technology, gen. Sylwestra Kaliskiego 2, 01-476 Warszawa, Poland

---

Abstract

This paper contains concepts of lightweight unmanned ground vehicle adapted to transport cargoes. It was assumed that platform should be able to moving in urbanized area. This type of area doesn't require developed low MMP indicators of the platform. It was assumed that the platform should be built from universally and commercial available elements. The drive system of the platform is the high efficiency hydrostatic drive system. The platform include active slip control which increasing the mobility on low adhesion coefficient areas.

© 2016 A. Bartnicki, S. Konopka, T. Muszyński, K. Spadło.
Peer-review under responsibility of the Kaunas University of Technology, Panevėžys Faculty of Technologies and Business

Keywords: unmaned ground vechilce; dynamic model; active slip control.

---

1. Introduction

Unmanned ground vehicles are used everywhere where they can support a man in a variety of activities or relieve it. A common type of platform is the logistics platform [4, 6]. Its main task is to transport goods or passengers. It was assumed that the proposed platform should be able to transport about 150kg cargo. This is equivalent to load which can be transported by 4 men. This payload will also provide the ability to transport injured or use platform as manned.

---

* Corresponding author. Tel.: 48 22 261 83 74 16
E-mail address: kacper.spadlo@wat.edu.pl

© 2016 A. Bartnicki, S. Konopka, T. Muszyński, K. Spadło.
Peer-review under responsibility of the Kaunas University of Technology, Panevėžys Faculty of Technologies and Business

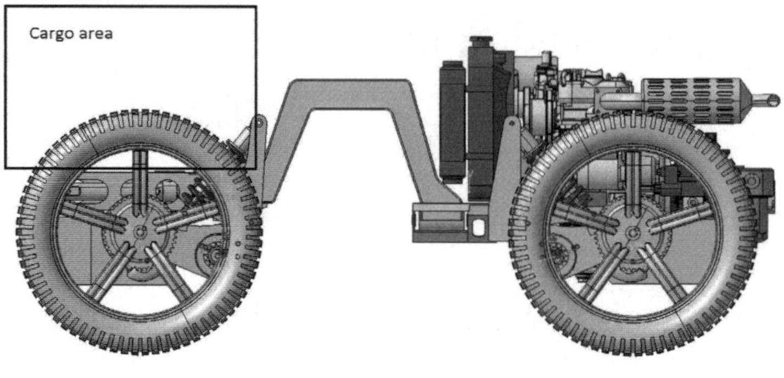

Fig. 1. Arrangement of elements in the platform.

2. Assumptions regarding the model

It was assumed that a vehicle able to provide support to man is one that can transport up to 150 kg of cargo. This is equivalent to the weight of equipment that can be transport by 4 people in rough terrain [8, 11]. A preliminary list of vehicle components - engine, hydraulics pumps, hydraulic motors and actuators, gears, frame, wheels, battery and oil and fuel tank - used in platforms of this type made it enabled to determine the approximate weight of the vehicle as 250 kg. The assumed width was 1 m, ensuing from the width of track necessary for two people to move freely when walking side by side. Since the vehicle is designed to move in urban, a high degree of maneuverability is required. It was assumed that the platform shout have wheel with 600mm diameter. It helps to negotiate obstacles which can be in urban areas [1, 5, 13, 14]. The system ensuring high maneuverability while featuring low steering resistance is an articulated steering system positioned above the front axle. This enables the handling of loads of a relatively large mass while maintaining high maneuverability and stability. It was assumed that the platform have elastic suspension. It was assumed that the front vehicle part would feature a steering system, fuel tank and the cargo space. Engine with equipment, pumps and oil tank would be on the rear platform part. It was assumed that the vehicle would be one with an articulated steering system [2, 7, 9, 10]. Their location is illustrated in Figure 1. Turning of the two vehicle part relative to one another would be achieved by a turning joint positioned above the front axle. Vertical motion of the front part would be provided by a thrust bearing located in rear part of platform.

Arrangement of component will allow the determination of the center of gravity of both parts. On this basis it can be build dynamic model.

2.1. Ability to move in terrain

In order to determine the ability to move in urban area a model of vehicle was built using a software environment for modelling multibody systems [6, 10, 13]. The model was built in MSC Adams software and it is shown in Figure 2.

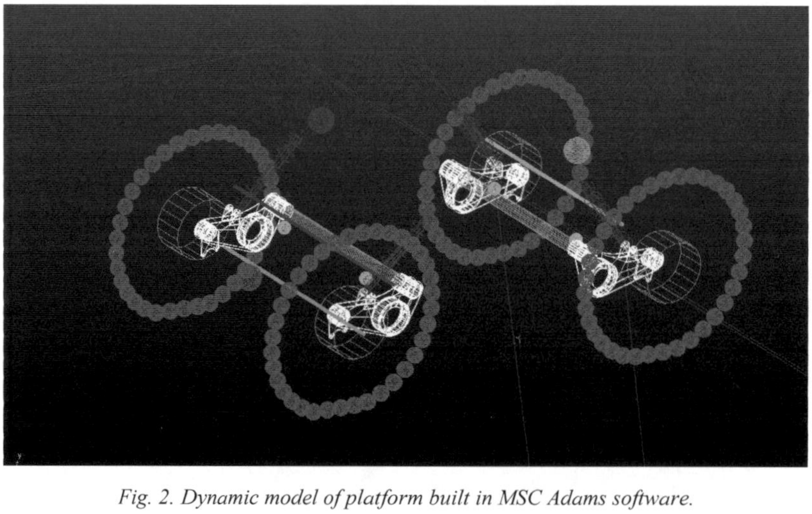

Fig. 2. Dynamic model of platform built in MSC Adams software.

This model featured a flexible tire model that takes into account shaped engagement with the ground and slippage [5, 10, 12]. In was assumed that model waist 250kg without load and 400kg with load. In order to simplify the model, it was assumed that the weights ensuing from the weight distribution of the front and rear part and the weights of loads would be treated as centred on the local centres of gravity.

The type of terrain obstacles adopted as representative were a two types of fog (radius 200 and 300mm), small curb – height 100mm and curb – height 200mm and 300mm height wall. The model overcame through the obstacle after another. The track is shown in figure 3.

Fig. 3. Track with obstacles.

It was assumed that the model will overcome obstacles when driving straight ahead and at an angle 15 °, 30 ° and 45°. The platform was running with speed of 1, 2, 5 and 10km/h. This is equivalent to the speed which develops man. The objective of research was calculation the load in points of mounting wheels. It helps to designed frame and rest of constructive elements in terms of strength. In figure 4 there are example of forces exerted on the front left wheel.

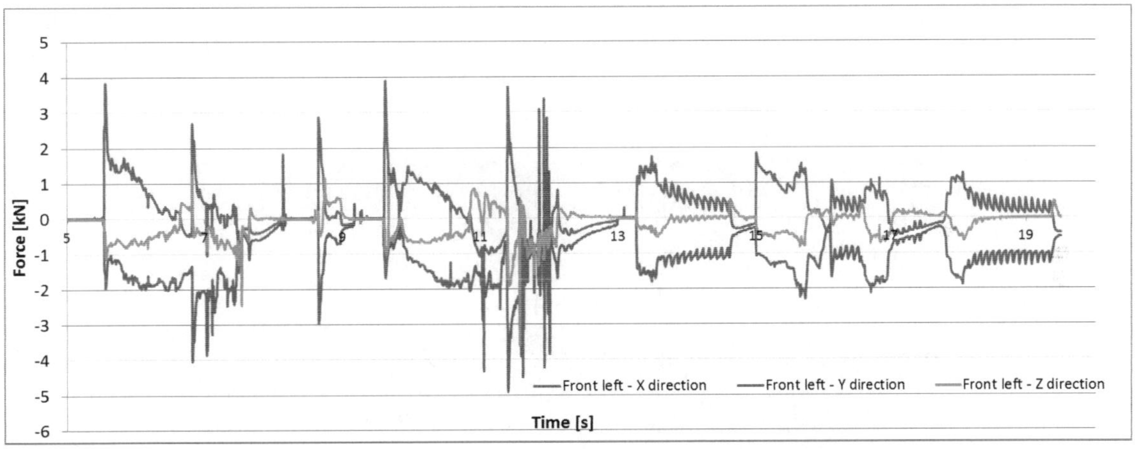

Fig. 4. Forces exerted on front left wheel during negotiate obstacles with speed of 5km/h.

Determined values forces allow the verification of the model frames in a later stage of the design

2.2. Concept of frame design

Accepted arrangement of the basic elements and the position of the bearings allows to design the shape of frame. The steering mechanism was situated in front part. Concept of platform was to build it from commercial parts. For this reason parts like wheels, brakes, suspensions, chain transmissions are used in motorcycles or quads. This allows to lowering construction costs and increases the availability of spare parts. Using articulated steering system positioned above the front axle require to designed high mounted frame.

Fig. 5. Visualization of high mobility platform.

Hydrostatic system is built on high speed, axial piston hydraulic motors. These type of motor is characterize by higher efficiency than rest of type of hydraulic motors. Due to high speed of motors it is necessary to use reduction gear. It was assumed that platform has elastic suspension. For this reason there was used two reduction gears. First is planetary gear and second is chain gear. Both of it is mounted in rocket arm – figure 6.

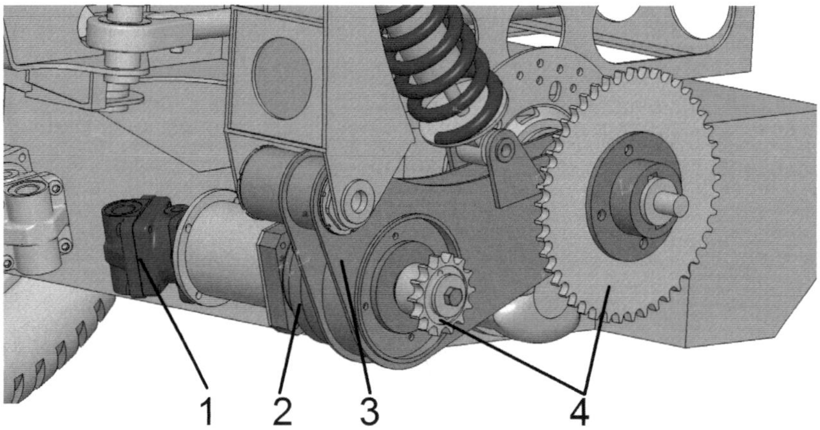

Fig. 6. Components mounted in rocket arm, where: 1 – hydrostatic motor, 2 – planetary gear, 3 – rocket arm, 4 – chain gear.

Proposed configuration allows to mounted fuel tank between rocket arms. Elastic suspension ensure ±80mm stroke.

Developed frame was strength analyzed in MESS module in MSC Catia software. Force values was determinate in chapter 2.1. It was assumed that material of frame is steel 18G2A which characterized yield strength about 335MPa. As permissible stresses it was assumed about 220MPa with is 0,65 of Re.

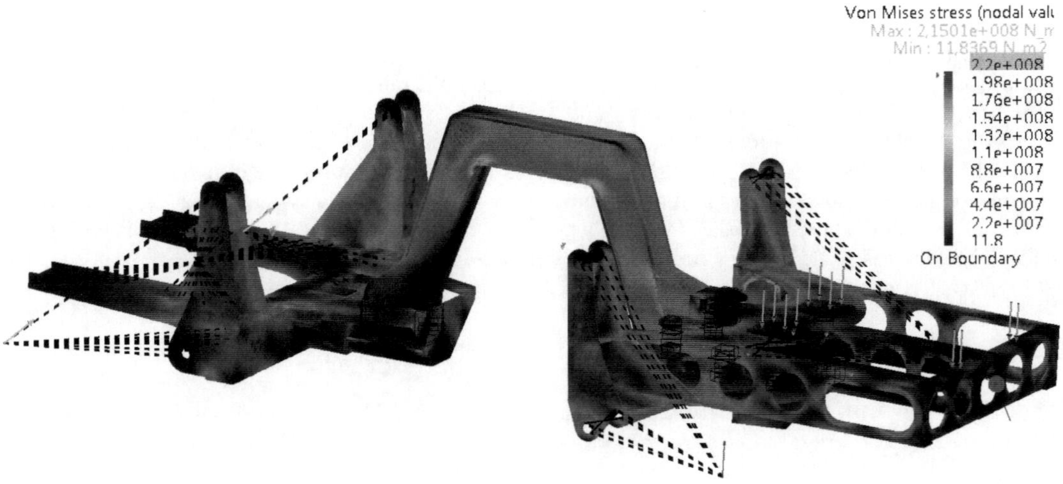

Fig. 7. Example of MESS analysis results.

In the figure 7 there is show example of MESS analysis results. Maximum tension in material is above 215MPa and it is located in bottom part of frame. It was analyzed several types of frames position include position of turning system and longitudinal coupler. In all cases stress do not exceed allowable stress level.

2.3. Active slip control

It was assumed that platform will be equipped with active slip control system. It will be composed of speed sensors mounted in hydrostatic motors, laser scanner, CPU unit and hydraulic brake system. Available analyzes [3, 11, 15, 16, 17,] show that the use of that system significantly improves the ability to overcome obstacles. The idea of system operation was shown in figure 8.

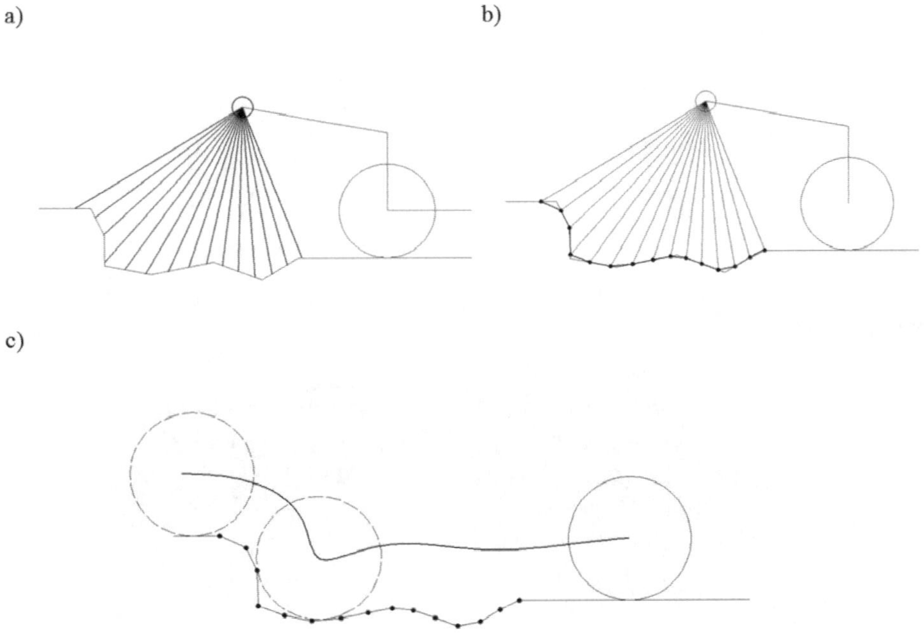

Fig. 8. The idea of active slip control system operation (a) scanning road, (b) creating a road profile, (c) determining the trajectory and speed of the wheel.

Slip control system will operate on the basis of the brake system powered by an electric brake pump. CPU unit will compare real platform's speed to wheel speed. On this basis it determinate slip level. If the slip level will be outside the specified range CPU will lunch brake on required wheel. Conducted analysis show that slip control system improves capacity to overcome obstacles. Example of system working is show in figure 9 and 10. In the first case platform model overcome log 300mm without anti slip system. In that case in 18s front wheel start to slip. Finally according to the operation of the parallel hydraulic system (hydraulic differential) front left wheel rotates 4x faster than start wheel speed. Result is the immobilization of the model – obstacles was not overcome. Figure 10 shows wheel speed of the same mode but with anti slip system. When front wheels starts to slip (about 18s) CPU lunch the brakes for front wheels. As a result front wheel speed decreases. Driving torque on rear wheel increases and finally model overcome obstacle.

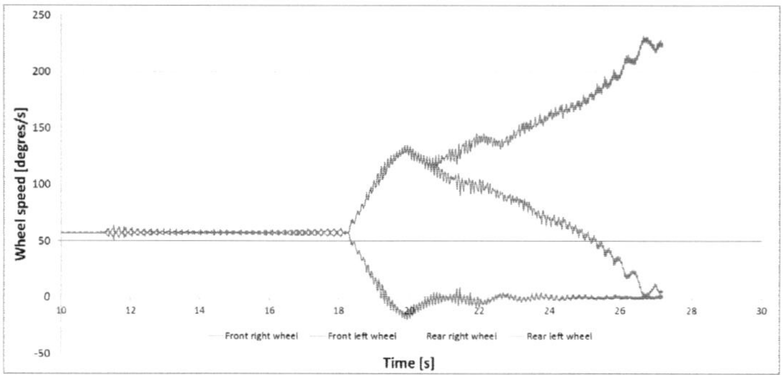

Fig. 9. Wheel speed during overcome log 300mm without anti slip system.

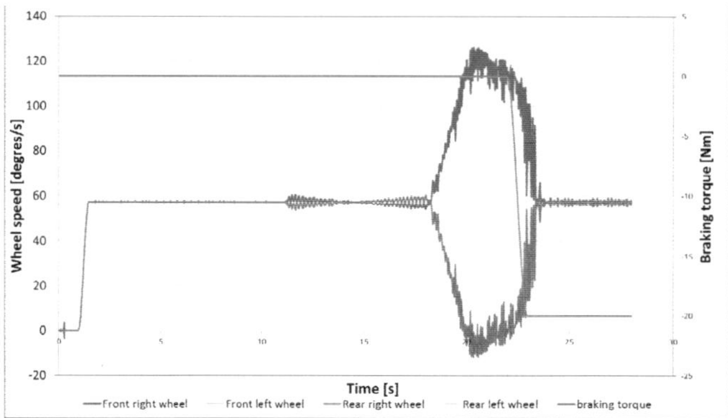

Fig. 10. Wheel speed and braking torque during overcome log 300mm with anti slip system.

2.4. Transport capacity

Designer platform is able to transport up to 150kg of loads. Dimensions of cargo space is 600mm length, 530mm width and 600mm height.

Fig. 11. Visualization of platform with cargo.

The specified load capacity allows to adapt the platform to transport man. It can be high mobility wheelchair.

Fig. 12. Visualization of platform as high mobility wheelchair.

Conclusion

The proposed concept of lightweight high mobility platform with hydrostatic drive system has been realized. The platform is characterized by mas about 250 kg and load capacity of 150 kg. The use of an articulated steering system provides high maneuverability and a small turning radius. Conducted analysis and dynamic simulation allowed to determine the loads acting on the frame and others elements of the platform. This allowed the verification of the frame structure in terms of strength. The use of an active slip control system provides the possibility of movement in the area with low coefficients of adhesion. It helps to overcome obstacles that may be found in urban areas. The dimensions of the cargo space provides the ability to transport all kinds of boxes, backpacks, etc.. Cargo space can be adopted to allow the transport of man co it cause that the platform can be used as a wheelchair high mobility.

References

[1] Andreev AF, Kabanau VI, Vantsevich VV. *Driveline Systems of Ground Vehicles – Theory and Design*. Taylor & Francis Group, USA; 2010.

[2] Bartnicki A, Dabrowska A, Łopatka M.J. Muszyński T. Experimental research on directional stability of articulated tractors. *Mechatronic systems, mechanics and materials II. Solid State Phenomena*, Trans Tech Publications Ltd; vol. 210. p. 77-86

[3] Castillo J, Cabrera J, Guerra A, Simon A. A Novel Electrohydraulic Brake System With Tire-Road Friction Estimation and Continuous Brake Pressure Control. *IEEE Transactions on industrial electronics*, USA; 2016. vol. 63.

[4] Dąbrowska A, Typiak R. Universal Robotic Control Station for a Grouping of Unmanned Engineering Machines. *Proceedings of 8th International Conference on Intelligent Technologies in Logistics and Mechatronics Systems ITELMS 2013*, Kaunas University Technology Press, Kaunas; 2013. p. 71-75

[5] Dabrowska A, Konopka S, Przybysz M. Ability to negotiate terrain obstacles by lightweight six-wheeled unmanned ground vehicles. *Proceedings of 10th International Conference on Intelligent Technologies in Logistics and Mechatronics Systems, ITELMS 2015*, May 2015;. p. 102-109

[6] Dąbrowska A, Jaskółowski M, Krogul P, Spadło K. Mobility evaluation of a lightweight four-wheel unmanned ground vehicle with articulated steering system. *Proceedings of 10th Intelligent Technologies in Logistics and Mechatronics Systems, ITELMS'2015*, Panevezys, Lithuania; 2015.

[7] Haueisen B. *Mobility analysis of small, lightweight robotic vehicles*; April 2003.

[8] Jaskolowski M, Konopka S, Krogul P, Spadlo K. Capabilities to negotiate terrain obstacles by articulated tracked platforms. *Proceedings of 10th International Conference on Intelligent Technologies in Logistics and Mechatronics Systems, ITELMS 2015*, Kaunas; 2015

[9] Konopka S, Krogul P, Łopatka MJ, Muszyński T. Symulacyjna analiza stateczności maszyn przegubowych na przykładzie ładowarki kołowej Ł34. *Przegląd Mechaniczny*, nr 11/2013, str. 13, Warszawa; 2013.

[10] Konopka S, Sprawka P, Muszyński T, Spadło K. Investigating resistance of turn six-wheel skid-steer vehicle. *Proceedings of 8th International Conference ITELMS'2013*; 2013.

[11] Konopka S, Łopatka M, Spadło K. Analiza możliwości poprawy zdolności pokonywania przeszkód przez pojazdy wysokiej mobilności z napędem hydrostatycznym, *Logistyka* 3/2015, Poznań; 2015.

[12] Konopka S, Łopatka MJ, Przybysz M. Kinematic discrepancy of hydrostatic drive of Unmanned Ground Vehicle. *Proceedings of 8th International Conference ITELMS'2013;* 2013

[13] Łopatka M J. *Analiza metod oceny zdolności pokonywania terenu o niskiej nośności*. Biuletyn WAT; 10/2004.

[14] Maclurin B. Comparing the NRMM (VCI), MMP and VLCI traction models. *Journal of Terramechanics* 2007; 44: 43-51

[15] Miller J, Cebon D. Tyre curve estimation in slip-controlled braking. *Proceedings of the institution of mechanical engineers part d-journal of engineering*; vol. 230, 2016.

[16] Moorehead S, Shamah B, Teza J, Wagner MD, Wettergreen D, Whittaker W. Steering and Control of Passively Articulated Robot. *SPIE. Sensor Fusion and Decentralized Control in Robotic Systems IV*; October 2001.
[17] Xu G, Xu K, Zheng C, Zahid T. *Optimal Operation Point Detection Based on Force Transmitting Behavior for Wheel SlipPrevention of Electric Vehicles. IEEE Transactions on intelligent transportation systems*, vol. 17, USA; 2016.

Forum on Innovative Technologies and Management for Sustainability (ITMS'2016)

panevėžys faculty of technologies and business

The 11th International Conference Intelligent Technologies in Logistics and Mechatronics Systems (ITELMS'2016), 28-29 April 2016, Panevėžys, Lithuania

Currently Concept of Risk and Application

Breznicka A.[a], Stodola P.[b], Stodola J.[b]*

[a] Alexander Dubcek University of Trencin, Pri parku Str.19, 911 06 Trenin, Slovak Republic
[b] Jiri Stodola, University of Defence Brno, Kounicova Str. 65, 662 10 Brno,Czech Republic

Abstract

The article deals with up-to-date concept of risk. Its advantages are discussed and perspectives of development of this concept are shown. The article presents some examples of its application in mechanics. When the fundamental notion of risk is used in engineering applications, it is natural to feel a desire to incorporate this risk indicator into e. g. material losses, damage to human health or life loss as a result of an emergency been assessed in a material equivalent.

© 2016 A. Breznicka, P. Stodola, J. Stodola.
Peer-review under responsibility of the Kaunas University of Technology, Panevėžys Faculty of Technologies and Business

Keywords: Risk; probability; safety; analysis.

1. Introduction

The notion is always subjective because the assessment of the identical material losses in different countries for instance will undoubtedly be substantially different. However one can risk in life and not only avoid material losses but on the contrary obtain significant material profit. Hence, it is clear that an objective risk assessment should not include material losses directly. It certainly can and should reveal a certain risk with them. The concept of risk disclosed below is a pioneering effort, as far as we know, it attempts to separate the notion of risk proper from material losses or benefits it causes.

2. Approaches and Criterions to Risk

When the practice of operating technical objects serving essential purposes demanded a quantitative assessment of risk, it was proposed to interpret safety as reliability in respect to human life and health and the condition of environment [1]. Thus, a real possibility was assured to perform numerical analyses of risk and safety using the known methods of the theory of reliability, the theory of probability and mathematical statistics. In fact, assume $S(t)$ is a safety function determined as a probability that an emergency does not occur within the time t, i. e. no failure of system with severe consequences occurs. Then the function of risk is addition to the function of safety to a unity, and the following applies

$$H(t) = 1 - S(t) \tag{2.1}$$

since the condition of safety and risk are treated antagonistic. It is apparent that the function of safety is

* Corresponding author. Tel.: +420 973 442 278; fax: +420 973 442 284.
E-mail address: jiri.stodola@unob.cz

equivalent to the probability of failure-free operation and the function of risk is identical to a probability of failure[2]. Than the rate of risk is determined similarly to the rate of failures andthe following applies

$$h(t) = \frac{H«(t)}{(1 - H«(t))} \approx -S«(t) \tag{2.2}$$

The advantages and significance of the disclosed approach that have gained wide recognition and achieved much progress, specifically when applied to technical systems, are well known and are not discussed here. The basic postulates are:

1. Risk is a significantly broader and deeper notion than probability. Hence, the idea of risk (2.1) as identical to the notion of probability can be considered insufficiently universal.
2. Since risk is a broader and deeper notion than probability, should it be acknowledged that the range of numerical variations of the magnitude is also much broader than the range

$$0 \le H(t) = P(t \le t_0) \le 1 . \tag{2.3}$$

3. The analysis of probability (2.1) is applicable exclusively to random events; also, if risk calculation is some fundamental phenomenon cannot fit into an extremely limited range (2.3), it implies that the risk indicator should be a real number.
4. If the idea of randomness of unfavourable events is entirely excluded from the notion of risk, we will face an obvious contradiction with reality. This contradiction is removed easily: we will assume that the risk indicator has a causal-consequential relation with the probability of occurrence of same or other random event.
5. The nature of risk in our analysis is double yet integral: though it is due to random events it is described by real numbers.

If the safety analysis assesses a probability that the system fails, the consequences are determined in the sense of damage to equipment and the consequences to humans around. In this case H_r (failures/time) is the function of risk; r_s (losses/time) is the risk and it is calculated as a product of the frequency v_r ; and I_c (losses/failures) is times the average rate of failures, i. e.

$$H_r \cdot I_c = r_s \tag{2.4}$$

The practical significance of the function the risk is in Fig. 1[3] as dependencies of v_r on I_c for traffic accidents per year (region 1 is all vehicles, region 2 is defective vehicles). If v_r is measured by the number of accidents per year and I_c represents the average number of fatal outcomes per accident, the risk$r_s$has the dimensionality of mortality per year. The data in Fig. 1 permits to draw the following conclusions:

- grave accidents happen rarer than milder accidents;
- the major share of risk is due to driving a faulty vehicle;
- in numbers the gravity of accidents distributes irrespectively whether the accidents are due to a faulty vehicle or to a driver´s fault.

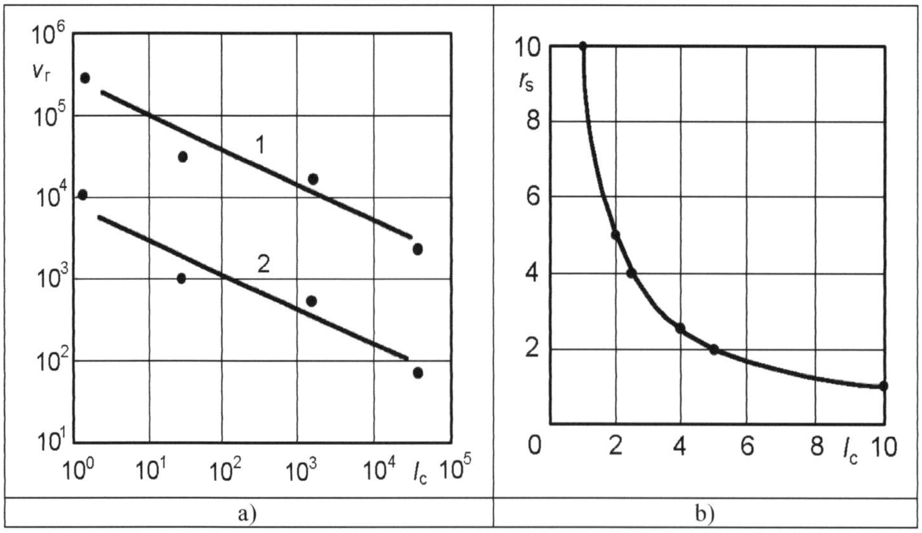

Fig. 1 Relation between frequency v_r and gravity I_c of traffic accidents per year a) and curve of identical risks b).

The possibility of obtaining such information is highly important, but it should be remarked that the aftermath of the accident is assessable more or less accurately only after it happens. The value r_s according to (2.4) is not probabilistic, hence the numerical values of risk go far beyond range (2.3), however remaining essentially positive. Though the idea of the r_s risk has definitely gained broad and useful application, it manifests a number of well-known drawbacks that should be analysed in brief:

1. According to (2.4) the risk due to a great number of minor failures may be equal to the risk of a single accident with grave consequences. It means that the case in (2.4) represents a measure of danger rather than a measure of risk or if to be more exact some intermingling of these two notions. In fact, danger rated subjectively should be ambiguous.
2. According to (2.4) there is a multitude of identical risks that do not imply identical consequences. The multitude of ambiguous assessments of the r_s risk hinders extremely the perception of this specific phenomenon.
3. Having assessment of (2.4) it is impossible to rank objectively the risk, at least, for instance, by the extent of tolerance. It can of course be done subjectively from this or that view of the danger of this or that risk in the sense of damage it entails; yet material losses are calculated my multiple variants.
4. Note that indicator (2.4) does not permit its improvement or development. The reason is that it is the indicator-individualist, it is considered self-sufficient; no relations have been traced with other different indicators of a variety of conditions of objects.

In order to avoid many shortcomings of risk determination, the idea of the risk proper should be separated from material aftermath of the accident [4]. Of course, it should be supplemented with a possibility of determination of risk relation to material losses when it is necessary and wise. The analysis of strength of materials, elements, systems, etc. of structures is based, as a rule, on an accepted criterion of ultimate state. From a great number of such known criteria we select just one because it becomes acceptable to describe the destruction of very unlike objects under essentially unlike condition.

3. Foundations of Risk

In relation to technical objects the idea of risk appears every time when any damage is detected in operation. The sooner the damage approaches a critical value the nearer the emergency situation is and correspondingly the stronger the riskis. In our view the risk induces the expectation of damage in objects, system processes; in generalized sense it is an expectation of any unfavourable phenomena, events or situations, whether social or natural. For example, if $P(A)$ is a probability of an unfavourable event A, then$Q(B)$ is the probability of occurrence of an opposite positive event B, then the risk indicator

$$\rho = P(A)/Q(B) \tag{3.1}$$

is determined by a simple relation between the probabilities of the above events and expresses the share of "bad" (P(A)) in "good" (Q(A)). Because, according to the theory of probabilities, the sum of probabilities of opposite events is equal to a unity, i. e.

$$P(A)+Q(B)=1, \tag{3.2}$$

the risk indicator (3.1) can be determined with the allowance for (3.2) either only by the probability of unfavourable events

$$\rho = (k_P - 1)^{-1} \text{ and } k_P = (P(A))^{-1}, \tag{3.3}$$

or only by the probability of favourable events

$$\rho = k_Q - 1 \text{ and } k_Q = (Q(B))^{-1}. \tag{3.4}$$

The risk interpreted in (3.1) implies no material damage. It can relate to one if the probability $P(A_1)$ of occurrence of a concrete unfavourable event A_1. The interpretedequation (3.1) does not contain any threat of life, health or the environment. It can relate to such threat if the probability $P(A_2)$ of occurrence of a concrete unfavourable situation A_2 entails relevant events. When there are no unfavourable events, or situations, or phenomena, then $P(A)=0$ and according to (3.1) the risk indicator $\rho = 0$. On the contrary, if there are no favourable events then $Q(B)=0$ and the risk is $\rho = \infty$. Hence, the numerical values of the risk indicator can vary within the interval

$$0 \le \rho \le \infty. \tag{3.5}$$

4. Operative Risk Characteristic

The dependence of the risk indicator ρ on the probabilities $P(A)$ and $Q(B)$ is termed by the operative risk characteristic. The Fig. 2presents it graphically for a limited interval

$$0 \le \rho \le 1 \ . \tag{4.1}$$

The theoretical data to plot the operative risk characteristic within interval (4.1) are listed in Tab. 1[3]. The lower border of risk $\rho = 0$ corresponds to the case when $P(A) = 0$. The upper analysed risk value in Fig. 2[3]and Tab. 1 when $\rho_k = 1$ is limited by the condition $P(\overline{A}) = Q(\overline{B}) = 0.5$.

Table 1 Data for plotting operative risk characteristic.

$P(A)$	$Q(B) = 1 - P(A)$	$\rho = P(A)/Q(B)$	$P(A)$	$Q(B) = 1 - P(A)$	$\rho = P(A)/Q(B)$
0.01	0.99	0.0101	0.60	0.40	0.6666
0.05	0.95	0.0526	0.70	0.30	0.4286
0.1	0.90	0.1111	0.80	0.20	0.2500
0.2	0.80	0.2500	0.90	0.10	0.1111
0.3	0.70	0.4286	0.95	0.05	0,0526
0.4	0.60	0.6666	0.99	0.01	0.101
$P(\overline{A}) = 0.5 = Q(\overline{B})$	0.5	$\rho_k = 1$	$Q(B)$	$P(A) = 1 - Q(B)$	$\rho = P(A)/Q(B)$

The value $\rho_k = 1$ is considered critical. The numerical value of the risk indicator can change by several orders of magnitude; the scale ρ of the operative characteristics of often assumed logarithmic, see Fig. 2a [4]. The current risk value indicates the operative characteristic by a full vertical line AB. The probability of favourable and unfavourable events relates to risk as

$$Q(B) = \frac{1}{1+\rho} \text{ and} \tag{4.2}$$

$$P(A) = \frac{1}{1+1/\rho}. \tag{4.3}$$

In principle, a risk normative value [ρ] can be established. It is justified by the relevant feasibility calculation with the account of an aftermath of occurrence of some unfavourable events. The risk normative value [ρ] is indicated by a vertical dashed line AB along the risk characteristic, see Fig. 2.

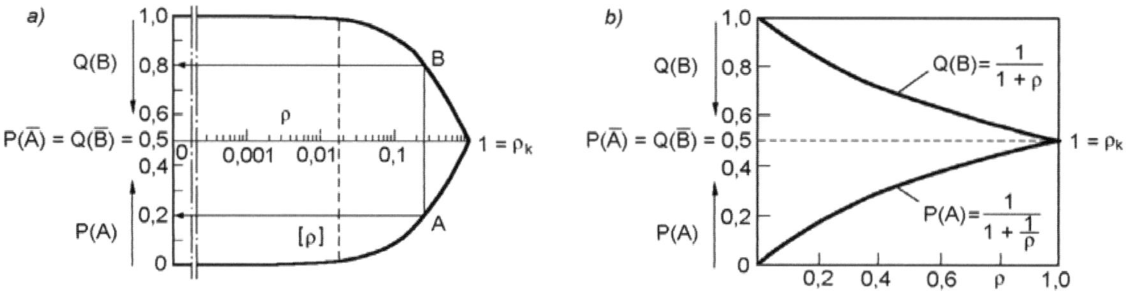

Fig. 2. Operative risk characteristic plotted in logarithmic (a) and regular (b) scale.

Any event or any situation for which

$$\rho \le [\rho], \tag{4.4}$$

is assumed acceptable or tolerable. The values

$$[\rho] < \rho < \rho_k \tag{4.5}$$

are assumed unacceptable or intolerable. When it is necessary to perform a comparative analysis of risk in many different events, a sector risk diagram is plotted. In the sector diagram: the critical risk $\rho_k = 1$ corresponds to the external circumference and the risk normative value [ρ] is reflected by the internal circumference.

5. Risk and Safety

It is established above the risk in an anticipation of any unfavourable phenomenon, event, or situation A_1, A_2,..., A_n. If the probability P of such events is

$$P(A_i) = P(A_1, A_2,..., A_n) \equiv P \tag{5.1}$$

and the probability Q of the opposite favourable events $B_1, B_2,..., B_n$ is

$$Q(B_i) = P(B_1, B_2,..., B_n) \equiv Q \ , \tag{5.2}$$

the risk determined by the share of a bad in goods is

$$0 \le \rho(A_i, B_i) = \frac{P(A_i)}{Q(B_i)} \Lambda(A_i) \le \infty \ , \tag{5.3}$$

providing a peculiar law of preservation of oppositions in the form

$$P(A_i) + Q(B_i) = const = 1 \ , \tag{5.4}$$

it applies that neither $P(A_i)$ nor $Q(B_i)$ disappear but interact transforming into one another

$$P(A_i) \Leftrightarrow P(B_i) \tag{5.5}$$

It means, for example, that in case the probability $P(A_i)$ grows the probability $Q(B_i)$ may not grow or remain constant; it diminishes following law (5.4). Thus, formulas (5.3) and (5.4) outline the procedure of risk analysis tor the case of a multitude of events A_i and B_i that dictate it in fact. The probabilities $P(A_i)$ and $P(B_i)$ are calculated by the methods of the theory of sets. To make the discussion simpler the indexes i are omitted from the calculation of the events A_i and B_i. Graphically interlinked variations of the probabilities $P(A_i)$ and $P(B_i)$ caused by $\rho(A,B)$ are depicted by the full operative risk characteristic. Unlike (3.2), the formula (5.3) contains a function $\Lambda(A_1)$ of function $A_i \Leftrightarrow A_j$ of events A_i and A_j, the numerical values can be

$$\Lambda(A_i \Leftrightarrow A_j) \le \text{or} \ge 1 . \tag{5.6}$$

If $\Lambda < 1$, the events $A_i \Leftrightarrow A_j$ interact so that the risk (5.3) diminishes. If $\Lambda > 1$,the events $A \Leftrightarrow B$ interact according to the equation (5.3), and conversely the risk increases. If $\Lambda = 1$, it means that the risk is independent of the processes of interaction between events A and B; formula (5.3) in this case transforms into (2.1). Assume that safety is an opposition to risk. Then

$$S_\rho + \rho = 1, \tag{5.7}$$

where S_ρ is the indicator of safety liking to the idea of risk in this relation (3.2) [5]. Considering that the idea of risk in (5.3) is broader than in (3.2), obtain

$$S_\rho + \frac{P(A)}{Q(B)} \Lambda(A_i) = 1 . \tag{5.8}$$

Tab. 2 [4] shows us the lists of quantitative parameters of risk prediction, hence the analysis of the conditions of safety, because the relation (5.7) combines the indicators of risk and safety together.

Table 2. Risk and safety analysis.

Risk condition	ρ	$P(A_i)$	$Q(B_i)$	S_p	Safety situations
Zero risk	0	0	1	1	Absolute safety
Scantly risk	$0 < \rho < 1$	$0 < P < 0.5$	$0.5 < Q < 1$	$1 > s_p > 0$	Scanty safety
Critical risk	$1 = \rho_k$	0.5	0.5	0	Zero safety
Transcritical risk	$1 < \rho < \infty$	$0.5 < P < 1$	$0 < Q < 0.5$	$0 > S_p > -\infty$	Negative safety
Infinite risk	∞	1	0	∞	Absolute danger

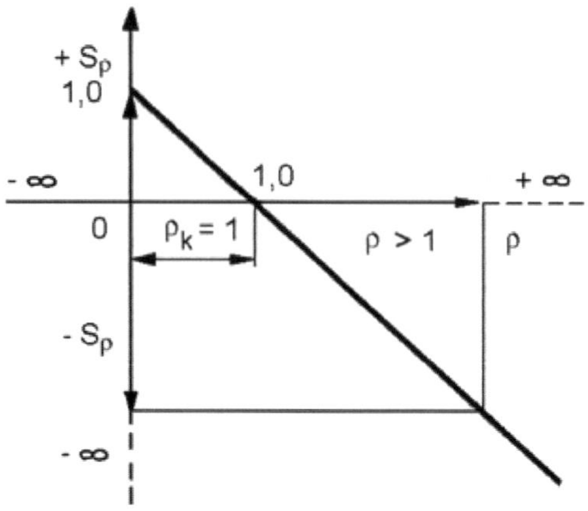

Fig. 3. Diagram of risk and safety.

According to this relation, if the risk diminishes, the safety augments correspondingly and vice versa. Yet, during any variations of the values S_ρ and ρ, their sum remains unchanged and equal to a utility [6]. Assuming for analysis of the interval specified in (5.3), it is established from (5.7) that the safety indicator can have both positive (at $\rho < 1$) and negative ($\rho > 1$) values [7]. According to the data in Tab. 2 and Fig.3[3]:

1. the absolute safety is identical to the zero risk; the safety is zero when the risk reaches a critical value $\rho_k = 1$;
2. if the risk $\rho_k > 1$, the safety becomes negative, i.e. a situation emerges that can be characterized as dangerous; the danger grows because the transcritical risk value augments;
3. the danger is absolute when the risk becomes infinitely large that corresponds to the safety $S_\rho = -\infty$.

Thus the risk manifests two oppositions, they are safety and danger[7]; the danger being interpreted as negative because the safety proper is positive; hence, in our view, danger is a peculiar opposition to safety; it is a negative form of safety[8]. The main conclusions are as follows:

1. Only undamaged objects ($\omega_\Sigma = 0$) are absolutely safe, naturally there is no risk operating them, it means $S_\rho = 1$ and $\rho = 0$. Damaged objects ($\omega_\Sigma < 1$) have typically a scanty safety, their risk of operation is within the interval $0 < \rho < 1$; any higher risk reduces their safety correspondingly $(1 > s_\rho > 0)$.
2. When the risk is critical $(\rho_k = 1)$ an object enters into the limiting state characterized by the limiting damage level $(\omega_c = 1)$; it is a break-down situation with a typically zero safety; negative safety is the danger without any guaranteed safety. A catastrophe is possible when a system reaches a relevant translimiting state ($\omega_\Sigma >> 1$) for which the risk is transcriticaltoo $(\rho >> 1)$ [9].
3. Absolute danger $(S_\rho = -\infty)$ is identical to the infinite risk $(\rho = \infty)$ [10]; apparently cataclysms occur in such conditions when objects are decomposed.

Conclusion

This article discloses just basic relation to risk and safety concept being developed. Both the risk indicator ρ and the interrelated safety indicator S_ρ do not include directly the material losses, they are alienated. Yet, an analytical procedure is advanced how the ρ and material losses interrelate. Finally, basic principles of designing of active systems of modern machinery and equipment have been recently using the criteria of risk and safety. In our view, it is specifically important to be able to analyze risk in translimiting or transcritical states of systems. From the available information the concept of risk and safety pose a now problem in the first approximation.

References

[1] Bolotin V. *Prediction of Life of Machines and Structures*. Moscow; 1984.

[2] Zio E, Sansavini G, Maja G, Marchionni G. Analysis of the Safety Efficiency of Road Network: a Real Case Study. *1st Summer Safety and Reliability Seminars;* 2007. p. 401 - 406

[3] Risk Control: Risk. Self-Sustainable Development, Synergy. In *Cybernetics: Unlimited Opportunities and Possible Restrictions*. Moscow, 2000.
[4] Henly AJ. *Reliability of Technical Systems*. Moscow; 1984.
[5] Zhmailik V, Sosnowskiy L. New Concept of Risk and its Technical Applications. *Proceedings of VI. International Symposium on Tribo-Fatigue*. Minsk 2010; 2: 15 – 24.
[6] Vintr Z, Koucky M, Valis D. Contribution to Modeling of Complex System Reliability. *9th International Conference PSAM.* Hong Kong; 2008.
[7] LiW. *Risk Assessment of Power Systems : Models, Methods and Applications*, Wiley-IEEE Pres; 2014.
[8] Flammine F. Railway Safety, Reliability and Security: Technologies and Systems Engineering. *IEEE Computer Society*; 2014.
[9] Breznická A, Chovanec A. Mathematic and simulation modelling for analysis prediction of risk, *Reliability: Theory & applications 2015;* 10 (4) (39):.29-34.
[10] Sinay J. et al .*Risk of technical equipment. Risk management*. OTA, Košice; 1997.

Forum on Innovative Technologies and Management for Sustainability (ITMS'2016)

panevėžys faculty of technologies and business

1922

The 11th International Conference Intelligent Technologies in Logistics and Mechatronics Systems (ITELMS'2016), 28-29 April 2016, Panevėžys, Lithuania

Inverse kinematics for robotic motion platform applied on an interdisciplinary performance involving dance, music and robotics

Česnulevičius A.[a], Sinkevičius V.[a], Urbanavičiūtė L.[a,c]*, Urbonas J.[b], Vaičiulis D.[a], Vileišis E.[d]

[a]Kaunas University of Technology, Nemuno str. 33, 35212 Panevėžys, Lithuania
[b]Vilnius Academy of Art, Maironio str. 6, 01124 Vilnius, Lithuania
[c]Panevėžys College, Laisves Sq. 23, 35200 Panevėžys, Lithuania
[d]UAB PROFIBUS, Kniaudiškių str. 12-3, LT-37104 Panevėžys, Lithuania

---

Abstract

This article is about the research of the robotic motion platform to be used for composing and experiencing kinetic journeys engaging the whole body. The motion platform must move the seats mounted on the platform in six degrees of freedom (6-DOF) that can be experienced by anybody free to move in space. These are three rotations Pitch, Roll and Yaw, and three linear movements Heave (up and down), Sway (side to side) and Surge (fore and aft). The jack layout used is generally that of the so-called Stewart platform. The analysis of parallel robot design and kinematics parameters will be presented in this paper in order to create the best option for such platform. The originality of this research is in floating tool centre point (TCP) of the platform. Calculating the kinematics parameters of the platform its location and orientation depends on the person sitting on the platform.

© 2016 A. Česnulevičius, V. Sinkevičius, L. Urbanavičiūtė, J. Urbonas, D. Vaičiulis, E. Vileišis.
Peer-review under responsibility of the Kaunas University of Technology, Panevėžys Faculty of Technologies and Business

Keywords: parallel robot; robotic platform; kinematics; floating tool center point (TCP); Stewart platform.

---

1. Introduction

The research key element, a motion platform, will be applied on an interdisciplinary performance involving dance, music and robotics. It is similar to the one used in advanced vehicle simulators, is capable of simulating wide range of movements: both found in real vehicles such as aircraft, ship or military tank and those produced by some

Nomenclature

$\boldsymbol{a}_{A\,(i)}$	linear acceleration of point A with respect to reference frame (i)
H_0	the minimal distance between base and platform when they are parallel
$\boldsymbol{l}_{j\,(i)}$	j^{th} robot leg position vector with respect to reference frame (i)
$\dot{l}_j$	j^{th} robot leg linear velocity
$\ddot{l}_j$	j^{th} robot leg acceleration

* Corresponding author. Tel.: +370-685-21564.
E-mail address: lina.urbanaviciute@ktu.lt

© 2016 A. Česnulevičius, V. Sinkevičius, L. Urbanavičiūtė, J. Urbonas, D. Vaičiulis, E. Vileišis. **T428F0006** . MEDIMOND
Peer-review under responsibility of the Kaunas University of Technology, Panevėžys Faculty of Technologies and Business

$\boldsymbol{r}_{AB\,(i)}$	position vector $\overrightarrow{AB}$ with respect to reference frame (i)
$\boldsymbol{R}_{(i,j)}$	rotation matrix from reference frame (i) to reference frame (j)
$\boldsymbol{R}(\boldsymbol{\theta}_{j\,(i)})$	rotation matrix when rotation angles vector $\theta_{j\,(i)}$ is given
$\boldsymbol{v}_{A\,(i)}$	linear velocity of point A with respect to reference frame (i)
$\boldsymbol{\varepsilon}_{j\,(i)}$	angular acceleration of jth frame (body) with respect to reference frame (i)
$\boldsymbol{\theta}_{j\,(i)}$	j^{th} frame (body) Euler rotation angles θ_x, θ_y, and θ_z vector with respect to reference frame (i)
$\boldsymbol{\omega}_{j\,(i)}$	angular velocity of jth frame (body) with respect to reference frame (i)

sort of fictitious vehicles from computer games or movies. But this platforms not concerned with the simulation of 'real' movements, it shifts its creative attention to rather the 'unreal', the hybrid or the imagined. In the performance, the public will be invited to sit strapped to the device, and ride it, while experiencing 'choreographed' multi-sensorial compositions of movements, sounds and dance imagery (Fig. 1). Part a kinetic sculpture, part an amusement ride, part a vehicle simulator, and even part a medical disorientation device, this platform will employ the experiences of dancers, acrobats, pilots, astronauts, transporting the 'riders' to unique aesthetic realms where dance and music penetrates the innards, and spectators are confused with performers. Thus the motion base will also serve as a collaborative discursive platform for contemplating and sharing diverse knowledge from a wide field of subjects such as motion aesthetics, spectatorship, technologies and arts, and bodily knowledge.

There are two main robot constructions: serial and parallel robots. The first one is modelled as a sequence of links connected by either sliding or hinged joints forming a serial chain, while the second is a closed-loop kinematics chain mechanism made up of a moving platform and a fixed base, linked together by several independent kinematics chains. For realization of this project a parallel robot was selected. The selection was conditioned by requirements for the platform and better characteristics of parallel robots in comparison to serial robots.

Dr. E. Gough established the priciples of a mechanism with a closed-loop kinematic chains in 1947 [1] and D. Stewart suggested the use of such structure for flight simulator in 1965 [2]. However, the first parallel robot was invented by James E. Gwinnett, who applied for a patent in 1928 [3]. J. E. Gwinnett suggested a platform for use in movie theaters.

The advantages of parallel robots [4]: high load-carrying capacity; the ratio of the mass of the payload over the mass of the robot can be larger than 10; very good positioning accuracy; low inertia; higher structural stiffness; they are almost insensitive to scaling; can be built using almost any type of actuator or transmission, for example, wire transmissions can be used. The drawbacks of parallel robots: small workspace; the singularities that can appear within the latter.

Today there are many application fields of parallel robots [4], such as medicine, industry, motion simulators, space exploration, vibration suppression or generation and also the art. For instance, the Pendulum Choir [5] shown in Fig. 2. This is a nine-person choir, where each singer stands on a tilting platform. The aim of the project was the research of the relationship between music and motion.

In comparison with platforms used for Pendulum Choir this project requires platform which can possess general spatial motions, i.e. three translation motions (Heave (up and down), Sway (side to side) and Surge (fore and aft)) and three rotation motions (Pitch, Roll and Yaw), plus additional continuous rotation around axis perpendicular to moving platform. In pursuance of generation of new emotional senses it is important to have the platform which could be programmed to rotate around any axis of rotation, i.e. the TCP of the platform should be easily relocated as it is done, for example, in human balance and postural researches [6, 7].

Fig. 1. The idea of the project.

Fig. 2. Pendulum Choir [5].

2. Requirements for the robot platform

The motion platform will be moved by six hydraulic jacks (as the majority of advanced motion platforms) that can move the seats mounted on the platform in any of the six degrees of freedom (6-DOF) that can be experienced by anybody free to move in space. Each leg is connected to the base and to the moving platform with universal joints at both ends, referred to 6-UPU. The letters U and P refer to type of joints or type of kinematics pairs [8]. U means universal joint and P means prismatic joint. The jack layout used is generally that of the so-called Stewart platform and on which the seats are mounted. The linear actuators will be used for the robot control. Other requirements for the platform are listed in Table 1.

Table 1. Requirements for the robot platform.

Parameter	Value
Lifting power	150 kg
Base / platform diameters	1500 / 1000 mm
Linear displacements: vertical (*have*) / horizontally (*surge, sway*)	+ 400 mm / ± 250 mm
Angular displacements: around a vertical Z-axis (*yaw*) / longitudinal turn around Y-axis (*pitch*) / lateral turn around X-axis (*roll*)	± 45° / ± 30° / ± 30°
Vertical (*have*) linear speed / Angular speed	0.005 – 0.6 m/s / 0.5 – 40 °/s
Maximum linear acceleration / Angular acceleration	9.8 m/s^2 ($\geqslant$ 1*g*) / 250 °/s^2

3. Inverse kinematics of parallel robot with floating tool centre point

Solving the inverse kinematics of the parallel robot the linear actuators displacements (robot legs lengths), velocities and accelerations will be determined. The analysis of works, related with inverse kinematics of parallel robots showed that TCP usually is attached to moving platform origin, i.e. points T and P are coincident (Fig. 3b) as well as axes of coordinate systems attached to these two points. There is impossible directly apply such solution if we want to get rotation around the free selected axis. Therefore, application the solutions of other authors is limited. The solving of inverse kinematics problem, when TCP could take any position and orientation, is presented in this work.

The solution can be found when $\boldsymbol{r}_{\mathrm{BT\,(B)}}$, $\boldsymbol{\theta}_{\mathrm{T\,(B)}}$, $\boldsymbol{v}_{\mathrm{T\,(B)}}$, $\boldsymbol{\omega}_{\mathrm{T\,(B)}}$, $\boldsymbol{a}_{\mathrm{T\,(B)}}$, $\boldsymbol{\varepsilon}_{\mathrm{T\,(B)}}$, $\boldsymbol{r}_{\mathrm{BB}i\,\mathrm{(B)}}$, $\boldsymbol{r}_{\mathrm{PP}i\,\mathrm{(P)}}$, $\boldsymbol{r}_{\mathrm{PT\,(P)}}$ and $\boldsymbol{\theta}_{\mathrm{T\,(P)}}$ are known; here $i = 1, 2, \ldots, 6$ is the number of robot leg.

Rotation matrix from reference frame P to reference frame B is expressed as:

$$\boldsymbol{R}_{(\mathrm{T,B})} = \boldsymbol{R}_{(\mathrm{P,B})}\,\boldsymbol{R}_{(\mathrm{T,P})} \Rightarrow \boldsymbol{R}_{(\mathrm{T,B})}\,\mathrm{inv}(\boldsymbol{R}_{(\mathrm{T,P})}) = \boldsymbol{R}_{(\mathrm{P,B})}\,\boldsymbol{R}_{(\mathrm{T,P})}\,\mathrm{inv}(\boldsymbol{R}_{(\mathrm{T,P})}) \;\Rightarrow \boldsymbol{R}_{(\mathrm{T,B})}\,\mathrm{inv}(\boldsymbol{R}_{(\mathrm{T,P})}) = \boldsymbol{R}_{(\mathrm{P,B})}; \tag{1}$$

here $\mathrm{inv}(\boldsymbol{R}_{(\mathrm{T,P})})$ – inverse matrix of $\boldsymbol{R}_{(\mathrm{T,P})}$.

Eq. (1) can be expressed as:

$$\boldsymbol{R}_{(\mathrm{P,B})} = \boldsymbol{R}(\boldsymbol{\theta}_{\mathrm{T\,(B)}})\,\mathrm{inv}(\boldsymbol{R}(-\boldsymbol{\theta}_{\mathrm{T\,(P)}})) = \boldsymbol{R}(\boldsymbol{\theta}_{\mathrm{T\,(B)}})\,\boldsymbol{R}(\boldsymbol{\theta}_{\mathrm{T\,(P)}}); \tag{2}$$

here $\boldsymbol{R}(\dot{\mathbf{e}}) = \begin{bmatrix} \mathrm{c}\theta_y\,\mathrm{c}\theta_z & \mathrm{c}\theta_z\,\mathrm{s}\theta_x\,\mathrm{s}\theta_y - \mathrm{c}\theta_x\,\mathrm{s}\theta_z & \mathrm{c}\theta_x\,\mathrm{c}\theta_z\,\mathrm{s}\theta_y + \mathrm{s}\theta_x\,\mathrm{s}\theta_z \\ \mathrm{c}\theta_y\,\mathrm{s}\theta_z & \mathrm{s}\theta_x\,\mathrm{s}\theta_y\,\mathrm{s}\theta_z + \mathrm{c}\theta_x\,\mathrm{c}\theta_z & \mathrm{c}\theta_x\,\mathrm{s}\theta_y\,\mathrm{s}\theta_z - \mathrm{c}\theta_z\,\mathrm{s}\theta_x \\ -\mathrm{s}\theta_y & \mathrm{c}\theta_y\,\mathrm{s}\theta_x & \mathrm{c}\theta_x\,\mathrm{c}\theta_y \end{bmatrix}$; $\mathrm{c}\theta \equiv \cos\theta$, $\mathrm{s}\theta \equiv \sin\theta$.

Location of TCP (point T) may be expressed as follows (Fig. 3b):

$$\begin{aligned} \boldsymbol{r}_{\mathrm{BT\,(B)}} &= \boldsymbol{r}_{\mathrm{BB}_i\,\mathrm{(B)}} + \boldsymbol{r}_{\mathrm{B}_i\mathrm{P}_i\,\mathrm{(B)}} + \boldsymbol{r}_{\mathrm{P}_i\mathrm{P}\,\mathrm{(B)}} + \boldsymbol{r}_{\mathrm{PT\,(B)}} = \boldsymbol{r}_{\mathrm{BB}_i\,\mathrm{(B)}} + \boldsymbol{l}_{i\,\mathrm{(B)}} - \boldsymbol{r}_{\mathrm{PP}_i\,\mathrm{(B)}} + \boldsymbol{r}_{\mathrm{PT\,(B)}} = \\ &= \boldsymbol{r}_{\mathrm{BB}_i\,\mathrm{(B)}} + \boldsymbol{l}_{i\,\mathrm{(B)}} + \boldsymbol{R}_{(\mathrm{P,B})}\left(\boldsymbol{r}_{\mathrm{PT\,(P)}} - \boldsymbol{r}_{\mathrm{PP}_i\,\mathrm{(P)}}\right) \end{aligned} \tag{3}$$

The length of each leg in reference frame B could be computed using Eq. (3):

$$\boldsymbol{l}_{i\,\mathrm{(B)}} = \boldsymbol{r}_{\mathrm{BT\,(B)}} - \boldsymbol{r}_{\mathrm{BB}_i\,\mathrm{(B)}} - \boldsymbol{R}_{(\mathrm{P,B})}\left(\boldsymbol{r}_{\mathrm{PT\,(P)}} - \boldsymbol{r}_{\mathrm{PP}_i\,\mathrm{(P)}}\right) = \boldsymbol{r}_{\mathrm{BT\,(B)}} - \boldsymbol{r}_{\mathrm{BB}_i\,\mathrm{(B)}} - \boldsymbol{r}_{\mathrm{P}_i\mathrm{T\,(B)}} = \boldsymbol{r}_{\mathrm{BT\,(B)}} - \boldsymbol{r}_{\mathrm{BB}_i\,\mathrm{(B)}} + \boldsymbol{r}_{\mathrm{TP}_i\,\mathrm{(B)}}\,. \tag{4}$$

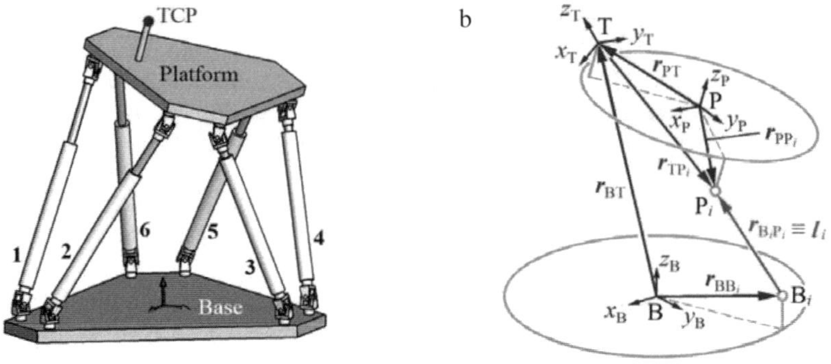

Fig. 3. Parallel robot: a – general view of robot with numbered legs; b –calculation scheme of the robot's i^{th} leg parameters.

To calculate the linear velocity and acceleration of the robot's legs first necessary to find the points P_i velocities and accelerations. Therefore, the velocity of the point P_i is:

$$\begin{aligned} \boldsymbol{v}_{P_i\,(B)} &= \boldsymbol{v}_{T\,(B)} + \quad {}_{T\,(B)} \times \boldsymbol{r}_{TP_i\,(B)} + \boldsymbol{v}_{P_i\,relative\,(B)} = \boldsymbol{v}_{T\,(B)} + \quad {}_{T\,(B)} \times \left(\boldsymbol{R}_{(P,B)}\,\boldsymbol{r}_{TP_i\,(P)}\right) = \\ &= \boldsymbol{v}_{T\,(B)} + \quad {}_{T\,(B)} \times \left(\boldsymbol{R}_{(P,B)}\left(\boldsymbol{r}_{PP_i\,(P)} - \boldsymbol{r}_{PT\,(P)}\right)\right) \end{aligned} \tag{5}$$

here $\boldsymbol{v}_{Pi\,relative} = 0$, because it is considered that points T and P_i don't move according to each other. Then linear velocities of robot legs are:

$$\dot{l}_i = \boldsymbol{v}_{P_i\,(B)} \circ \frac{\boldsymbol{l}_{i\,(B)}}{\left\|\boldsymbol{l}_{i\,(B)}\right\|}; \tag{6}$$

here the symbol " ∘ " means vectors dot product.

The acceleration of the point P_i is:

$$\boldsymbol{a}_{P_i\,(B)} = \boldsymbol{a}_{T\,(B)} + \mathring{\boldsymbol{a}}_{T\,(B)} \times \boldsymbol{r}_{TP_i\,(B)} + \dot{\boldsymbol{u}}_{T\,(B)} \times \left(\dot{\boldsymbol{u}}_{T\,(B)} \times \boldsymbol{r}_{TP_i\,(B)}\right) + \boldsymbol{a}_{P_i\,relative\,(B)} + 2\,\dot{\boldsymbol{u}}_{T\,(B)} \times \boldsymbol{v}_{P_i\,relative\,(B)}. \tag{7}$$

Taking into account that $\boldsymbol{v}_{Pi\,relative} = 0$ and $\boldsymbol{a}_{Pi\,relative} = 0$, we obtain:

$$\boldsymbol{a}_{P_i\,(B)} = \boldsymbol{a}_{T\,(B)} + \mathring{\boldsymbol{a}}_{T\,(B)} \times \left(\boldsymbol{R}_{(P,B)}\left(\boldsymbol{r}_{PP_i\,(P)} - \boldsymbol{r}_{PT\,(P)}\right)\right) + \dot{\boldsymbol{u}}_{T\,(B)} \times \left(\dot{\boldsymbol{u}}_{T\,(B)} \times \left(\boldsymbol{R}_{(P,B)}\left(\boldsymbol{r}_{PP_i\,(P)} - \boldsymbol{r}_{PT\,(P)}\right)\right)\right). \tag{8}$$

Then linear accelerations of robot legs are:

$$\ddot{l}_i = \boldsymbol{a}_{P_i\,(B)} \circ \frac{\boldsymbol{l}_{i\,(B)}}{\left\|\boldsymbol{l}_{i\,(B)}\right\|}. \tag{9}$$

The geometric parameters of robot model, listed in Table 2, were used for validation of given method. Let say that TCP moves from point $\{0, 0, 1100\}^T$ to point $\{100, 180, 1350\}^T$. It takes 0.6 seconds. Motion is linear and vectors of velocity and acceleration are parallel to trajectory of TCP. Platform rotates around all axes. First, using inverse kinematics solving method described above, linear velocity of all robot legs are calculated. The latter data are used for solving the direct displacement, velocity and acceleration kinematics problem using SolidWorks Motion programme. The comparison of calculations are displayed in Fig. 4. The TCP displacements (rotation) vary less than 14 %, velocities – less than 8 % and accelerations – less than 12 % (Fig. 4).

Table 2. Geometric parameters of robot (rounded up to 1 mm accuracy), mm.

$\boldsymbol{r}_{BB_1\,(B)}$	$\boldsymbol{r}_{BB_2\,(B)}$	$\boldsymbol{r}_{BB_3\,(B)}$	$\boldsymbol{r}_{BB_4\,(B)}$	$\boldsymbol{r}_{BB_5\,(B)}$	$\boldsymbol{r}_{BB_6\,(B)}$	$\boldsymbol{r}_{PP_1\,(P)}$	$\boldsymbol{r}_{PP_2\,(P)}$	$\boldsymbol{r}_{PP_3\,(P)}$	$\boldsymbol{r}_{PP_4\,(P)}$	$\boldsymbol{r}_{PP_5\,(P)}$	$\boldsymbol{r}_{PP_6\,(P)}$	$\boldsymbol{r}_{PT\,(P)}$
$\{739, -130, 70\}$	$\{739, -130, 70\}$	$\{-257, 705, 70\}$	$\{-482, 575, 70\}$	$\{-482, -575, 70\}$	$\{-257, -705, 70\}$	$\{354, -354, -70\}$	$\{354, 354, -70\}$	$\{129, 483, -70\}$	$\{483, 129, -70\}$	$\{-483, -129, -70\}$	$\{129, -483, -70\}$	$\{-150, -100, 200\}$

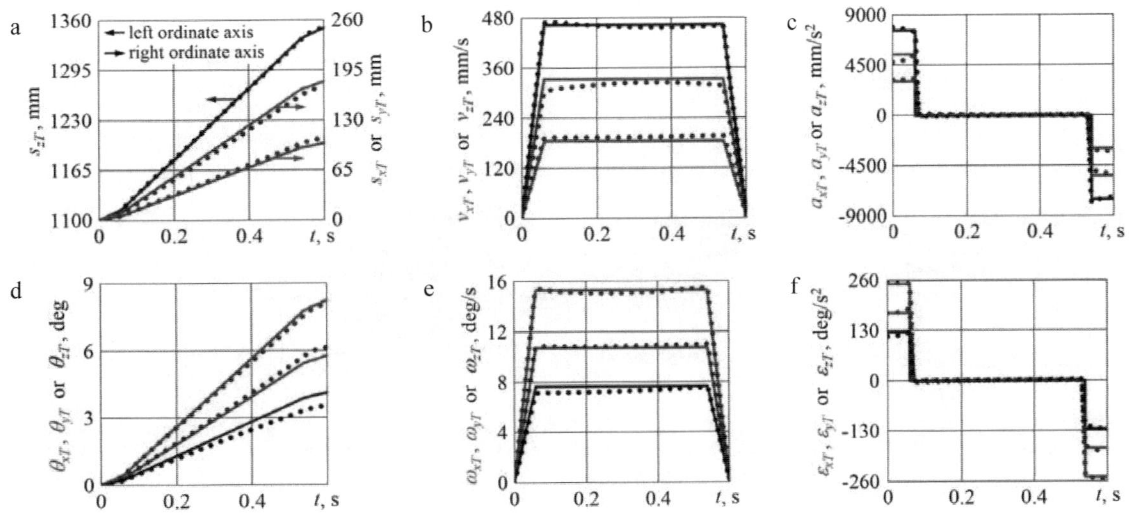

Fig. 4. Kinematics parameters of TCP, when it moves from point $\{0, 0, 1100\}^T$ to point $\{100, 180, 1350\}^T$ in 0.6 seconds: (———) – original values; (● ● ●) – values calculated solving direct kinematics problem with SolidWorks Motion, when linear velocities of robot legs are calculated using method presented in this section; a – displacements; b – linear velocities; c – linear accelerations; d – angles of rotation; e – angular velocities; f – angular accelerations.

4. Research of robot legs parameters

For the purpose of kinematics analysis, the strokes of hydraulic cylinders should be known. We do premise that one leg of the robot is loaded with 1500 N force; the length of retracted hydraulic cylinder is ≈ *Stroke* + (200 … 350) mm; a step of stroke variation is 50 mm; base and platform are parallel when TCP moves in effective workspace (Table 1); $\boldsymbol{r}_{PT\,(P)} = \{0, 0, 0\}^T$; $\boldsymbol{\theta}_{T\,(P)} = \{0, 0, 0\}^T$. Fig. 5 shows dependency of robot leg length and hydraulic cylinder stroke upon minimum distance between robot base and platform H_0. Hydraulic cylinder should satisfy these conditions:

Length of retracted cylinder≤Calculated minimal length of robot leg

Maximum length of hydraulic cylinder≥Calculated maximumlength of robot leg

The parameters of hydraulic cylinders suitable for robot with chosen configuration should be selected from the green zone (Fig. 5). As we can see the stroke of hydraulic cylinder should be 650 mm and minimal distance H_0 between base and platform when they are parallel can vary from ≈ 1080 mm to ≈ 1130 mm. For further research the following parameters were chosen: minimum distance between base and platform – 1100 mm; length of retracted cylinder– 950 mm. The TCP, when base and platform are parallel, will reach any point of workspace (Table 1) if leg length vary from 950 mm to 1600 mm. Increasing of robot legs is not rational decision because it leads to increase of robot mass and height and decrease of legs stiffness.

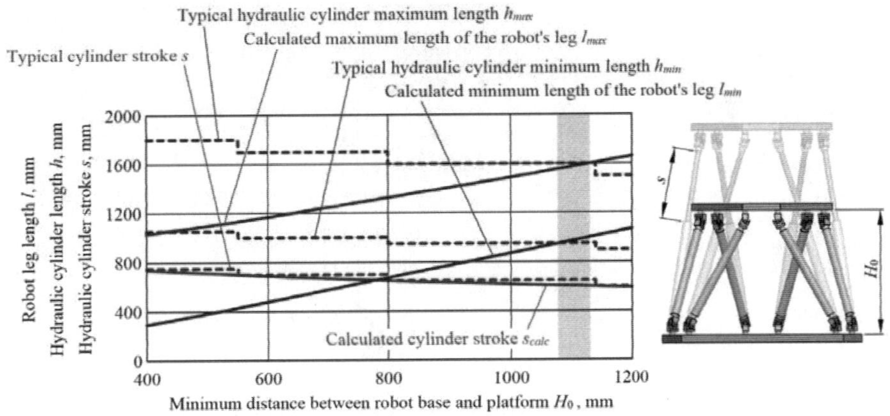

Fig. 5. Dependency of robot leg length and hydraulic cylinder stroke upon minimum distance between robot base and platform when base and platform are parallel.

Fig. 6 shows the workspaces of robot when platform rotates around different axes of TCP. The results of calculations shows that simultaneous rotation of platform around all three axes is impossible. There are two possible solutions for this problem: selection of new cylinders with longer strokes or additional degree of freedom –continuous rotation around z_T axis.

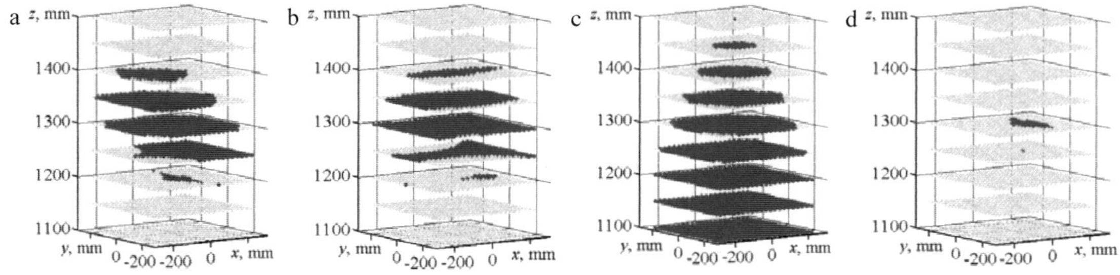

Fig. 6. Workspaces of robot when: a – platform rotates ±30° only around x_T axis of TCP; b – platform rotates ±30° only around y_T axis of TCP; c – platform rotates ±45° only around z_T axis of TCP; d – platform rotates ±30° simultaneously around x_T and y_T axes of TCP.

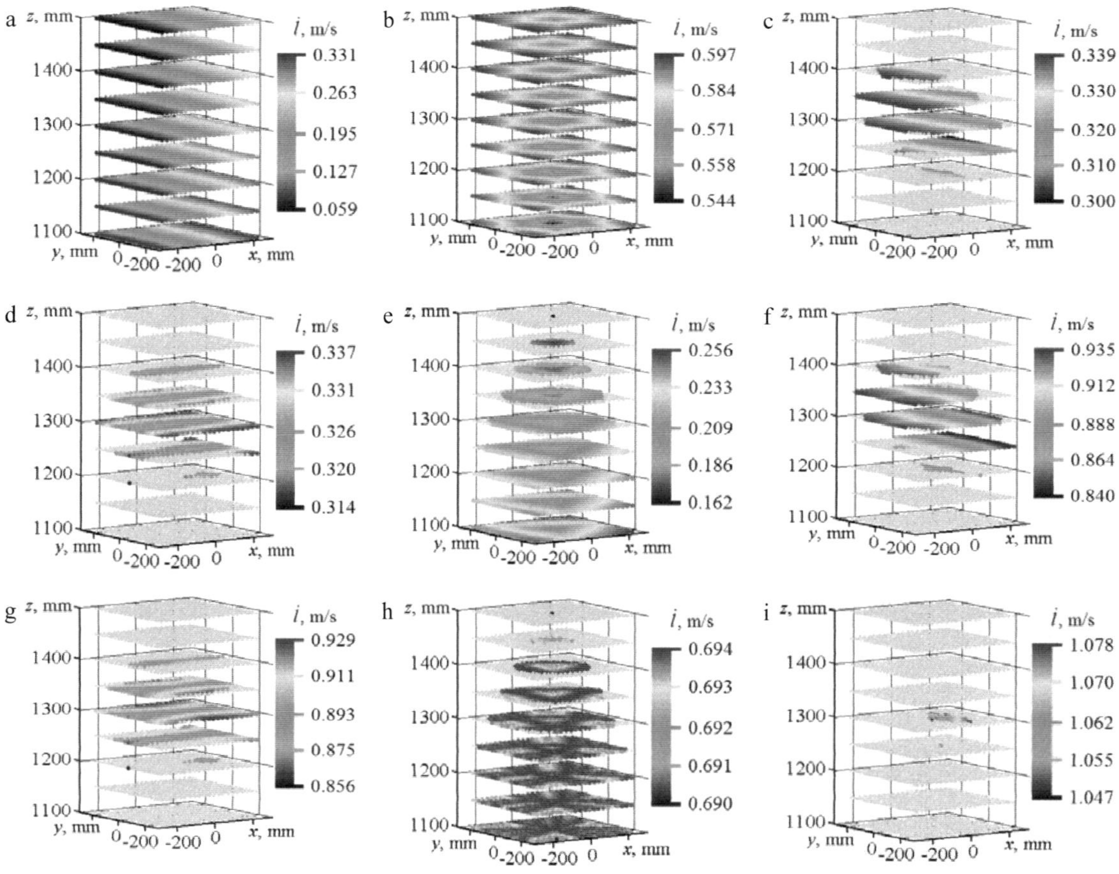

Fig. 7. Maximum velocities of the legs: a – platform moves with 0.6 m/s velocity along x_T axis without rotation; b – platform moves with 0.6 m/s velocity along z_T axis without rotation; c – platform rotates ±30° with 40°/s angular velocity around x_T axis of TCP; d – platform rotates ±30° with 40°/s angular velocity around y_T axis of TCP; e – platform rotates±45° with 40°/s angular velocity around z_T axis of TCP; f – platform moves with 0.6 m/s velocity along z_T axis and rotates ±30° with 40°/s angular velocity around x_T axis of TCP; g – platform moves with 0.6 m/s velocity along z_T axis and rotates ±30° with 40°/s angular velocity around y_T axis of TCP; h – platform moves with 0.6 m/s velocity along z_T axis and rotates ±45° with 40°/s angular velocity around z_T axis of TCP; i –platform moves with 0,6 m/s velocity along z_T axis and rotates ±30° with 40°/s angular velocity around x_T and y_T axes of TCP.

The analysis of velocities presented in Fig. 7 showed that the maximum linear velocity of robot leg is reached when platform moves along z_t axis which is almost twice larger than linear velocity of robot leg when platform only rotates and velocity exceeds 1 m/s in case of synchronous rotation and linear movement along z_t axis.

Conclusions

- The minimal required stroke of hydraulic cylinder taking into account the workspace of TCP and considering that movement of the platform is linear and axes of base and platform coordinate systems are parallel is equal to 650 mm.
- The workspace of robot, when stroke of hydraulic cylinder is equal to 650 mm, decreases in case of platform rotation. Taking into account the selected lengths and strokes of hydraulic cylinders the simultaneous rotation of platform with maximum angular displacements around all three TCP axes is impossible. Therefore, additional degree of freedom – continuous rotation around z_t axis is required.
- The maximum linear velocity of robot leg when platform moves without rotation is equal to 0.6 m/s and in case of synchronous rotation and linear movement of platform the velocity exceeds 1 m/s.

Acknowledgment

This research was made possible under grant No. MIP-098/2013 "Research and development of an interdisciplinary performance involving dance, music and robotics" from Agency for Science, Innovation and Technology (MITA).

References

[1] Gough VE, Whitehall SG. Universal tyre test machine. *Proceedings of the FISITA Ninth International Technical Congress*; May; 1962. p. 117-137

[2] Stewart D. A. platform with six degrees of freedom. *Proceedings of the Institution Mechanical Engineers*; Vol. 180, No. 1,1965. p. 371–386

[3] Gwinnett JE. Amusement devices. *US Patent No. 1,789,680*, January 20; 1931.

[4] Patel YD, George PM. Parallel Manipulators Applications - A Survey. *Modern Mechanical Engineering*; Vol. 2, 2012. p. 57-64

[5] *Cod. Act – Pendulum Choir*. http://codact.ch/gb/pendugb.html.

[6] Barton GJ, Vanrenterghem J, Lees A, Lake M. A method for manipulating a movable platform's axes of rotation: *A novel use of the CAREN system. Gait & Posture*; Vol. 24, 2006. p. 510-514

[7] Commissaris DACM, Nieuwenhuijzen PHJA, Overeem S, de Vos A, Duysens JEJ, Bloem BR. Dynamic posturography using a new movable multidirectional platform driven by gravity. *Journal of Neuroscience Methods*; Vol. 113, 2002. p. 73-84.

[8] Shen H, Yang T, Ma L. Synthesis and structure analysis of kinematic structures of 6-dof parallel robotic mechanisms. *Mechanism and Machine Theory*; Vol. 40, 2005. p. 1164-1180

Forum on Innovative Technologies and Management for Sustainability (ITMS'2016)

panevėžys faculty of technologies and business

The 11th International Conference Intelligent Technologies in Logistics and Mechatronics Systems (ITELMS'2016), 28-29 April 2016, Panevėžys, Lithuania

Strain Field Evolution on the Surface of Stainless Sheet Steel 12Cr17 Exposed to a Specific Impact with Oscillation Loading

Chausov M.[a], Hutsaylyuk V.[b]*, Sniezek L.[b], Pylypenko A.[a], Sobchak A.[c]

[a]Department of Mechanics, Technical Education and Research Institute, National University of Life and Environmental Sciences of Ukraine, 12 Geroiiv Oborony, Kyiv, 03041, Ukraine
[b]Institute Design Machine, Faculty of Mechanical Engineering, Military University of Technology, 2 Gen. S Kaliskiego str., Warsaw, 00-908, Poland
[c]Department of Structural Mechanics, Faculty of Mechanical Engineering, Ternopil Ivan Pul'uj National Technical University,56 Ruska str., Ternopil, 46001, Ukraine

Abstract

Strain field evolution on the surface of stainless sheet steel 12Cr17 has been analyzed during static deformation and during a specific high-speed and oscillation loading-dynamical non-equilibrium process. Experiments have been conducted by using an original mechanical testing technique and specially developed software solution for non-contact study of strain field, using digital image correlation fitted with a high-speed camera. It has been established that deformation process kinetics for tested stainless sheet steel after a dynamical non-equilibrium process during further static deformation changes as compared with a process of static deformation mainly manifests itself in significant delay in process "neck formation". It was found that relaxation processes do make influence on kinetics of process deformation steel after dynamic non-equilibrium process.

© 2016 M. Chausov, V. Hutsaylyuk, L. Śnieżek, A. Pylypenko, A. Sobchak.
Peer-review under responsibility of the Kaunas University of Technology, Panevėžys Faculty of Technologies and Business

Keywords: strain field; dynamics, impact testing; digital correlation (2D); stainless sheet steel 12Cr17; dynamic non-equilibrium process (DNP).

1. Introduction

In our previous works it has been shown that shock-vibration load of plastic materials in various stages prior to static tension, which the authors call a dynamic non-equilibrium process (DNP) that leads to a significant changes in the materials structure and in turn leads to changes of mechanical properties with a further static tension [1-4].

The primary question is, what triggers a new deformation mechanism associated with the structural change? It is impossible to answer this question without additional detailed physical tests. On the other hand, another prospective research line aimed at understanding the physical processes, that take place in material during DNP may be connected with the evolution of strain field, evolution on the surface of ductile sheet materials.

It has been established that deformation process evolution for aluminum alloys D16ChATV and 2024-T3 after DNP changes significantly as compared with a standard process of static tension. This phenomenon refers to the formation of a new specific structure as a result of DNP effect [5].

Under the static loading on stainless steels usually as a rule the uniform, monotonous process of static deformation is being registered, which is not accompanied by wave-like change in shape of the strain field. This

* Corresponding author. Tel +48 22 261 83-92-45; fax: +48 261-837-366.
E-mail address: volodymyr.hutsaylyuk@wat.edu.pl

deformation behavior of stainless steels differs significantly from that of the aluminum alloys. Therefore is would be a very interesting to investigate deformation of the material features of this class directly at the realization of DNP and also the effect of this short-term process on a repeated static tension of the steel.

The aim of this study is the investigation of strain field evolution on the surface of stainless sheet 12Cr17 during DNP by using digital image correlation (DIC) to follow the dynamic strain distributions with a high-speed Phantom v711 camera (frame rate up to 1,400,000).

2. Test procedure and material

A new type of mechanical testing for DNP implementation was developed by Professor M. Chausov et al. and has been applied in this study and described in various works [6, 7].

The study was conducted on a modified hydraulic installation for static tests ZD-100Pu. Modification of this installation was carried out to obtain a complex mode loading type of static tension-high-frequency, high-oscillatory process (dynamic non-equilibrium process).

A modified version of the installation consists of two circuits-external (load frame of the testing machine) and internal see Fig. 1. The internal circuit is a simple statically indeterminate mechanical system in the form of three parallel elements, that are loaded simultaneously-the central specimen and two symmetrical specimens satellite ("brittle samples") of a different cross-section, made of hardened steel 65G or U8-U12. When implementing the loading of this installation the satellites specimens are destroyed (under given loads or deformation) and impulse input of load energy into the material sample is being performed.

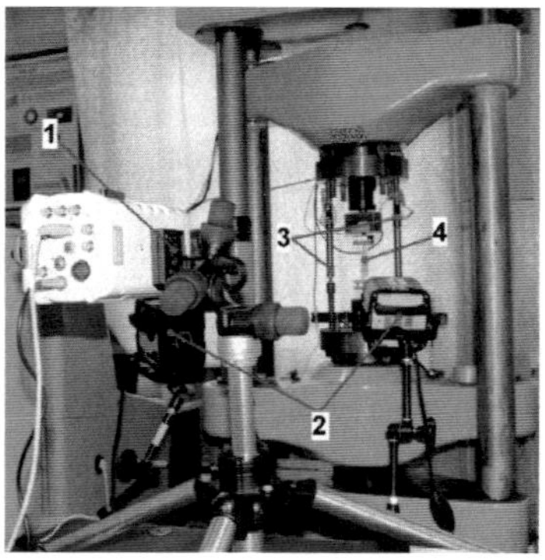

Fig. 1. A test installation with equipment to determine the kinetic field deformation at DNP: 1 – high speed camera Phantom v711; 2 – highly stable halogen source of light; 3 – bars with tension meters; 4 – testing specimen.

As it has been already noted, the essence of DNP is in a high-speed tensile material with overlay onto it of oscillatory process with a high frequency (several kilohertz). This mode of loading is achieved due to free vibrations of mechanical system and excess of pressure in the hydraulic system. Thus, this process can be controlled by changing the ratio of stiffness/mass of outer contour system and size inertia displacement of servo hydro cylinder, which is defined by failure loading of brittle specimens.

During the experiments effort on the specimen and brittle samples is registered by the independent extensometers.

During the process of static tensile strain and realization of DNP deformation on the specimens flat surfaces deformation recorded by the standard extensometer based on 11 mm, and by the correlation method of digital images over all working field of specimens by using specially developed software system and high-speed cameras.

However, when testing specimens with high speed deformation (DNP realization in steel) registration of the deformation field carried out by high-speed camera Phantom v711 with registering a speed of 15,000 f/s to 680 000 f/s. (Maximum registration speed 1400000 f/s), Fig. 1.

At the time of static tests of specimens a registration of deformation field was carried out with assistance of Casio Exilim Pro EX-F1 camera (maximum speed registration 1200 f/s).

Table. 1 shows the mechanical properties of the investigated steel.

Table 1. Mechanical properties of stainless steel 12Cr17

Material	Yield stress, MPa	Ultimate stress, MPa	δ, %	ψ*,%
12Cr17	280	455	60	31
* based on of cross deformation on the surface while maintaining the volume				

Samples of stainless steel 12Cr17 were made by laser cutting.

The flat form of specimens was specially selected for the testing, since it allows to capture deformation field by using only one camera, in this case the resulting deformation field can be considered two-dimensional. The form of a sample for the experiments is shown on Fig. 2.

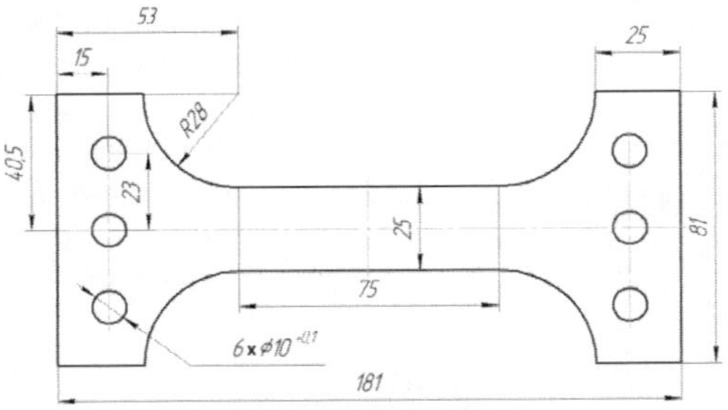

Fig. 2. Specimen for testing.

In all testing after realization of DNP fracture efforts of brittle samples were within: -60...160 kN. Impulse uploading at the static tensile of stainless steel was superimposed at deformation of 0,05 ... 8 %. A total of 18 specimens with thickness of 3 mm were tested.

In testing and the implementation of DNP captured the whole working part of the specimen, with the camera mounted at a distance of 2 m, thus giving an error of 0,05% value of longitudinal deformation at the value of the transverse deformation of 30%. This accuracy is more than sufficient to analyze the kinetics of deformation field.

Since all impulse uploading in this testing was performed on the ascending section of deformation diagram, then it is important that the residual strain after a difficult load mode "static stretching-DNP" did not exceed the deformation of neck forming steel 12Cr17 with standard static tension (20 %).

In this testing realization of DNP in steel 12Cr17 carried out at various levels of the previous static deformation. Realization of DNP at low initial static strain was specifically chosen to reduce the impact of material damage on the character of the process. For example, the Fig. 3 presents the experimental results in the case of when uploading impulse F_{imp} = 94,1 kN carried almost on elastic deformation area. For clarity, the same figure shows the curve of deformation under a static strain (curve 2).

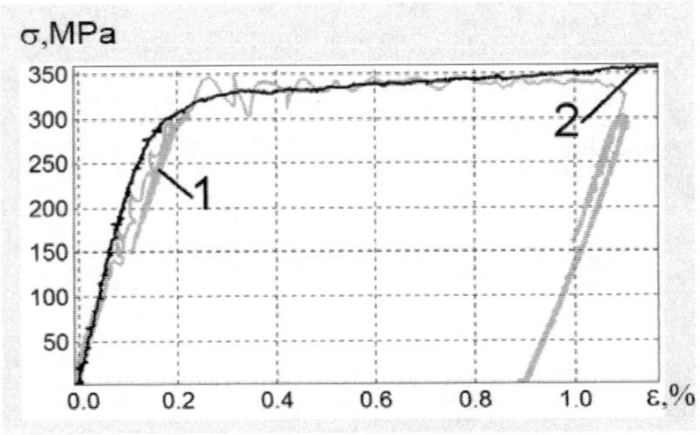

Fig. 3. A process of static strain and DNP of stainless steel 12Cr17 (value of impulse uploading 94,1 kN): 1 – DNP curve; 2 – static strain curve.

It is noted that strain curves under static and DNP are largely the same, and a significant deviation is observed only in the late implementation of DNP. This may indicate the fact that the dissipative structure is formed mainly at the final stage of the specific speed oscillating deformation. One should also pay attention to the fact that the implementation DNP does not make steel practically strengthened on macro level.

Unlike from the previously tested aluminum alloys [5] graphs of local deformation of steel at DNP showing evidence of significant deformation field uniformity (in points) (Figure 4). Due to such homogeneity of deformation field it is possible to reliably measure a high-frequency vibrations of steel at DNP (1,2 ... 2,4 kHz), which depends on random factors associated with fluctuations in the spatial frame testing machine and the rate of deformation of steel during the process of DNP realization (2…10 s^{-1}).

The character of minimum main strain virtually unchanged, when compared to static tension.

When analyzing Fig. 4 one should pay attention to the kind of wave that is on the surface of the specimen at DNP, which illustrating the field of maximum main deformation. The front of this "wave" is moving at a low speed of 0.3-0.7 m / s. Such difference between the field maximum and minimum deformation leads to the conclusion, that structural changes may be coefficient of transverse strain, but to verify it, a detailed study of metal-physical research in real time DNP (10-20 ms) required, which unfortunately at present is not possible. Also Fig. 4 clearly shows oscillatory loading process, especially in the initial part of the process. These fluctuations are 1,2-2 kHz frequency caused by external circuit loading machines, since this frequency range corresponds to its own frequency. Their spatial variation caused by fluctuations frames, which are the main element of randomness at these trials. During the DNP maximum deformation is sent to the central part of the specimen, but after unloading and subsequent deformation it is shifted to the periphery. Its movement is not homogeneous and trajectory usually has a crescent shape. Except for these mentioned features a process of deformation during the DNP especially no is different from the static deformation process.

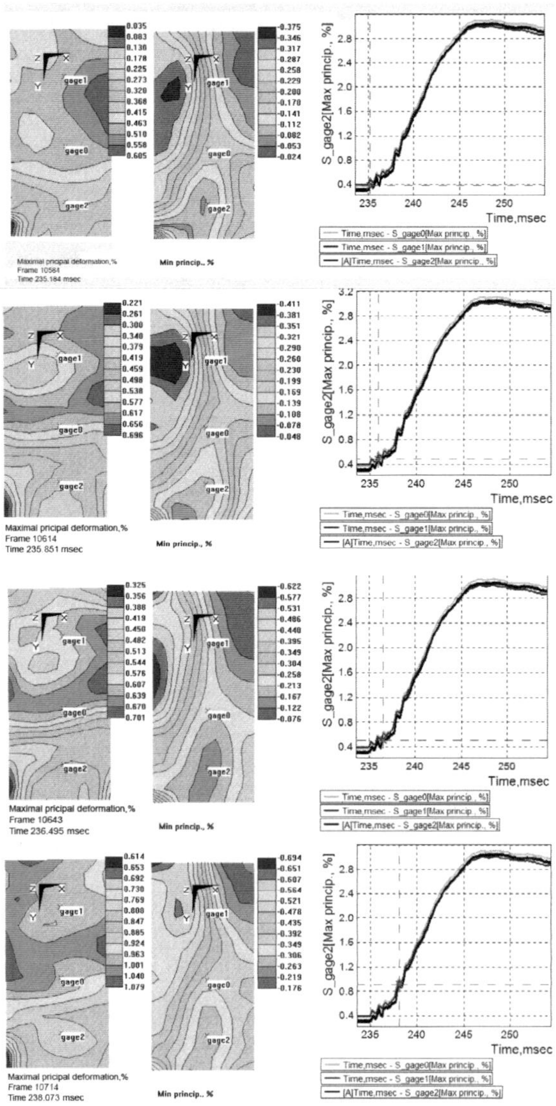

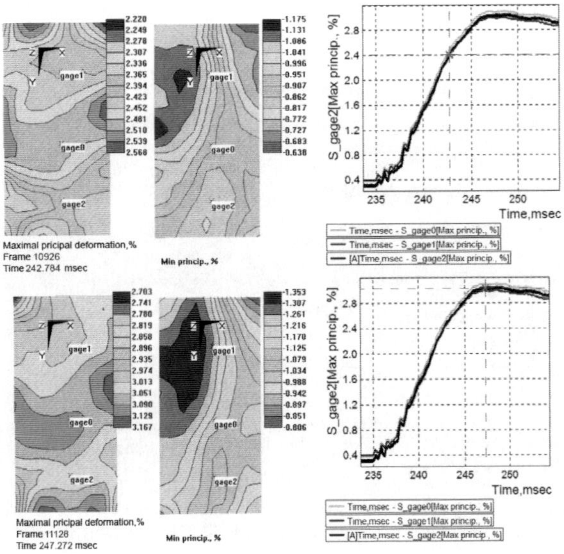

Fig. 4. A typical deformation kinetics of stainless steel under DNP, at a base of zero frame.

This paper evaluated the impact of DNP at the beginning of neck formation and overall ductility of steel 12Cr17. The beginning moment of neck was rated by a classic method-determined maximum deformation on the diagram deformation-effort, and by a specific method of digital image correlation, which used the whole strains field. Between the values obtained by different methods, it was found a clear correspondence that proved the effectiveness of the use of information about the field of deformation.

Figure 5 shows the dependence of neck beginning from local fracture deformation, this graph clearly shows the delay in process of the neck formation caused by the action of DNP.

Such delay may be caused by accumulation of dislocations in the band structure, so that one band passes dislocation (band corridor) and the other resists their motion (barrier band). The resistance of this structure dislocation motion during the subsequent deformation after DNP is likely to lead to reduction of their concentration in the band-corridors. At the same time, band barriers are saturating with dislocations, which provides the time delay in neck creation to the moment of penetration of barrier bands by dislocations.

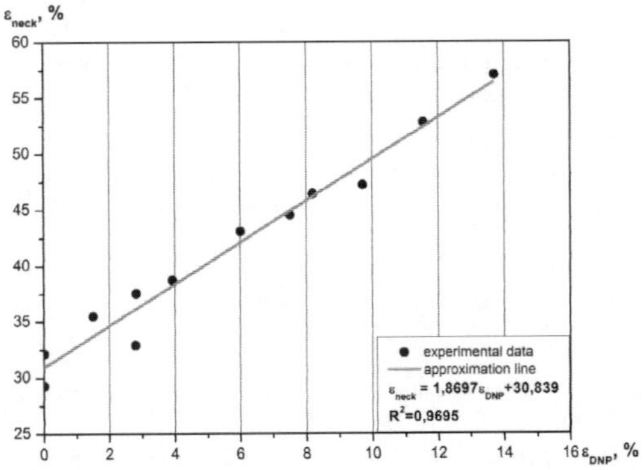

Fig. 5. Dependence of early neck creation in percentage to the local fracture deformation in regards to of strain values obtained as a result of DNP (the difference between the initial and final values) as compared to static.

On Figure 6 it is clearly shown decrease in plasticity after of DNP for this material.

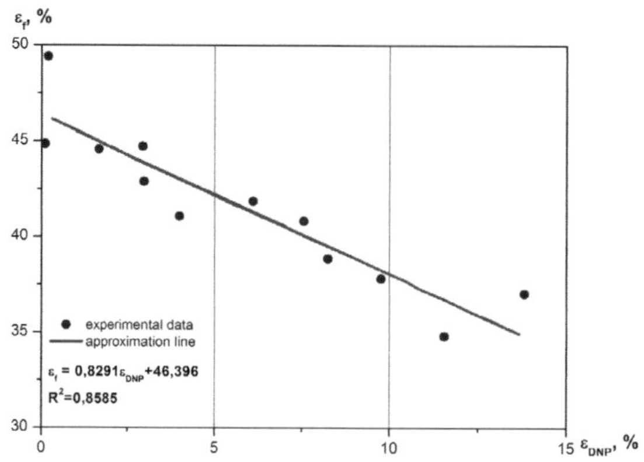

Fig. 6. Dependence of fracture strain on a virtual basis of 2 mm deformation values obtained as a result of DNP (the difference between the initial and final values) as compared to static.

This phenomenon can have such explanation-in the corridors bands a process of porosity formation is occurring much more intense than in the base material, which is caused by the reduced density of dislocations, because formed pores are larger and their fusion processes are accelerated.

Preliminary tests showed a significant sensitivity of the effects caused by DNP to a time elapsed since its realization. In order to assess this initial effect, DNP was implemented on 3 pairs of specimens with endurance 0-7 days and 3 months. A pair of test specimens were subjected to the same deformation prior to DNP, the blow was inflicted with effort within 123...147 kN.

Overall during the short exposure growth of 5-11 % recorded in strength and significant drop in ductility of 20-30%. With a long exposure a significant increase in strength is not recorded, only drop inductility of 15-35 % was noted. This fact indirectly indicates the presence of relaxation processes at the macro level, the effects of which should be considered when constructing material behaviors models under these complex load modes and strength factor coefficient estimates.

3. Conclusions

- Patterns of field deformation kinetics steel 12Cr17 at DNP are not significantly different from the kinetics fields of deformation of steel at static tensile. There was shown a typical kinetics of deformation field and the presence of "wave" that moves on the specimen at a speed of 0,3-0,7 m/s.
- The movement was found of a most intensely deformed part of the strain field in the12Cr17 steel to the center after DNP - the appearance of clearly pronounced kinetics in the deformation process.
- The effect of DNP on the kinetics of the field strain of steel 12 Cr17 during the subsequent static deformation. In particular, it was found the cause of the delay in neck creation associated with the movement of the most intensely deformed part of a specimen to the center.
- The influence of relaxation processes on the process of steel 12Cr17 deformation after DNP has been determined. Their influence leads to changes in the kinetics of deformation.

References

[1] Chausov M, Maruschak P, Pylypenko A. et al. Effect of high - force impulse loads on the modification of mechanical properties of heat - resistant steel after service. *Est. J of Eng* 2012; 18:3, 251 - 258.

[2] Hutsaylyuk V, Chausov M, Berezin V, Pylypenko A. Strength analysis of mechanical systems at dynamic non - equilibrium processes. *Eng. Failure Analysis* 2013; 35, 636 - 644.

[3] Zasimchuk E, Markashova L, Baskova O, Turchak T, Chausov N.et al. Influence of combined loading on microstructure and properties of aluminum alloy 2024 -T3.*J of Mat Eng. and Perf* 2013; 22, 3421-3429.

[4] Chausov N, Zasimchuk E, Markashova L, et al. Peculiarities of plastic materials deformation during dynamical non-equilibrium processes. *Fact. Lab. Mater. Diagn* 2009; 75, 52-59.

[5] Chausov M. B Berezin V, Pylypenko A and Hutsaylyuk V. Strain field evolution on the surface of aluminum sheet alloys exposed to specific impact with oscillation loading . *J. Strain Analysis* 2014; 50:1, 61 -72.

[6] Chausov N, Voityuk D, Pilipenko A et al. Setup for testing materials with plotting complete stress-strain diagrams. *Strength Mater* 2004; 36, 532-537.

[7] Chausov M and Pylypenko A. Laws of deformation processes and fracture of plastic steel from the point of view of dynamic overloading. *Mechanika* 2005; 54, 24-29.

Forum on Innovative Technologies and Management for Sustainability (ITMS'2016)

panevėžys faculty of technologies and business

The 11th International Conference Intelligent Technologies in Logistics and Mechatronics Systems (ITELMS'2016), 28-29 April 2016, Panevėžys, Lithuania

Stiffness identification of hydraulic hoses

Cieślik K.[a]*, Łopatka M.J.[a], Muszyński T.[a]

[a]*Military Univercity of Technology, Kaliskiego 2 street, 00-908 Warsaw, Poland*

Abstract

In this paper was presented how to determine the stiffness of flexible hydraulic hoses. The stiffness of the hoses was determined based on a research changes of the pressure of oil in the hydraulic hoses. There was presented the measuring station used for this, and described the method to realization of this research. In the research the hydraulic hoses worked in the system with the hydraulic actuator. The obtained results of measurements of the pressure changes relate to the actuator and the hydraulic hoses together. The research was made for the hydraulic hoses of the inter diameter 5 and 6 mm. The length of the hoses was 1, 3 and 6 m. Moreover, the investigated hydraulic hoses were different material which it was made and the number and type of the reinforcement.

© 2016 K. Cieślik, M. J. Łopatka, T. Muszyński.
Peer-review under responsibility of the Kaunas University of Technology, Panevėžys Faculty of Technologies and Business

Keywords: hydraulic hoses; stiffness of flexible hydraulic hoses; hydraulic actuator.

---

1. Introduction

The stiffness of the hydraulic elements is a very important parameter in the modeling and designing of the manipulators with hydrostatic drive system [1, 2, 3]. The working tools of the manipulators was moved with different velocity. The time needed to stop the working tool of the manipulator in a specific position is different. Moreover, the precision of the motion of the manipulator is different for different speeds of movement it. On the precession and time of this movement was affected by many factors [4]. One of them is a stiffness of the hydraulic hoses. In the modeling of steel structures using the hydrostatic drive system usually skips the stiffness of steel and focusing on the stiffness of the hydraulic fluid (oil).

The elasticity module of steel with a value 220 x E^9 N/m^2 is nearly 200 times larger than the elasticity of oil. Therefore, the omission the elasticity module of steel in the modeling is not introduce a big error.

In the literature can be found many item about how to appoint a bulk module of hydraulic oil and how equal to its value for the deferent pressures. The value of the bulk module of hydraulic oil is between 1 and 2.1E9 N/m^2. The value of the bulk module depends on the percent of the air – containing and it is less for the oil with a higher air content [5, 6, 7]. In the literature there are very few items about the stiffness of the flexible hydraulic hoses connected with hydraulic actuator or only about the hoses or the actuator [8, 9]. Based on the literature there was determined the formula which allowing of the analytical calculation of the stiffness of the hydraulic hoses with oil - k_{pp} [1, 2].

$$k_{pp} = \pi \cdot d_p^2 \cdot L_p - 4 \cdot \square E_{zc}. \tag{1}$$

---

* Corresponding author. Tel.: +48 22 261 837 416.
E-mail address: karol.cieslik@wat.edu.pl

© 2016 K. Cieślik, M. J. Łopatka, T. Muszyński.
Peer-review under responsibility of the Kaunas University of Technology, Panevėžys Faculty of Technologies and Business

where: d_p – hydraulic hose inside diameter, L_p – hydraulic hose length, E_{sc} – reduced bulk module of elasticity of the hydraulic fluid.

The reduced bulk module of elasticity of the hydraulic fluid – E_{sc} was determined from formula [2]:

$$E_{sc} = \frac{E_{oc}}{1 + \frac{d_p \cdot E_{oc}}{\delta_p \cdot E_c}} \tag{2}$$

where: E_{oc} – bulk module of the hydraulic fluid, δ_p – thickness of the wall of hydraulic hose, E_c - module of elasticity of the material which a hydraulic hose was made.

On the market there are many kind of hydraulic hoses made with different materials. Calculation the stiffness of the hydraulic hoses with oil on a base of formula 1 and 2 is very difficult. The E_c parameter is different for every type of the hydraulic hoses. Determined of the E_c parameter requires a research on a testing machines for every kind of the hoses.

The stiffness of hydraulic hoses has determined on the basis of analysis of the oil pressure changes in time. The change can be caused by oscillate mass, which indirectly influences on the oil in hoses. The intermediate element can be hydraulic actuator. The oscillate mass influences on the actuator, it is compresses the oil in an actuator and in the connected hoses to it. Exemplary chart of change pressure in the hose was presented on figure 1.

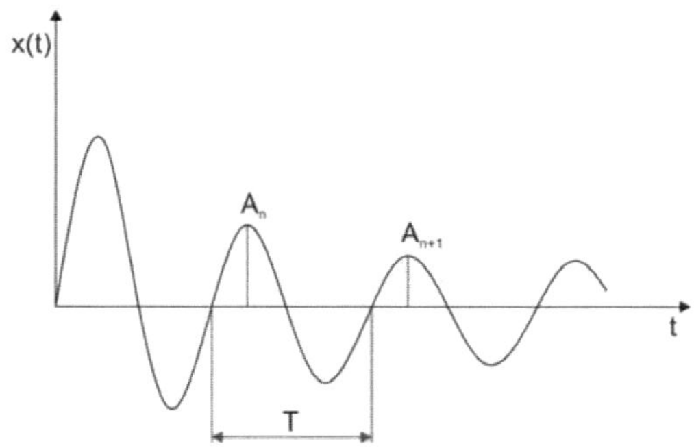

Fig. 1. Exemplary chart of change pressure in the hose.

One a base of the graph (fig. 1) can calculate the stiffness of hydraulic hoses connected with hydraulic actuator by formula [1, 2]:

$$k = \omega_0^2 \cdot m_L \tag{3}$$

where: ω_0^2 – natural frequency, m_L – oscillate mass.

The natural frequency given by formula [1, 2]:

$$\omega_0 = \frac{2\pi}{T} \tag{4}$$

where: T - the oscillation period.

Because the hoses were connected with hydraulic actuators the stiffness calculate on base of the formula 3 concern stiffness of this elements together. Assumed that the hoses and the actuator was serial connected, so:

$$k = \frac{k_H \cdot k_A}{k_H + k_A} \tag{5}$$

where: k_H - the stiffness of the hydraulic hoses, k_A – the stiffness of the hydraulic actuator.

On the stiffness of the hydraulic actuator consist three kind of elements. On a figure 2 was presented the scheme of the hydraulic actuator and the physical model of it. In model on the figure 2 the stiffness of the steel was omitted.

a)

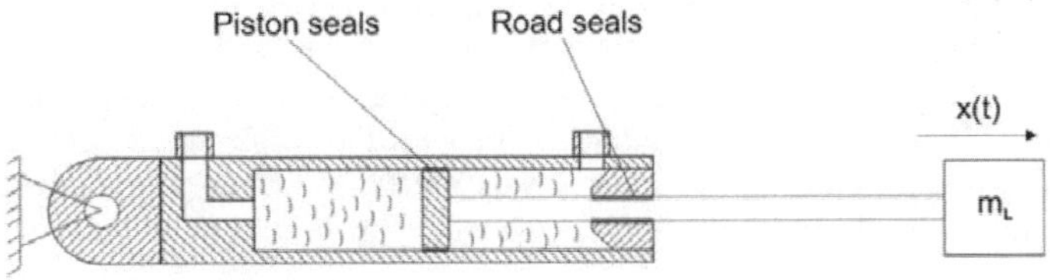

b)

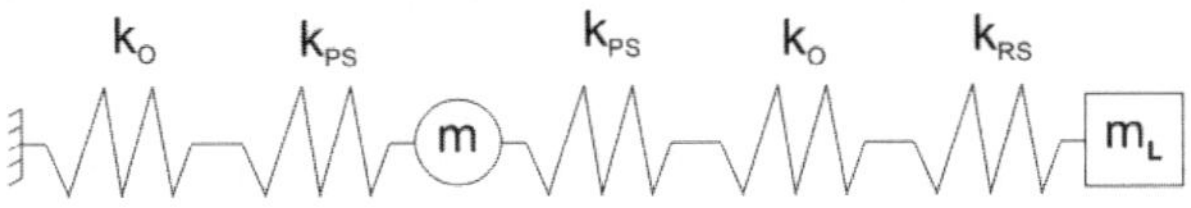

Fig. 2. Scheme of the hydraulic actuator (a); physical model the hydraulic actuator (b); k_{PS} – stiffness of the piston seals, k_O – stiffness of the hydraulic oil, k_{RS} – stiffness of the road seals, m – piston and piston road mass.

On a basis of figure 2 the equation was given:

$$k_A = \frac{k_O \cdot k_{PS}}{k_O + k_{PS}} + \frac{k_{PS} \cdot k_O \cdot k_{RS}}{k_{PS} \cdot k_O + k_{PS} \cdot k_{RS} + k_O \cdot k_{RS}} \tag{6}$$

The formula (6) shows that the elasticity of seals of the actuator have a large part of the total stiffness of this hydraulic component.

One a basis a graph on figure 1 can calculate also two parameters which are useful to compare hydraulic hoses and actuator with another. The parameters are the damping coefficient and logarithmic damping decrement. The logarithmic damping decrement is defined by [1, 2]:

$$\delta = \ln\left(\frac{A_n}{A_{n+1}}\right) \tag{7}$$

where: A_n – the amplitude at time t, A_{n+1} – the amplitude of the peak 1 period away.

The logarithmic decrement δ is used to find the damping ratio of an under damped system in the time domain. The damping coefficient is defined by [1, 2]:

$$c = 2 \cdot k \cdot \zeta \cdot \sqrt{\frac{m_L}{k}} \tag{8}$$

where: ζ – the dimensionless damping coefficient.

The dimensionless damping coefficient is given by the equation [10, 11]:

$$\zeta = \frac{\delta}{\sqrt{4 \cdot \pi^2 + \delta^2}} \tag{9}$$

All of presented equation was used to description researched hydraulic hoses and actuator.

2. Research of stiffness of the hydraulic hoses

2.1. Methodology

Objective of the research was determined the stiffness of the hydraulic hoses with connected actuator to it. The research was carried for two types of hoses and for differed length of it. The research relied on the registration of the pressure change in hydraulic actuator and in connected hoses to it. The hoses were connected to both of flow port of the actuator, in order to remapped the real operating conditions. The pressure was measured in both of hoses connected to the actuator. Moreover, the displacement of the piston road of the actuator was measured. It was changed when the actuator was loaded. The value of the displacement was helped to calculated the bulk module of elasticity for the actuator, oil and hoses. The bulk module was given by equation [6]:

$$K = \frac{\Delta p}{\frac{\Delta V}{V}} \tag{10}$$

where: Δ p – pressure increase, Δ V – change in volume, V – original volume.

The research related on loaded the hydraulic actuator with connected one type and length hose on a special measure station. The piston road displacement and the pressure change was registration. The next step was set in motion the load and saved the data. This action was repeated when the structure stopped oscillate. The study was carried out for three different strokes of the piston road, therefore the compressed volume of hydraulic oil was differed. During the research the actuator was tensile, the volume on the piston road area was compressed. Obtained with the research graph was used to calculate the summary stiffness of the hydraulic hose, actuators and oil, bulk module of it, damping coefficient and logarithmic damping decrement for every with investigated hoses.

2.2. The measure station and elements

To carried out the research the measure station was made. The scheme of it was presented on figure 3.

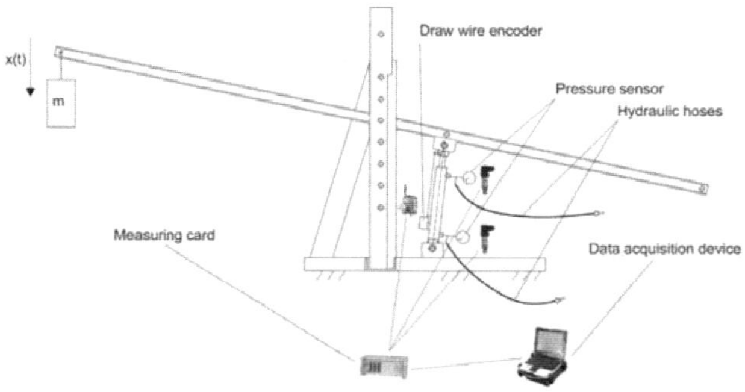

Fig. 3. The scheme of the measure station

Basic data of the measuring devices used in a measure station was presented in table 1. The mass - m (fig. 3) which was set in motion had 180 kg.

Table 1. The basic data of the measuring devices.

O.N.	Name of element	Basic data
1	Pressure sensor	Measurement Range 0 – 600 bar Accuracy 0,5%
2	Draw wire encoder	Measurement Range 0 – 1250 mm Accuracy 0,05%
3	Measuring card	Gantner q.brixx station Module: 3xA101, 4xA107, 2xD101
4	Data acquisition device	Durabook DPS-U12Ci

The basic data of the hydraulic actuator which was used in the research was presented in table 2.

Table 2. The basic date of the hydraulic actuator.

Piston Diameter, mm	Piston road diameter, mm	Stroke, mm	Type of seals	Working pressure, bar
40	22	150	Heavy duty seals	250

The basic data of the hydraulic hose was presented in table 3.

Table 3. The basic date of the hydraulic hose.

Image of the hydraulic hose		
Hose type	KP206 EN857 2SC 06	NY104 DN5
Inside diameter, mm	6	5
Working pressure, bar	400	300
Inner layer	Oil resistant synthetic rubber	Polyester elastomer
Insert	Two high tensile steel wire braided inserts	One high tensile steel wire braided insert
Outer layer	Oil resistant and weatherproof synthetic rubber	Polyurethane

2.3. The results of the research

The first of the research was concern of the single actuator without the hoses. The research was made for three piston position: 10 mm higher of the minimum limit stroke, a half of the stroke and 10 mm less than the maximum stork is. Then to the actuator was connected hoses and the research was made again. The time function of the pressure change for two position of the actuator and connected to it different length a type hose was presented in Figure 4.

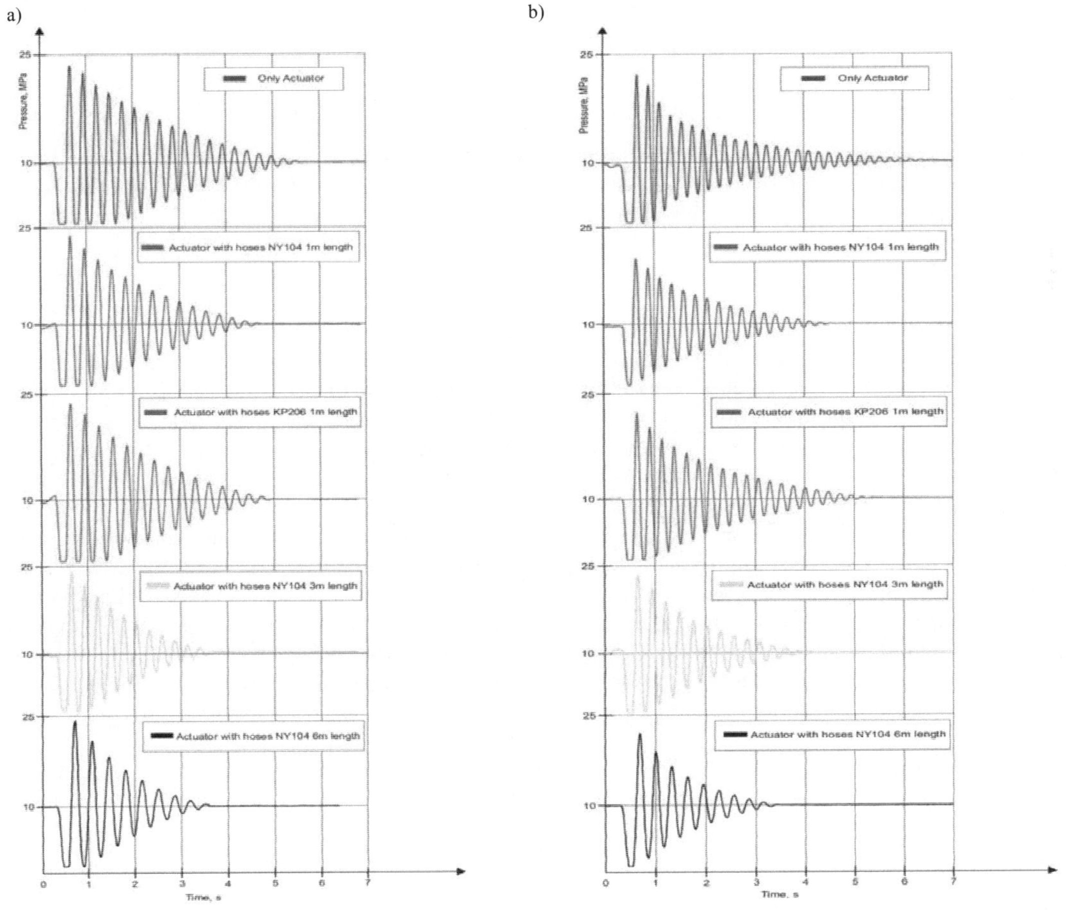

Fig. 4. The time function of pressure change of the actuator with hoses; (a) pressure change for a minimum stroke of the actuator; (b) pressure change for a maximum stroke of the actuator.

The pressure change in time function (figure 4a, b) was present only for the piston rod side because the actuator was tensile. Pressure on the piston side was value close 0, so was omitted.

On a base the research results the stiffness of the hydraulic hose connected with actuator was calculate. The results of it was presented in table 4. In a table also given a period, damping coefficient and logarithmic damping decrement for every time function which was realized. The reduced mass which acted on the actuator was calculate on a basis the piston road side pressure. The mass was change because the stroke of the actuator was change. The mas was 839 kg to minim stork and 853 for maximum stroke of the actuator.

Table 4. The research results.

Parameter name	Actuator stroke	Only actuator	Actuator with hose NY104 1m	Actuator with hose KP206 1m	Actuator with hose NY104 3m	Actuator with hose NY104 6m
Period, s	min	0,27	0,29	0,3	0,33	0,36
Damping coefficient, Ns/m		626	820	892	919	1086
Logarithmic damping decrement		0,1006	0,1260	0,1541	0,1806	0,2330
Stiffness, N/m		454826	394255	368408	304470	255839
Period, s	half	0,24	0,26	0,27	0,31	0,34
Damping coefficient, Ns/m		795	891	1189	1016	1300
Logarithmic damping decrement		0,1124	0,1319	0,1823	0,1857	0,2607
Stiffness, N/m		581763	495704	459664	348694	289874
Period, s	max	0,21	0,23	0,24	0,28	0,41
Damping coefficient, Ns/m		1072	1245	1357	1312	1783
Logarithmic damping decrement		0,1320	0,2155	0,1830	0,2258	0,3244
Stiffness, N/m		763853	636785	584845	429667	350530

Presented in table 4 stiffness was concerned actuator with hydraulic hoses. On a base a formula (6) can calculate the stiffness of hoses and actuator with seals separately. But in modeling of the hydraulic drive system usually the actuators occurred with hoses, so volume of stiffness concerned such type system. During of the investigate the air content in oil did not exceed 5%.

3. Conclusions

Investigated stiffness of two type of the hydraulic hoses KP206 and NY105. The hoses were connected to hydraulic actuator. The value of stiffness concerned summary stiffness of oil, actuator seals and hoses. The stiffness of hoses was depended of the period of the investigated system. The period value was increased when the length of the hoses also did it. The compress volume of the hydraulic fluid and pressure also affected on period value. For increased volume value the period also increased. Observed that when the length of the hose increased the time of the oscillation was decreased. For single actuator the oscillation time was 6.2 s and for the actuator which 6 m hoses was 3.1s. It was indicated that the length of the hoses has large influence on increased a dumping coefficient. Difference of the period value for hoses KP206 and NY105 was 0.1 s. Structural difference of those hoses was significant, but large stiffness almost 10% has hose NY105, which has less reinforcement number. The research was indicated that the actuator seals was significant part with summary stiffness of investigated elements. The stiffness of the hoses should be taken into modeling hydrostatic drive systems. It not included the results of the simulation research are inconsistent with reality.

Acknowledgements

We would like to thank of National Centre for Research and Development for founding grant number UOD-DEM-1-145/001 within have been conducted investigation described in this paper.

References

[1] Tomczyk J. *Modelowanie dynamiczne elementów i układów napędów hydrostatycznych*. WNT, Warsaw, Poland; 1999.

[2] Konopka S, Łopatka M J. *Podstawy konstrukcji maszyn z CAD Modelownie ruchu maszyn*. MUT, Warsaw, Poland; 2005.

[3] Will D, Gebhardt T. *Hydraulic Grundlagen, Komponenten. Schaltungen*. Berlin: Springer, 2008.

[4] Konopka S, Lopaka MJ, Krogul P. *Simulation Research of Hydrostatic Power System Control of Engineer Robot Manipulator*.

[5] Yang H, Feng B, Gong G. Measurement of Effective Fluid Bulk Modulus in Hydraulic System. *Journal of Dynamic Systems Measurement and Control Transactions of*, v. 133, Nov 2011.

[6] Hruzik L, Vasina M, Burecek A. Evaluation of Bulk Modulus of Oil System with Hydraulic Line. *7th International Conference on Experimental Fluid Mechanics (EFM)*, Tech Univ Liberec, Hradec Kralove, Czech Republic; NOV 20-23. 2012.

[7] Sunghun K, Hubertus M. Measurement of Effective Bulk Modulus for Hydraulic Oil at Low Pressure *Journal of Fluids Engineering*, v. 134, 6 MAR. 2012.

[8] Vandenhorn B, Kuipers M. *Strenght and stiffness of a reinforced hose*. Kluwer Academic Publ, Netherlands; 25 JAN 1988.

[9] Sunghun K, Hubertus M. Measurement of Effective Bulk Modulus for Hydraulic Oil at Low Pressure. *Journal of Fluids Engineering*, v. 134, 6 MAR 2012.

[10] Gutowski R, Swietlicki WA. *Dynamika i drgania układów mechanicznych*. PWN. Warszawa; 1986.

[11] Szulej J. *Wyznaczenie ekwiwalentnego wiskotycznego tłumienia drgań w konstrukcjach wielomateriałowych*. Rozprawa doktorska. Wydawnictwo Politechniki Lubelskiej. Lublin; 2010.

Forum on Innovative Technologies and Management for Sustainability (ITMS'2016)

ktu 1922 panevėžys faculty of technologies and business

The 11th International Conference Intelligent Technologies in Logistics and Mechatronics Systems (ITELMS'2016), 28-29 April 2016, Panevėžys, Lithuania

Using of FEM in Assessment of Shooting Resistance of Vehicles

Čornak S.[a], Novak L.[a]*

[a]*University of Defence, Kounicova 65, Brno, 662 10, Czech Republic*

Abstract

Article authors deal with an issue of ultimate strength changes of ARMOX 500T passive armor in dependence on a temperature. As an introduction there is the issue analysis. In further parts of the paper a mathematical model of dependence between a strength of passive armor and a temperature is suggested. The paper is complemented by results of numerical simulations in ANSYS AUTODYN software and results of shooting experiment. The results have confirmed an original hypothesis that a strength decreases with rising temperature.

© 2016 Š. Čornak, L. Novak.
Peer-review under responsibility of the Kaunas University of Technology, Panevėžys Faculty of Technologies and Business

Keywords: FEM; strength; temperature; ansys; experiment.

---

1. Introduction

The issue of a ballistic resistance assessment of military vehicles has outstanding areas of interests nowadays, when a very accurate dimension of a ballistic protection is needed [2] [5]. One of these areas is a temperature effect on basic material properties of a passive armor, especially strength, hardness and toughness. Researches were interested in these effects in a past, but it has never been formed any mathematical descriptions of dependences between listed material properties and a temperature. The paper is going to solve an issue of strength of an armor material.

One of the current trends of a military vehicles design is low-weight vehicles with a high mobility. A conclusion can be made that temperature differences affect listed material properties and a ballistic resistance in association with various operating conditions (Iraq, Afghanistan, etc.). An accurate suggestion and verification by fast empiric method should be applied during the design of military vehicles, instead of using difficult systems and expensive tests, only.

All armies are affected by a current trend of budgets reductions, therefore simulations by a finite element method (FEM) have increasing role during the ballistic protection research and practically in every technical discipline. The method can substitute expensive shooting tests and allows obtaining much data, which cannot be measured in experiments.

---

* Corresponding author. Tel.: +420-608-166117.
E-mail address: lukas.novak3@unob.cz

2. Model of the Problem and Basic Derivation

The problem is simplified for the need of a mathematical description derivation, FEM simulations and experiments.

A steel ball impacts an armor target. The projectile is the bearing ball made of a *STANDARD 14109* material and the cartridge of the ball was created especially for the experiment (figure 1). A material of the target is *ARMOX 500T*, which is the basic material of a ballistic protection of the Czech army's PANDUR II 8x8 vehicle. Detected values are a depth of penetration H and an impact velocity of a projectile v_{proof}.

Fig. 1. Cartridge with steel ball (diameter 4.5 mm).

The hypothesis, that tensile test is standard static test to obtain an ultimate tensile strength R_m, was chosen as a basic starting point. The relation for determination of an ultimate tensile strength is [3]:

$$R_m = \frac{F_m}{S_0} \tag{1}$$

where F_m is a maximal force, which is achieved by a tensile test and S_0 is an original cross-section of a test specimen.

The premise that kinetic energy E_k changes completely on a work W, is a basic assumption of a force F_m determination. The complete derivation of this relation was published in [1]. The relation of R_m after the preliminary derivation is:

$$R_m = \frac{m_p \cdot v_{proof}^2}{2\pi H \cdot \left[\dfrac{r-H}{tg\left(arcsin \dfrac{r-H}{r} \right)} \right]} \tag{2}$$

where m_p is a mass of projectile, v_{proof} is an impact velocity, H is a depth of a penetration and r is a radius of a projectile (ball).

3. Shooting Experiment

Shooting experiment was performed to verify of modelled results. There were compiled the measuring chain, which consists of a gun, optical gates (for an impact velocity measuring) and holder with a target. Sheets from the ARMOX 500T material were used as targets (figure 2). Widths of sheets were 5 mm (real width 5,4 mm) and 7 mm (real width 7,6 mm) [1]. Shootings were performed in two regimes:

- Constant (laboratory) temperature T = 17.1 °C;
- After heating of material in a converter (interval of a temperature ranging from 100 °C to 17.1 °C).

A constant temperature experiment was conducted by v_{50} method. First, an impact velocity v_{proof} was step by step increased and then the mean value of the armor penetration was determined. Next, the target was heated up to 100 °C and placed to a holder. Than the decreasing temperature in a relation with time was monitored by a thermal camera and during the measuring were projectiles (balls) fired. Following, times of impacts were recorded. An accurate time of impact was necessary for the assignment of an actual temperature.

Results of shooting experiment are the bases for finding the relation between static and dynamic (terminal ballistic) tests for a determination of basic properties of an armor material.

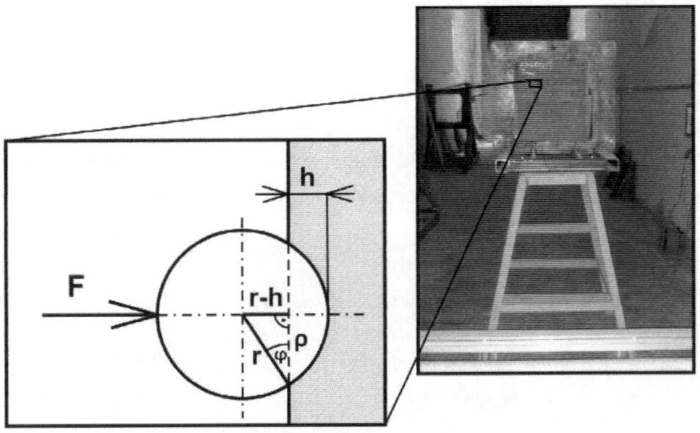

Fig. 2. Target (Sample of ARMOX 500T material) with detail of impacting ball.

4. FEM Simulation

Simulations by software, which are based on the solving of construction problem by finite element method (FEM), divide the continuous model by the mesh into the finite number of elements. Elements are joined by nodes. Every node has 3 degrees of freedom (DoF), which are described by 3 equations. The equations are included in a matrix and solved using the basic equation of motion [4]:

$$[M]\{\ddot{U}\} + [C]\{\dot{U}\} + [K]\{U\} = \{F(t)\} \tag{3}$$

where *[M]* is a matrix of mass, *[C]* is a matrix of damping, *[K]* is a matrix of toughness, *{U}* is a column vector of model transposition and *{F(t)}* is a column vector of external loads.

The bases for correct results are especially appropriately created mesh and accurate definition of marginal conditions. In case of a terminal ballistic task is the mesh creation quite simple. A major part of an armor target is created by a basic mesh. More quality mesh is used for the projectile and near the impact side, where is need for exact results. An example of a mesh is in a figure 3.

In case of a simplified terminal ballistic model, the definition of marginal conditions is restricted to a boundary of target edges and an entering of impact velocity of a ball. The results of FEM simulations are the depth of penetrations and deformed shapes of targets.

The FEM simulation is used to verify model accuracy for its other use with more widths of targets. More data for derivation of the most accurate mathematical description are going to be obtained by this way.

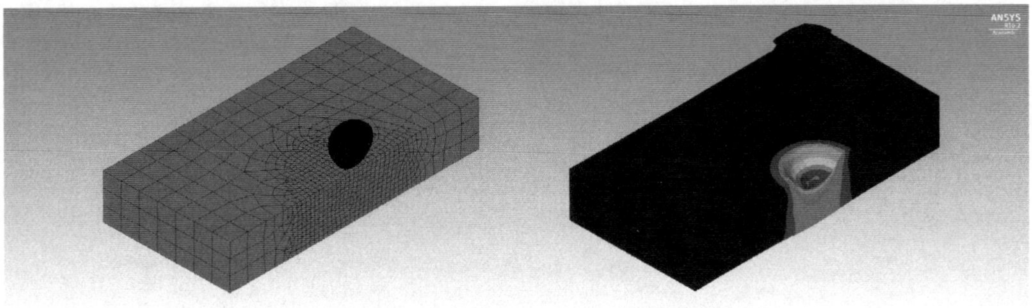

Fig. 3. (left) Simplified model pro problem with mesh; (right) finally deformation.

5. Evaluation and Final Mathematical Model

Shooting experiments and FEM simulations in Ansys Autodyn software were performed by the description in chapter 3 and 4. In the figure 3, there is seen the crater (imprint) with a depth *H* created by the ball in the target. Impact velocities (kinetic energies) of the projectile were randomly chosen in the experiment. A measurement of a depth of penetrations was performed using a *SOMET XYZ 643C* machine with a measurement accuracy of thousandths of millimeter. FEM simulations with the same impact velocities were performed afterwards. Differences between experimental results and FEM simulations results were in hundredths of millimeters. These differences can be explain such as deflection of a projectile during firing. The ball impacts the target exactly perpendicular to its surface in FEM simulations. The results are shown in the table 1.

Table 1. Depth of penetration in dependence on impact velocity and temperature of armor target.

w	T	v_{proof}	H_{EXP}	H_{MKP}	ΔH
	17.1	1379.3	1.772	1.657	0.115
	54.5	926.8	0.858	0.769	0.089
	67.5	1087.2	1.174	1.092	0.082
7,6	69.4	1099.2	1.158	1.213	0.055
	74.5	1312.1	1.196	1.196	0.495
	78.8	1389.4	1.843	1.791	0.052
	90.5	1273.4	1.695	1.613	0.082
	17.1	1326.0	2.125	2.026	0.098
	60.5	1203.9	1.679	1.636	0.043
	65.1	1201.5	1.686	1.855	0.168
5,4	70.1	1071.6	1.324	1.371	0.047
	75.7	1096.5	1.574	1.552	0.022
	85.2	914.9	0.917	0.869	0.067
	95.6	918.7	0.976	0.781	0.195

Table legend: w [mm] is a real width of the armor target, T [°C] is an actual temperature, v_{proof} [$m \cdot s^{-1}$] is an impact velocity, H_{EXP} [mm] is a depth of the imprint obtained from the experiment, H_{FEM} [mm] is a depth of the imprint obtained from FEM simulations and $\Delta H = |H_{EXP} - H_{FEM}|$ is the deference between obtained depths.

The equation, which describes changes of an ultimate tensile strength in dependence on a temperature of an armor and a kinetic energy of a ball, was derived based on the obtained data:

$$R_m = \alpha \cdot \psi \cdot \left(\frac{0.5 \cdot m_p \cdot v^2_{proof,REF}}{2\pi \cdot H_{REF} \cdot \rho^2_{REF}} \right)^{\frac{V_{v,REF}}{V_{z,REF}}} \quad (4)$$

where m_p is a mass of a projectile, v_{proof} is an impact velocity of a projectile, H is a depth of an imprint in a target, ρ is a radius of an imprint, V_v is a volume of an imprint and V_z is a volume of a plug in a target. A lower index REF indicates values, which were measured at a reference (laboratory) temperature. A function ψ describes a relation between a static and dynamic (terminal ballistic) test of an ultimate tensile strength and α is a function, which includes an effect of temperature difference between a laboratory temperature and an actual testing temperature. Functions ψ and α are expressed as follow:

$$\psi = 1636.1 \cdot e^{-9.203 \frac{V_{v,REF}}{V_{z,REF}}} \quad (5)$$

$$\alpha = -5 \cdot 10^{-6} \cdot \Delta T^2 + 10^{-4} \cdot \Delta T + 0.9999 \quad (6)$$

Values of an ultimate tensile strength of an ARMOX 500T material in dependence on a temperature were calculated based on equations (4), (5) and (6). Results are shown in a figure 4.

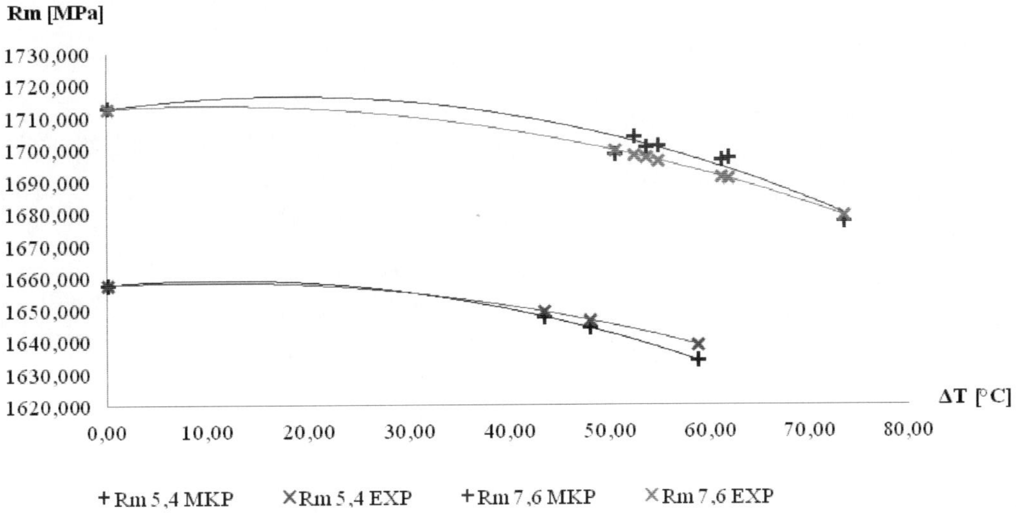

Fig. 4. Values of ultimate tensile strength of ARMOX 500T material with widths 5.4 mm and 7.6 mm, dependence on temperature changes.

Conclusion

A strength of a material is one of basic properties, which are provided for a passive armor. An evaluation and descriptions of temperature effects, which offers this paper, are based on results of shooting experiments and FEM simulations. Differences between data values from experiments and simulations are only in hundredths of millimeters, so differences could be considered insignificant and both methods may be considered applicable for solving the problem.

Useful data were selected after an evaluation of measured and simulated values. Marginal conditions were defined for the most accurate evaluation of an armor strength. The measuring must be performed with a steel hardened ball with the diameter 4.5 mm, in the interval of impact velocities from 950 $m \cdot s^{-1}$ to 1300 $m \cdot s^{-1}$. The best results are obtained in impact velocities around 1200 $m \cdot s^{-1}$. An important condition is the depth of penetration *H*, it has not to be higher than a radius of a ball *r*. Results are going to be incorrect, if the conditions are not comply.

Results of measurements and simulations confirm the hypothesis that a strength of passive armor decreases with an increasing temperature. In case of evaluated ARMOX 500T material the decrease of the strength is $\Delta R_m \approx 35$ MPa in maximal temperature difference $\Delta T_{max} = 73.4$ °C.

The measurement dealt with the specific material; therefore derived mathematical descriptions are limited to described case. In case of using different materials further experiments and simulations have to be performed. We can presume different course of ψ and α functions here, in connection with the specific materials and therefore different temperature effects.

Acknowledgements

Presented work has been prepared with the support of the Ministry of Defence of the Czech Republic, Methods Development for Increasing of Military Vehicles' Mobility, K-202, Department of Combat and Special Vehicles, University of Defence, Brno.

References

[1] Čorňák Š, Novák L. A Contribution to Shooting Resistance Evaluation of Military Vehicles. *2015 International Conference on Military Technologies*, Brno: University of Defence; 2015. p.135-138

[2] Droppa P, Vančo M, Ferencey V. Response of structure to ballistic load. *Advances in Military Technology* 2012; 7 (2): p. 93-110.

[3] ISO 6892-1:2009. Metallic materials-Tensile testing: Part 1: Method of test at room. *International Organization for Standardization*; 2009.

[4] Juříček L, Novotný P. Modeling of Small Arm Projectile Penetration Through a Steel Plate. *Military Medicine Science Letters* 2004; 3: 97-106.

[5] Koucký M, Vališ D. Reliability of sequential system with restricted number of renewals. In *Risk, Reliability and Social Safety ESREL 2007*. London: Taylor & Francis; 2007.

Forum on Innovative Technologies and Management for Sustainability (ITMS'2016)

ktu 1922 panevėžys faculty of technologies and business

The 11th International Conference Intelligent Technologies in Logistics and Mechatronics Systems (ITELMS'2016), 28-29 April 2016, Panevėžys, Lithuania

Efficiency of hydrostatic drive system applied in Unmanned Ground Vehicle

Dąbrowska A.[a]*, Łopatka M.J.[a], Typiak A.[a]

[a]*Military University of Technology, Gen. Sylwestra Kaliskiego Str. 2, 00-908, Warsaw, Poland*

Abstract

The main purpose of this paper was to present problems connected with designing of hydrostatic drive system applied in Unmanned Ground Vehicles in order to increase its total efficiency. The introduction describes the advantages, the disadvantages and the development trends of the hydraulic systems. Then, the hydrostatic drive, the principle of its operation and the basic group of its components were briefly characterized. In order to determine the efficiency of the entire system, power and pressure losses for its individual components, such as the main pump, installation, charge pump and motor were defined. The great variability of the efficiency of the individual components depends on their operating conditions. Its influence on the total efficiency of hydrostatic drive system used in Unmanned Ground Vehicles was described based on example of pump and motor.

© 2016 A. Dąbrowska, M. J. Łopatka, A. Typiak.
Peer-review under responsibility of the Kaunas University of Technology, Panevėžys Faculty of Technologies and Business

Keywords: hydrostatic drive system; unmanned ground vehicle; efficiency; hydraulic system; hydraulic pump; hydraulic motor;

1. Introduction

The robotization is one of the main directions of development of today's technology. Its aim is to replace the man in different situations. One of them is to support him during performing many dirty and dangerous missions – from unmanned explorers of the moon and ocean floor, to industry applications in difficult and dangerous for human operator tasks, to transport of materials and equipment, to terrain reconnaissance in remote control mode, the importance of Unmanned Ground Vehicles (UGV) is bigger and bigger. It is especially significant in military applications where new dangers such as chemical and biological warfare or mine traps pose a serious threat to soldiers' health or life. Unmanned systems enhance the soldiers' protection considerably by enabling them to operate at a safe distance from a threat. Such support can provide only the UGV able to follow the man and to recognize the potential hazards on the explored area [1, 2, 3, 4].

The possibility of widespread use of the UGVs as a direct human support in areas difficult to access or affected by natural disasters, require from them the ability to:

- recognize such an area from the human's security point of view,
- support man in situations when he is not able to complete the mission alone,
- reduce the risk of secondary damages and losses,
- accelerate the pace of operations in order to minimize the number of victims,
- perform technological actions (collecting samples, manipulating of lying objects, etc.).

* Corresponding author. Tel.: +48-261-837-107; fax: +48-261-839-616.
E-mail address: agnieszka.dabrowska@wat.edu.pl

© 2016 A. Dąbrowska, M. J. Łopatka, A. Typiak
Peer-review under responsibility of the Kaunas University of Technology, Panevėžys Faculty of Technologies and Business

The proper execution of these tasks requires from the UGV the ability to quick access, return or change the area of action. The specificity of the tasks expected to be accomplished by these vehicles, necessary mobility level, aforementioned requirements and the specifics of the area put very high demands on their drive system.

Increasingly often, in this type of vehicle the hydrostatic drive systems are used, mainly because of their advantages:

- easy transfer of energy and its distribution to several receivers,
- ease of operation and high susceptibility to automation,
- low inertia and high precision of operation,
- low weight per unit of transmitted power,
- reliability

and progressive development of hydraulic components – development trends concern the minimizing of the dimensions and acoustic emission, easy operation and above all, energy efficiency. The last aspect is significantly important because the principal disadvantage of this type of drive systems is not very high efficient – less than a mechanical drive [5, 6, 7].

Despite the growing popularity of hydrostatic drive systems there are still relatively not many research papers concerning modelling them from increasing their efficiency point of view. In literature one can find examples of modelling different power transmission systems [8, 9, 10] – a few of them relate to hydrostatic drive systems [11, 12], and even fewer are those that are dedicated to Unmanned Ground Vehicles with high mobility [13].

Nomenclature

d	diameter of pipe
E_M	mechanical energy
E_{NM}	non-mechanical energy
E_P	pressurized hydraulic fluid energy
F_D	driving force
l	length of pipe
P_1	input power
P_2	output power
Δp_1	linear pressure losses in pipeline
Δp_2	local pressure losses in pipeline
Δp_{CHP}	charge pump pressure
Δp_{INST}	pressure losses in installation
Δp_M	motor pressure drop
Δp_P	pump pressure rise
ΔP_{CHP}	power losses of charge pump
ΔP_{INST}	power losses in installation
ΔP_M	motor power losses
ΔP_P	pump power losses
q_M	motor displacement per revolution
q_P	pump displacement per revolution
Q	volumetric flow rate
Q_{CHP}	displacement of charge pump
R_W	wheel radius
T_M	torque on the shaft of the hydraulic motor
T_P	torque on the shaft of the pump
v	velocity of flow
v_D	velocity of drive
η_M	motor efficiency
η_{MHM}	motor mechanical-hydraulic efficiency
η_{VM}	motor volumetric efficiency
η_P	pump efficiency
λ	linear losses coefficient
ξ	local losses coefficient
ρ	fluid density
ω_M	angular speed of the hydraulic motor shaft
ω_P	angular speed of the pump shaft

2. Characteristic of hydrostatic drive system

The main task of the drive system is the distribution of the mechanical power generated at the transmission shaft of the engine to all operating mechanisms (energy receivers) according to their current needs. This can be done with or without the participation of an intermediate form of energy (Fig. 1).

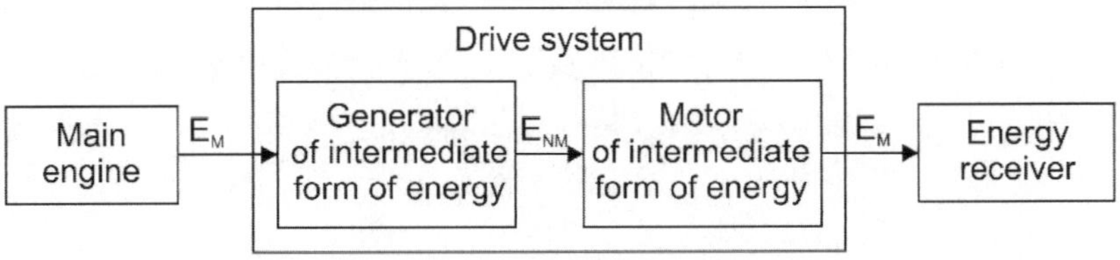

Fig. 1. Scheme of non-mechanical drive system [6].

The hydrostatic drive system operates according to the above scheme using the pressurized hydraulic fluid energy as an intermediate form of energy $E_{NM} = E_P$. The mechanical energy from the main engine is transferred through the transmission shaft to the positive displacement pump and consumed to perform mechanical work to trap a certain volume of fluid – pump displacement per revolution q_P – and to displace it into the discharge pipe. The pump forces the hydraulic fluid into the system with actual displacement Q_P resulting from theoretical displacement Q_{PT} and volumetric losses ΔQ_P. Then, the pressurized hydraulic fluid stream is supplied to energy receiver in form of rotary or linear hydraulic motor (hydraulic cylinder). The pressurized hydraulic fluid flows through the hydraulic motor and its pressure and volumetric flow are converted into torque T_M and angular speed of motor shaft ω_M or force F_C and linear velocity of cylinder piston v_C [6, 14].

In the general case, the hydrostatic drive consists of three main groups of components [6]:

- components converting one form of energy into another one (hydraulic pumps, motors and cylinders),
- components controlling the flow of energy from hydraulic fluid stream (valves),
- components gathering and conducting hydraulic fluid (tanks and hydraulic pipelines) as well as components allowing to maintain and control the hydraulic fluid properties (filters, coolers, manometers, thermometers).

An exemplary diagram of a closed hydrostatic drive system is shown in Figure 2.

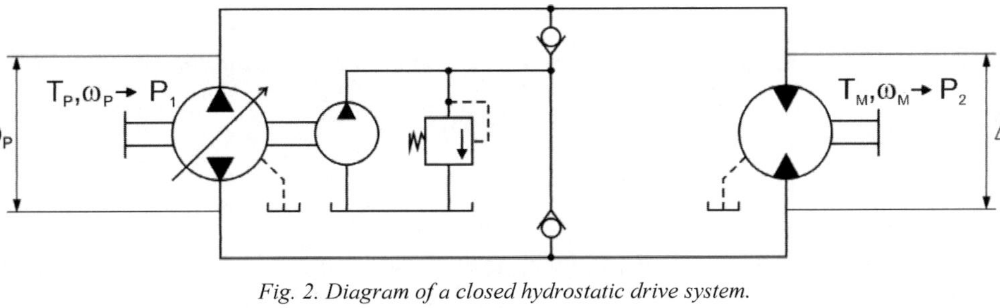

Fig. 2. Diagram of a closed hydrostatic drive system.

3. The efficiency of hydrostatic drive system

In order to estimate the total efficiency of the hydrostatic drive system, the power and the losses for individual components should be determined. The input power for pump can be described as:

$$P_1 = T_P \cdot \quad {}_P \tag{1}$$

Then, the pump power losses can be expressed as:

$$\Delta P_P = P_1 (1 - \quad {}_P) = T_P \cdot \quad {}_P - \Delta p_P \cdot Q \tag{2}$$

where: $Q = Q_P = Q_M$ – flow rate in system is equal to pump and motor displacement [l/min].

The efficiency η_P has no constant value but depends on the operating parameters such as pump pressure rise Δp_P, pump displacement Q_P and angular speed of the pump driving shaft ω_P. It is illustrated in efficiency map from Figure 3, which presents the overall pump efficiency depending on working pressure and varying rotational speed in percents of rated continuous speed.

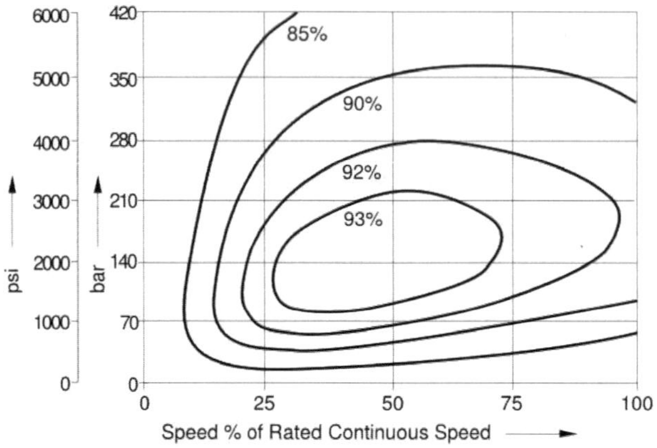

Fig. 3. The overall efficiency map for variable displacement pump (Sauer Danfoss) depending on system pressure and varying speed corresponding to rated continuous speed [15].

It is noteworthy that operating with nominal values of parameters do not guarantee the highest efficiency because the chart area with highest efficiency is achieved for speed range of 25-75 % of rated continuous speed and for operating pressure range of 7-23 MPa (at the nominal pressure equal 42 MPa).

Moreover, the pump pressure rise Δp_P should be enough to cover the motor pressure drop Δp_M and pressure losses in installation Δp_{INST}:

$$\Delta p_P = \Delta p_M + \Delta p_{INST} \tag{3}$$

The power losses in installation can be expressed as:

$$\Delta P_{INST} = \Delta p_{INST} \cdot Q \tag{4}$$

where pressure losses in installation Δp_{INST} result from linear and local losses in pipeline what can be described as:

$$\Delta P_{INST} = \Delta p_1 + \Delta p_2 = \left(\sum {}^{\text{'}} \frac{l}{d} + \sum {}^{\text{''}} \right) \frac{-v^2}{2} \tag{5}$$

The pressure drop in pipeline and components controlling the energy flow of fluid stream in correctly designed hydrostatic drive system should not exceed 0,5 MPa. Therefore its importance for the total system efficiency is marginal.

The similar situation is with charge pump. It is responsible for exchange of working fluid (filtration and cooling) and secures from cavity in low-pressure part of the system. The power losses of charge pump are following:

$$\Delta P_{CHP} = \Delta p_{CHP} \cdot Q_{CHP} \tag{6}$$

The charge pump pressure Δp_{CHP} equals usually about 2,5 MPa and its displacement Q_{CHP} is about 10 % of the main pump displacement. Hence, when working pressure is 25 MPa the power losses of charge pump equals about 1-2 % of input power. That makes it has no significant influence on total efficiency of hydrostatic drive system.

The component of hydrostatic drive system that is very important from total system efficiency point of view is hydraulic motor. Its output power can be represented by following equation:

$$P_2 = T_M \cdot \;\; {}_{M} \qquad (7)$$

The motor power losses can be described as:

$$\Delta P_M = \Delta p_M \cdot Q(1 - \;\; {}_{M}) = \Delta p_M \cdot Q - T_M \cdot \;\; {}_{M} \qquad (8)$$

It is worth to emphasize that hydraulic motors have similar changeability of efficiency characteristics as pumps. The motor efficiency has no constant value but depends on operating conditions. If the device driven by hydrostatic drive system is the UGV very important are terrain conditions where such a vehicle operates – type of ground, slope, terrain irregularities and different obstacles – they determine the operating parameters of hydraulic motor and the whole hydraulic drive system as well. It results from bellowed equations:

$$\begin{aligned} {}_{M} &= \frac{v_D}{2¶ \cdot R_W} = \frac{2¶ \cdot Q_M}{q_M} \;\; {}_{VM} \\ T_M &= F_D \cdot R_W = \frac{\Delta p_M \cdot q_M}{2¶} \;\; {}_{MHM} \end{aligned} \qquad (9)$$

In order to estimate how important is to know the operational parameters of hydraulic motor, from achieved efficiency and correct selection of particular component point of view, an example of torque T_M and rotational speed n_M for the moving UGV were generated (Figure 4). Red-coloured was marked the medium values of torque and rotational speed in particular time intervals, thereby forming a spectrum of load.

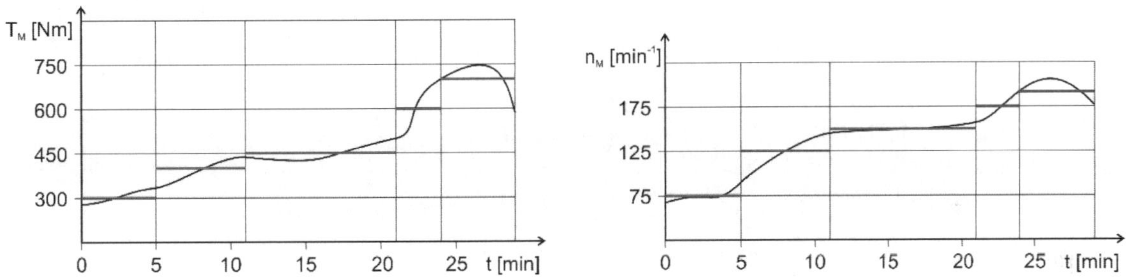

Fig. 4. The example example of torque T_M and rotational speed n_M for the moving UGV.

In the next step of analysis, estimated values of the load spectrum were placed in efficiency map of two different motor models from Sauer Danfoss Company – OMT 200 and OMT 400. For the motor OMT 400 efficiency range is about 87-89 % (Figure 5b) and for the motor OMT 200 is about 5-10 % lower (Figure 5a). Very important is as well the time duration of operating with particular values of torque and rotational speed (illustrated by different lengths of time intervals in Figure 4) and in consequence the time of operating with specific efficiency.

The conducted analysis shows how significant in process of components selection is to compare the efficiency map of particular element with required operating parameters in order to perform with the highest achievable efficiency.

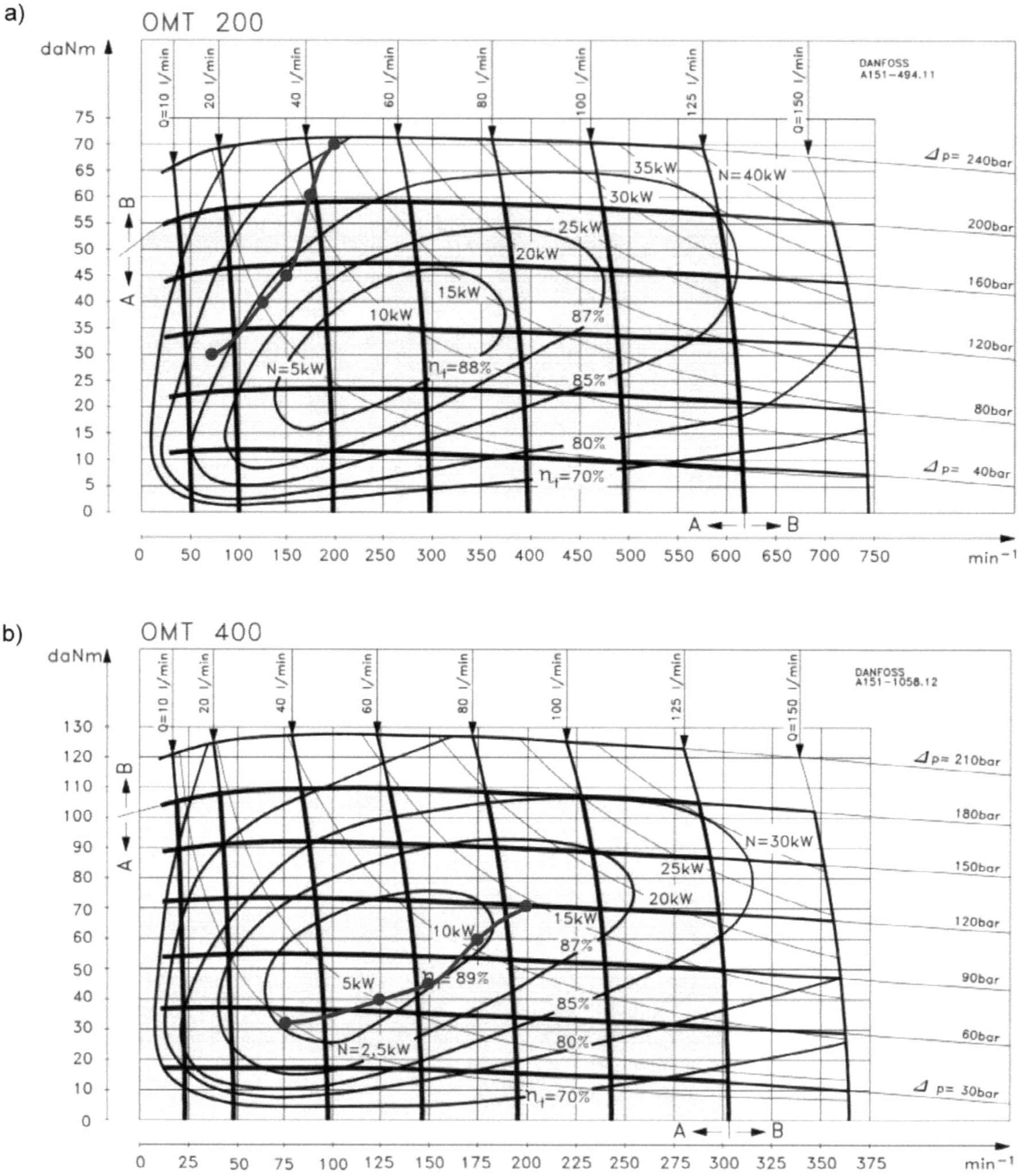

Fig. 5. The efficiency maps of two hydrostatic motors from Sauer Danfoss Company with exemplary efficiency: a) motor OMT 200, b) OMT 400 [15].

The total efficiency of hydrostatic drive system η_{HDS} can be expressed by following equation:

$$ {}_{\mathrm{HDS}} = \frac{P_2}{P_1} = \frac{P_1 - \Delta P_{\mathrm{P}} - \Delta P_{\mathrm{INST}} - \Delta P_{\mathrm{M}} - \Delta P_{\mathrm{CHP}}}{P_1} \tag{10} $$

Making an assumption that ΔP_{INST} and ΔP_{CHP} represent together less than 5 % of input power shows that the biggest influence on total system efficiency have the power losses in pump and motor. Hence, in the process of designing the hydrostatic drive system the special attention should be paid on these components selection. The crucial issue is to analyse the pump and motor efficiency maps in comparison to changing performance parameters during whole operating cycle.

Conclusion

Physical quantities necessary to determine the power of hydrostatic pump and motor are quite easy to define. The difficulty with designing of hydrostatic drive system concerns their changeability when system operates and that they depend on each other. The change of operating conditions and occurring loads result in significant change of pump, motor and whole installation efficiency. Therefore, the total system efficiency also has no constant value and it is strictly connected with operating parameters of individual components and the load of device it drives.

The conducted analysis shows that selection of hydrostatic components has extremely essential influence on transmission efficiency in hydraulic drive systems applied in UGV. The transmission efficiency depends on power losses in pump, installation and motor. The power losses in correctly designed installation should not exceed 3-5 % of input power. Therefore so important for the total system efficiency is the correct selection of hydrostatic components. Depending on the operating load (pressure and rotational speed) they operate with efficiency range from 60 to 90 % (piston units) or from 50 to 83 % (gear and vane units). The disadvantageous selection of hydrostatic components may result in total system efficiency at the level of 30-40 %. Whereas, carrying out the selection taking into consideration the characteristics of the operating load can allow to raise the average efficiency in operating cycle to 70-80 % for piston components and 60-70 % for gear and vane components. The efficiency increase reduces power requirement, heat dissipation and intensity of aging process, what results in decreasing of operating costs. Hence, the fundamental during selection of hydrostatic units is to rely on load characteristics and not on the nominal values of parameters characterising each component – especially when we want to raise the efficiency of the UGV drive system.

Moreover, carried out analysis identified a need to develop efficiency models of components of hydrostatic drive system applied in the UGV. Existing models are not sufficiently precise and do not take into consideration the variability of the operating parameters during the working cycle.

References

[1] Cieslik K, Przybysz M, Rubiec A, Typiak R. Concept for a vision system for an extreme mobility unmanned ground vehicle, INTELLIGENT TECHNOLOGIES IN LOGISTICS AND MECHATRONICS SYSTEMS - ITELMS'2015. *Intelligent Technologies in Logistics and Mechatronics Systems*, Kaunas: Univ Technology Press; 2015. p. 75-79

[2] Typiak R. Universal Control System for Managing Multiple Unmanned Engineering Machines, RECENT ADVANCES IN AUTOMATION, ROBOTICS AND MEASURING TECHNIQUES, *Advances in Intelligent Systems and Computing*, SPRINGER-VERLAG BERLIN. Berlin. 2014; 267: 523-531.

[3] Jaskolowski M, Konopka S, Krogul P, Spadlo K. Capabilities to negotiate terrain obstacles by articulated tracked platforms. *Proceedings of 10th International Conference on Intelligent Technologies in Logistics and Mechatronics Systems ITELMS 2015.* Kaunas: University Technology Press; 2015. p. 158 – 165

[4] Typiak R. Multiple Unmanned Engineering Machine Management System, MECHANICS AND MATERIALS II. *Solid State Phenomena. 210, Trans Tech Publications Ltd.* Zurich 2014; 210: 287-293.

[5] Kollek W, Palczak E. Optymalizacja elementów układów hydraulicznych. *Zakład Narodowy im. Ossolińskich – Wydawnictwo.* Wrocław; 1994.

[6] Konopka S. i inni: *Podstawy budowy i eksploatacji maszyn inżynieryjno-budowlanych.* WAT. Warszawa; 2002.

[7] Praca zbiorowa pod redakcją Garbacika A. Kierunki rozwoju napędów hydraulicznych i konstrukcji maszyn roboczych. *Fluid Power Net Publication.* Kraków; 1999.

[8] Jianjun Y, Yingce S, Diqing H, Chenggang L. A modularized design for multi-drivers off-road vehicle driving-line and its performance assessment. *International Journal of Computer and Information Engineering*; 2007.

[9] Kugi A, Schlachera K, Aitzetmullerb H, Hirmannb G. Modeling and simulation of a hydrostatic transmission with variable-displacement pump. *Mathematics and Computers in Simulation* 2000; 53:4-6, 409-414.

[10] Linares P, Mendez V, Catalan H. Design parameters for continuously variable power-split transmissions using planetaries with 3 active shafts. *Journal of Terramechanics* 2010; 47: 323-335.

[11] Czyński M. Energy efficiency of hydrostatic transmission. Comparing results of laboratory and simulation test. *Scientific Problems of Machines Operation and Maintenance* 2 (154); 2008.

[12] Jędrzykiewicz Z, Pluta J, Stojek J. Research on the properties of a hydrostatic transmission for different efficiency models of its elements. *Acta Montanistica Slovaca 2* 1997; 4: 373-380.

[13] Comellas M, Pijuan J, Potau X, Nogues M, Roca J. Analysis of a hydrostatic transmission driveline for its use in off-road multiple axle vehicles. *Journal of Terramechanics* 2012;49: 245-254.

[14] Tomczyk J. Modele dynamiczne elementów i układów napędów hydrostatycznych. *Wydawnictwa Naukowo-Techniczne.* Warszawa; 1999.

[15] Product catalogues of Sauer Danfoss Company.

Forum on Innovative Technologies and Management for Sustainability (ITMS'2016)

panevėžys faculty of technologies and business

The 11th International Conference Intelligent Technologies in Logistics and Mechatronics Systems (ITELMS'2016), 28-29 April 2016, Panevėžys, Lithuania

Detection of Technical Condition of Roller Bearings Using Vibrodiagnostics

Furch J.[a]*, Nguyen T.T.[a]

[a]University of Defence, Kounicova 65, Brno 662 10, Czech Republic

Abstract

This paper focuses mainly on the vibrodiagnostics of rolling bearings and the specification of their damage, and illustrates how to determine specific failure frequencies. Following the analysis of the measured signals we try to find causes of the damage occurrence. For the analysis the methods of a Fast Fourier Transform and envelope detection have been chosen. The paper addresses the aspect of the effective root-mean-square value of a signal (RMS) which might be used for evaluating the state of a bearing during longer-term observation. The practical part of the research was carried out in a laboratory where a place for measuring vibrations including relevant software equipment had been established for this purpose. Vibration sensors are one of the main components of a measuring device which has been described in the paper and modelled in Autodesk Inventor to have a better idea of it. For the measuring itself the equipment DEWE 50-USB-8 of the DEWETRON company as well as DEWEsoft were used.

© 2016 J. Furch, T. T. Nguyen.
Peer-review under responsibility of the Kaunas University of Technology, Panevėžys Faculty of Technologies and Business

Keywords: vibrodiagnostics; vibrations; roller bearing; root mean square (RMS).

1. Introduction

Vibrodiagnostics is one among many methods of technical diagnostics which has been continuously monitoring a technical state of a device by observing the level of a mechanical oscillation in a real time. The mechanical oscillation is the manifestation of a device during its operation. Certain parts become vibration exciters, others, depending on excitation, react specifically. Vibrations on machines are in general periodic in nature, although some unsteadiness in the signal is to be expected. This means that the signal repeats itself after a definite time interval.

The level of a device mechanical oscillation can be described by a bathtub curve. The bathtub curve describes single operating states in which a device can be during its technical life. These states correlate with arising vibration strokes which cause oscillations of different amplitude, velocity and acceleration.

Therefore the vibrodiagnostics is one of the most important methods used in technical diagnostics for identifying a technical state. With the use of vibration diagnostics we are able to detect an incipient failure, locate the place of an incipient failure and predict the length of time during which a device is going to work before a failure occurs or a preventive action is performed. The proper application of vibrodiagnostics can prevent from undesirable damaging machines, thereby saving money spent on a repair and a device in a disabled state.

* Corresponding author. Tel.: +0-420-608-618283.
E-mail address: jan.furch@unob.cz

© 2016 J. Furch, T. T. Nguyen.

Peer-review under responsibility of the Kaunas University of Technology, Panevėžys Faculty of Technologies and Business

Experimental vibrodiagnostics measurements have been performed while obtaining primary signals. Primary results of measurement are systematized and its analysis is done. Dependence of work characteristics between rotary systems with bearings of sliding friction and of roll is ascertained. Generalization of research results is done and conclusions are formulated [1], [2], [3]. A lot of companies spend a big amount of money on the preventive maintenance of machines and technical equipment despite the fact that at that particular moment there is no need to perform a preventive action. This leads to big financial losses mainly during the machine unavailability. Therefore it is desirable to perform preventive maintenance at the time when it is really necessary. Determining a preventive action or replacing a part based on a real technical state with the use of a mathematical model describing a machine technical state during its life leads to proactive maintenance [4], [5]. The aim of our article is to describe the technical state of a vehicle bearing box using vibrodiagnostics.

2. Description of frequencies typical bearing fault

Due to the surface fatigue of the material of antifriction bearing elements different kinds of damage occur, e.g. surface layer material pitting, abrasion resulting in play increase, corrosion, grooving, the snap ring cavities and failures. In bearings the damage of single elements can be localized on the basis of typical bearing frequencies. For the kinematic frequencies of impulses, providing that there is a regular rolling motion, the following equations apply [6], [7]:

a) BPFO - Ball Pass Frequency – Outer Race (outer ring defect)

$$f = \frac{n}{2} f_r \left(1 - \frac{BD}{PD} \cos\beta\right) \tag{1}$$

b) BPFI - Ball Passing Frequency Inner Race (inner ring defect)

$$f = \frac{n}{2} f_r \left(1 + \frac{BD}{PD} \cos\beta\right) \tag{2}$$

c) BSF - Ball Spin Frequency (bearing defect – of a bearing ball or a bearing roller)

$$f = \frac{PD}{2BD} f_r \left(1 - \left(\frac{BD}{PD} \cos\beta\right)^2\right) \tag{3}$$

d) FTF - Fundamental Train Frequency (snap ring defect)

$$f = \frac{1}{2} f_r \left(1 - \frac{BD}{PD} \cos\beta\right) \tag{4}$$

where n is the number of bearing balls or rollers, f_r frequency given by the relative revolution (speed) of an inner and outer ring, BP the diameter of a bearing ball or roller, PD pitch diameter, β contact angle [7].

Bearing wear might be generally divided into four basic stages. At the first stage initial bearing wear occurs. During the contact of a bearing with single rings acoustic emission spreads at the frequencies of up to a few MHz. At the second stage of bearing wear single elements are damaged. Diagnosing the bearing at this damage stage is performed in a frequency supersonic range from 20 kHz to 60 kHz. For measuring there are used special acceleration meters with a high value of their resonance frequency. During the third stage of the damage of bearing elements conventional acceleration meters might be used. A damaged element when touching another element causes mechanical impact during which kinetic energy is transferred into a bearing body. After the impact the bearing body is vibrated with its own frequency ranged from 5 kHz to 20 kHz, the vibrations are damped and linger quickly. During the last stage rapid wear occurs resulting in critical damage of bearing elements. Vibration spectral elements might be noticed directly in the area of low frequencies. Moreover, there is an increase in revolution spectral components. The wear stages introduced above are described in Fig. 1.

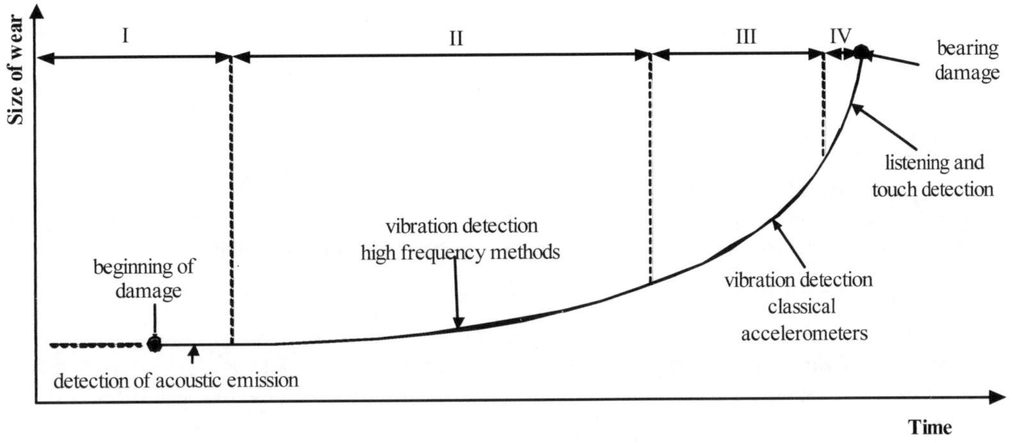

Fig. 1. Time course of bearing damage and possibility of identification [8].

The International Standards Organization (ISO), who establishes internationally acceptable units for measurement of machinery vibration, suggested the velocity – root mean square (RMS) as the standard unit of measurement. This was decided in an attempt to derive criteria that would determine an effective value for the varying function of velocity. Velocity RMS tends to provide the energy content in the vibration signal, whereas the velocity peak correlated better with the intensity of vibration. Higher velocity RMS is generally more damaging than a similar magnitude of velocity peak [6].

We calculated the root mean square (RMS) acceleration, so the root mean square indicates gRMS. For the calculation used the following relations. The mean absolute value A_{ave} can be expressed as follows

$$A_{ave} = \frac{1}{T}\int_0^T |a| dt \tag{5}$$

where T – a period expressed by the formula.
The root of mean square can be calculated by the equation below [3]

$$gRMS = \sqrt{\frac{1}{T}\int_0^T a^2 dt} \; . \tag{6}$$

3. Measuring workplace

The practical part was performed in a laboratory where a location for measuring vibrations was set. In order to measure vibrations the diagnosed object should be set into operation with the help of an outer source which is in our case a three-phase asynchronous electromotor. It is possible to regulate continuously engine revolutions and monitor their magnitude which is a lot easier with a display located inside a control box. We have made a simplified model of the working site using Autodesk Inventor 2012 program to have an idea what it looks like.

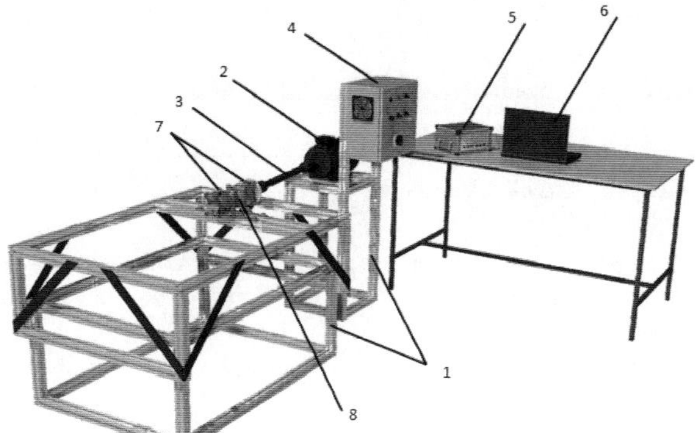

1 Adjustable aluminium chucking bench.
2 Asynchronous motor.
3 Joint shaft.
4 Frequency inverter MOVITRAC.
5 Measuring device from DEWETRON company.
6 Notebook with software DEWEsoft.
7 Two triaxial accelerometers S943B.100.
8 Bearing box of vehicle.

Fig. 2. Model of a measuring workplace [8], [9].

The measurement was carried out by the measuring device DEWE-50-USB-8 from DEWETRON company complemented by six pieces of accelerometers. These were in our case two triaxial accelerometers KS943B.100 which enabled us to measure two objects at the same time. This device also consists of software DEWEsoft including variance analysis. The software is used for processing data.

The sensors were located on the body of the bearing box directly above the bearings. The surfaces for accelerometers fixation were adapted to be fastened according to the set requirements. The fastening of sensors was performed in two ways. First, it was done with a precious rock magnet and then with bolts.

4. Measurement

For the evaluation of vibrations we used the root mean square for detection. This method is based on the Fast Fourier Transform. The following measurements were made to determine the technical condition of rolling bearings.

4.1. Long-term measurement

The observation described below was not based on a one-time measurement. This time we performed several measurements during long operation. Measurements were carried out during the operational period of 3168 hours, which it corresponds to the distance travelled 122500 km. For the analysis we could make a selection out of three possible axes which recorded vibrations in the bearing. We have chosen the axis "Z" on the bearing number 2.

The RMS measurements are very stable and therefore reproducible very well, the time trends are easy to read. In case this method is used, it is not necessary to know the design of a measured object and no considerable demands are made on the operating staff handling measuring devices. The disadvantage of this method is the impossibility of identifying a failure location. When having the increase in bearing wear, the values go up slower than during the PEAK measurement.

To put it simply, the RMS is an averaged value of all the obtained vibrations values during a certain period of time. If in this period of time a single maximum value occurs being even a few times higher than all the others, it will be lost in the final recalculation of all values. In practice it means that when the bearing defect increases causing a shock, the effective RMS value will grow only slowly.

The following figure 3 are presented the results of measurements that indicate the RMS value in the Z2 axis at 1600 RPM at the different times in the life of the rolling bearing. In Graph 1 we present the results of RMS axis Z2, when the bearing is in perfect operating condition. This means that the bearing clearance had an initial 0.2 mm, was lubricated and it was after running up, which was 20000 km.

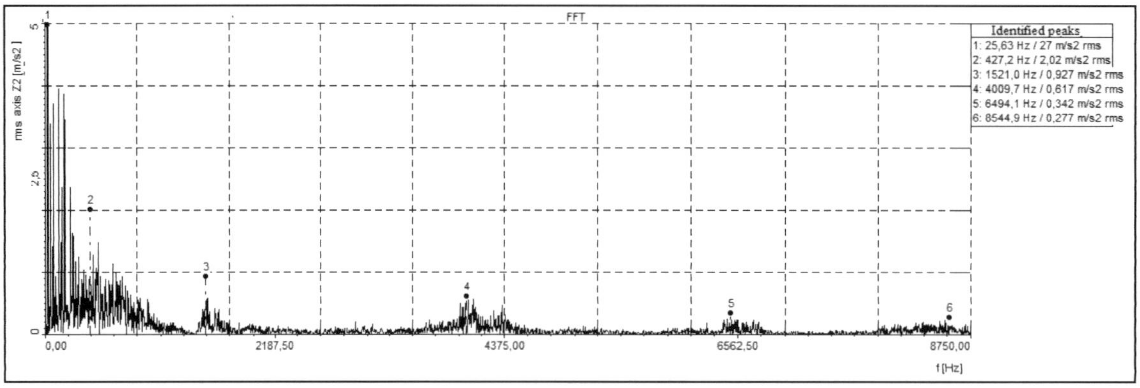

Fig. 3. Root mean square for a bearing in an ideal operating condition (clearance 0.3 mm) - after running up.

In figure 4 shows the results RMS for the same bearing which was operated 122500 km. The bearing was lubricated regularly, including his replacement at prescribed intervals. In the above bearing increased the operational clearance to 0.4 mm during operation. This over-limit clearance has probably result reduction RMS in the frequency range 3.5 kHz to 5 kHz. This result corresponds well with the measured values, which are shown in figure 6.

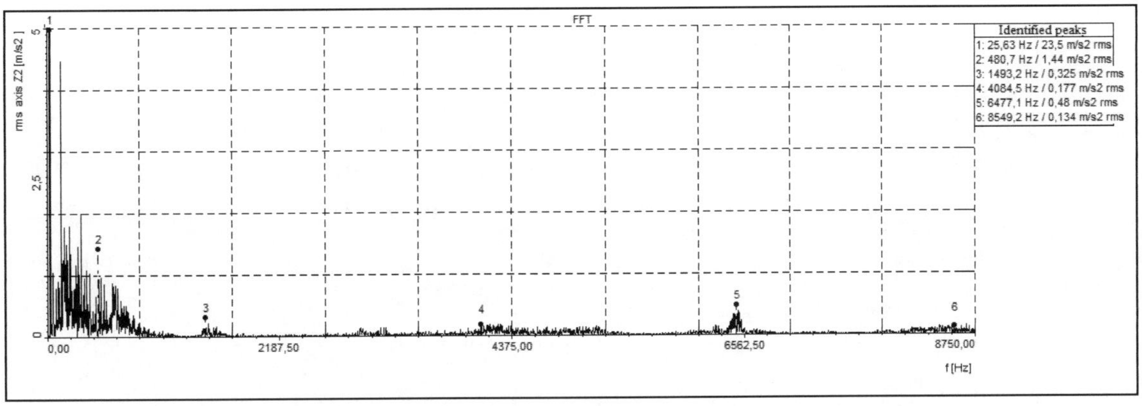

Fig. 4 Root mean square for a bearing lifetime 122 5000 km

In figure 5 are presented the results of RMS for the same bearing which was artificially damaged outer ring.

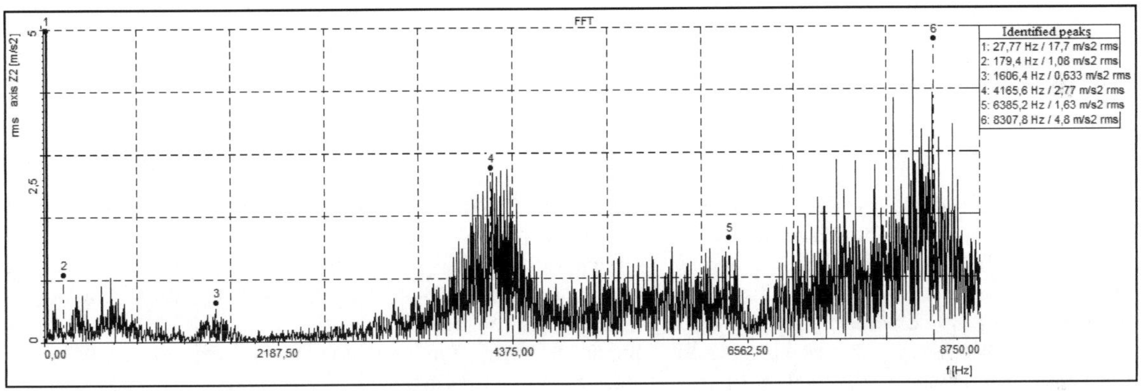

Fig. 5. Root mean square for bearing with damaged outer ring.

RMS curve fig. 6 grows not only in the range 3.5 kHz to 5 kHz, but also increasing the RMS throughout the illustrated frequency range.

During all this experiment we took 24 measurements which we had to evaluate. The records of the selected measurements were processed in DIAdem 2014 software from the company National Instruments. We selected only 5 measurements to make the experiment easy to follow and make clear what happened with the bearing during the experiment.

For the analysis we could make a selection out of three possible axes which recorded vibrations in the bearing. We have chosen the axis "Z" on the bearing number 2. After going over all the records this one seemed to be the most interesting.

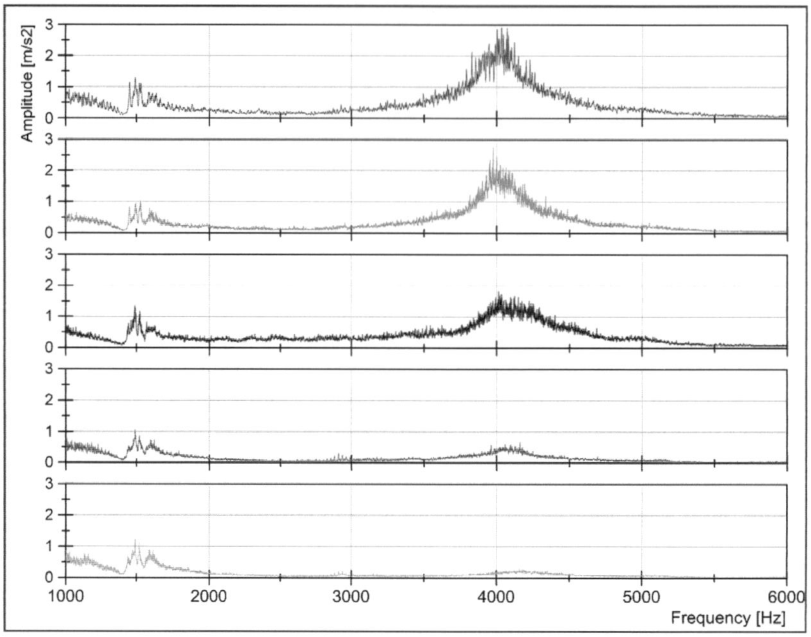

Fig. 6. Record of five frequency spectra processed with DIAdem 2014 program.

For the analysis we selected the range from 3.5 kHz to 5 kHz where we can see the effects of the bearing frequency. The graphs show how the values of amplitudes were gradually decreasing right in the spectrum of the bearing frequency due to running-in after disassembly. However, in no graph there were any of the specific failure frequencies. So we could make an obvious conclusion that although the bearing was in continuous operation for a long time, its operation life did not seem to go to an end.

After the bearing was disassembled, single components were degreased and put under the microscope where we performed a visual examination of their state. On the outer ring there was a visible trace left after the circulation of bearing elements which was, however, evenly coloured. After the examination of the outer ring we found also a few scratches which could be a result of disassembly and then assembly.

We checked the bearing elements too which also showed the signs of wear. However, there was no need to worry about it even if this damage might seem to be very serious at first sight. It is nothing more than local damage which appeared only at the edge of some bearing elements. On the other side, it is necessary to focus on the groove in the middle along the perimeter of all the rollers which could cause the increase in friction in this area.

4.2. Artificially created defect - Defect in an outer ring

In order to verify whether it is possible to use frequency values for specific failures calculated in advance, we gradually created defect in the second bearing.

As for the first mechanical damage to the bearing, we chose an outer ring in which a groove parallel with a rotation axis and perpendicular to the route of rolling elements was cut. This ensured that the rollers would be made to roll over the groove.

Table 1. The values calculated for bearing fault frequencies labelled FAG 31308A.

Designations of fault frequency	f_r frequency 20 [Hz]	f_r frequency 26,66 [Hz]	f_r frequency 36,66 [Hz]	f_r frequency 50 [Hz]
Snap ring defect - FTF	8.42	11.22	15.43	21.05
Bearing defect - BPF	107.22	143.72	194.63	269.55
Outer race - BPFO	134.64	*179.48*	246.80	336.6
Inner race - BPFI	185.36	247.08	339.76	463.4

For the analysis we selected the measurement 1600 RPM on a shaft and the axis "Z2" record, but also the same can be observed in the other axes. In the right upper corner we can see that these values are really within a given range (fig. 7). Minimum deviations can be caused by the inaccurate measurement of bearing parameters.

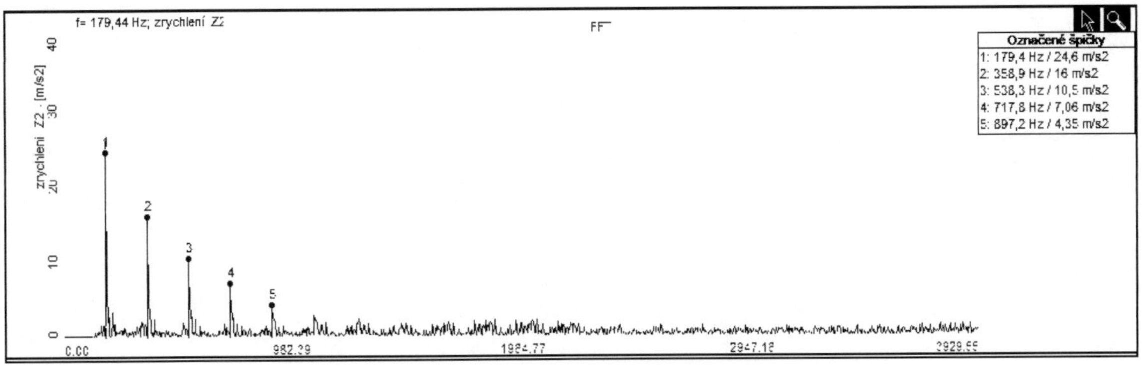

Fig. 7. Frequency spectrum record of the damaged outer ring.

In order to illustrate the frequencies of our interest better, we used a broadband filter where the lower filter limit was six times the amount of rotation frequency which was in our case 160 Hz, and also we applied the envelope. After the application of the filter and envelope we achieved a clear spectrum with the required frequencies.

5. Conclusion

The aim of the experiment was to verify whether the vibrodiagnostics is the right method of determining the technical condition of rotating elements. The attention was paid to the roller bearings since they are one of the most important parts of a machine transmission device.

Based on the performed measurements we can state that the vibration signal is a key signal when it comes to determining the technical condition of a rotating part of a vehicle. When processing the signal, it is necessary to take into account aspects which may create certain difficulties, e.g. determining a proper place for putting sensors, selecting right sensors, proper setting of software used for the evaluation, presenting the measurement results correctly, etc. The tools we selected for the measurement and the analysis were the Dewesoft company´s software and hardware which we considered to be very suitable. Along with the DEWEsoftX2 software we also used the DIAdem software product which enabled us to put the results in graphs (see fig. 6).

In the next step we verified the theoretically calculated value of damage to the outer ring, see Table 1 and performed the experiment. The results which are shown in figure 7 were obtained by using the envelope detection.

Acknowledgements

The presented work has been prepared with the support of the Ministry of Defence of the Czech Republic, Partial Project for Institutional Development, K-202, Department of Combat and Special Vehicles, University of Defence, Brno.

References

[1] Cong F, Chen J, Dong G, Pecht M. Vibration model of rolling element bearings in a rotor-bearing system for fault diagnosis. *Journal of Sound and Vibration* [online]; 2013, vol. 332, no. 8, p. 2081-2097

[2] Tsyfanskii S, Beresnevich V. Vibrodiagnostics of fatigue cracks in flexible, geometrically nonlinear, beam-type structural elements. *Russian Journal of Nondestructive Testing;* 1994. p 427-433

[3] Vekteris V, Jurevicius M, Cereska A. Vibrodiagnostics of the components of technological machines. *8th International Conference on Vibration Measurements by Laser Techniques Location*. Ancona; 2008.

[4] Vaško M., LeitneR B, Sága M. Computational fatigue damage prediction of the lorry frames under random excitation. In: *Communications: scientific letters of the University of Žilina*; 2010. vol. 12, no. 4, p. 62-67

[5] Leitner B. *A new approach to identification and modelling of machines dynamic systems behaviour*. Kaunas; 2010, p. 17-20.

[6] Girdhar P, Scheffer C. *Practical machinery vibration analysis and predictive maintenance.* Burlington: Newnes; 2004, 255 p.
[7] Reeves Ch. W. *Vibration monitoring handbook.* Oxford: Information Press, 1998.
[8] Furch J, Glos J, Blecha J. Identifying the Technical Condition of Rotating Parts by Means of Vibrodiagnostics. *Transaction of Famena;* 2015. vol. 39, no. 2, p. 75 – 86
[9] Furch J, Glos J. *Identifying a Roller Bearing Technical State Using Root Mean Square Acceleration Vibrodiagnostics.* Kaunas, 2015; p. 227-230

Forum on Innovative Technologies and Management for Sustainability (ITMS'2016)

ktu 1922 panevėžys faculty of technologies and business

The 11th International Conference Intelligent Technologies in Logistics and Mechatronics Systems (ITELMS'2016), 28-29 April 2016, Panevėžys, Lithuania

Tribo-diagnostics as an indicator and input for the optimization of vehicles preventive maintenance

Glos J.[a]*, Sejkorová M.[b]

[a]*Department of Combat and Special Vehicles, Faculty of Military Technology, University of Defence, Kounicova 65, 662 10 Brno, Czech Republic*

[b]*Department of Transport Means and Diagnostics, Jan Perner Transport Faculty, University of Pardubice, Studentská 95, 532 10 Pardubice, Czech Republic*

Abstract

The importance of using proper oil analysis is not always entirely clear to all users. There is a strategy that shows when it is necessary to change the oil. The oil might be used to indicate when and where the failure occurs in a vehicle. In other words, an oil analysis can reduce the severity of failures. Most often, oil analysis program combines the continuous attention to oil and the vehicles condition. By combining the instrumental methods of tribotechnical diagnostics (FTIR spectrometry, atomic emission spectrometry, particles analysis with the use of a laser analyzer) it was found that engine oil change intervals should be optimized to maximize the efficiency of engines and extend their lifetime. Changing the engine oil specified by the manufacturer of the buses is relatively conservative, does not reflect the operating motor conditions, fuel quality and individual driving habits and ambient conditions.

© 2016 J. Glos, M. Sejkorová.
Peer-review under responsibility of the Kaunas University of Technology, Panevėžys Faculty of Technologies and Business

Keywords: tribotechnical diagnostics; engine oil; instrumental methods; FTIR spectrometry; atomic emission spectrometry; laser analyzer.

---

1. Introduction

Ideally, modern maintenance would be based on the real condition of machines. The condition monitoring of machinery can be anything from simple visual inspection to continuously functioning real-time condition monitoring systems.

* Corresponding author. Tel.: +420 973 443 440;
E-mail address: Josef.Glos@unob.cz

Oil analysis is a significant part of a condition-based maintenance, because the oil tells us about the history of the machine since the last oil change.

Lubrication oils are during their operation in the lubrication system of an engine exposed to high temperatures, pressures and contamination by undesired materials from outside. These factors cause troubles with the operational quality of oils and materials of mechanical components of a lubrication system. The origin of degradation products and the contamination of oil from outside cause the additives to decrease [1, 2]. The resilience of oil to these processes defines its degree of life-cycle.

One of the methods of the non-dismantling technical diagnostics which uses the lubrication oil as a source of information about the processes and the mechanical changes in machinery mechanisms, is a tribotechnical diagnostics (TTD). Its function is to discover, evaluate and identify the occurrence of foreign substances in the lubricant and further monitor the effects and the results of lubricant degradation process during the operation. For the TTD to be valid, it is essential to perform the lubricant analysis fast, correctly and accurately [3, 4, 5, 6].

The tribotechnical diagnostics fulfils the following objectives:

© 2016 J. Glos, M. Sejkorová.
Peer-review under responsibility of the Kaunas University of Technology, Panevėžys Faculty of Technologies and Business

- to observe conditions and wornness of machines and equipment in order to determine the occurrence of metal particles in lubricants where the magnitude of measured values is important,
- to determine lubricants useful lifetime by specifying the level of their devaluation, the products of thermal and oxidation processes as well as external contaminants. The increased amount of contaminants in oil not only means that the lubricated parts are worn off more, but also that the sediments which may clog oil holes and grooves of the machines have been formed,
- to determine optimum regular intervals for oil exchange [7].

An efficient oil analysis programme is often based on an off-line oil analysis, performed in laboratories where all oil properties are analyzed. However, there are more and more requirements for real-time oil analysis data to support the maintenance decisions. Numerous continuously measuring on-line/in-line oil analysis sensors are commercially available. Conductometric sensors have been fabricated by applying imprinted polymers as receptors for monitoring engine oil quality in [8]. When evaluating the condition of automotive engine oil, the oil's viscosity is one of the most important parameters. Using microacoustic viscosity sensors, an oil-viscosity measurement can be performed on-board [9]. Wang [10] examined road tests of an oil condition sensor and sensing technique. The newly designed sensors were installed in three vehicles and tested on the road. The test results indicated that the degradation of engine oil could be divided into three stages: (1) good, (2) rapid increase in the total acid number (TAN), and (3) rapid increase in viscosity [11]. Various sensor principles of the on-line monitoring of thermal aging of engine oils are considered in Agoston et al [12]. Capone et al [13] deal with the application of a gas sensors array to detect defects in fuel system and contamination in engine oil. Zhan et al [14] made the operational approach to a bar-type engine oil cooler. However, these sensors often measure one property only.

Physico-chemical instrumental control methods have been developed extensively in recent years. These methods enable us to determine not only the actual state of an operated oil charge, but also determine the type of friction and wear of machinery lubricated by oil. This is performed on the basis of abrasion particle characteristics (the number, chemical composition, a structure, surface texture, morphological characteristics and so on). The methods used for this purpose are the infrared spectroscopy with Fourier transform (FTIR spectroscopy) [15-18] and electrochemical methods [19].

With the development of the FTIR spectrometry, software equipment of spectrometers got improved by adding highly demanding mathematic-statistical (chemo-metric) methods that allow to compress the extensive matrix of spectral data, to create regression calibration models for predicting the quality parameters of lubricants and fuel and, as the case may be, to classify the petrochemical samples into one of several (two or more) groups based on the similarity of their variables. A lot of consideration is given to the possibility of using the FTIR spectrometry as an alternative method for checking the oils and fuel in literature [20-23].

Another group of methods allows us to monitor the process and the amount of wear of components that are lubricated by the appropriate lubricant. These methods are suitable for determining the concentration of abrasion metals, describing their morphology and dividing into classification classes according to size, and also for determining e.g. contaminants from filtration materials (they are namely emission spectrometry, analyzer of particles and ferrography) [3, 4, 5, 24-30]. The obtained analytical data provide both the diagnostic information and the prognostic information – i.e. they allow us to predict disrepair situations and prevent their occurrence.

Kumbár et al say [2] that the current evaluation of the quality of automotive engine oil can be sufficiently expressed by dynamic viscosity oils, the analysis of wear metals by emission spectrometry and detection and monitoring of the quantity, type and size of wear debris by analyzer of particles. Similar conclusions may be found in [31] and [32].

2. Experimental

2.1. Diagnosed objects

The samples of engine lubrication oil were continuously collected from various types of buses which vary according to a type and a model year (Table 1). New engine oil was added when necessary to the original charge during the operation of the buses. Sampling intervals ranged from about 5 000 to 10 000 km. Elementary operating methods were performed in the laboratory of a transport company to determine the presence of water, dirt and relative dynamic viscosity.

Table 1. The technical specification of the buses and the operated oil chargers.

Bus marking	SOR B 9,5	IrisBus CityBus 18M	Karosa Axer
Type of engine		diesel - supercharged	
The number of cylinders	6	6	26
Year of manufacture	2000	2002	2003
Oil charger volume [l]	12	25	38
Recommended interval of oil change [km]	15	40	40
Mileage of engine with oil charge [Mth]	14	50	50
Engine oil marking	OMW Truck LD SAE 15W/40	OMW Truck FE Plus SAE 10W/40	OMW Truck FE Plus SAE 10W/40

By combining the instrumental methods of tribotechnical diagnostics (FTIR spectrometry, atomic emission spectrometry, analysing the particles using a laser analyzer), we analysed the relations between the degradation processes which occur during the motor oil operation and the characteristics of the process of operational wear of engine parts where this oil is applied.

2.2. Experimental methods used

FTIR spectrometry

The infra-red spectrometry is one of the optical methods of molecular spectrometry that uses the spectral range delimited by wavelengths of 0.78–1,000 μm, i.e. wavenumbers from 12,800 to 10 cm^{-1}. The most commonly used one is the middle infra-red (MIR) area of 4,000–400 cm^{-1}. The principle of the method is the absorption of infra-red radiation when passing through a sample where the vibration or vibration-rotation energy states of a molecule change depending on changes of its dipole moment. The analytical output is the band absorption spectrum which is a graphical representation of a functional dependency of energy, usually expressed in percent of transmittance or in units of absorbance.

The level of the engine oils wear can be determined using the differential spectroscopy for the spectra of the original oil and the degraded one. Therefore, it is possible to determine the relative content of oxidizing, nitrating and sulphatising products, the content of water, fuel and cooling liquid, the level of anti-oxidizing additives reduction, and so on [16].

Infrared spectra were recorded on a Nicolet IS10 spectrometer (Thermo Scientific) in the spectral range of 650–4000 cm^{-1}, with the resolution of 4 cm^{-1} and with the scan number 32 by means of the ATR technique (ZnSe crystal).

Atomic Emission Spectrometry

The emission spectrometry is one of the optical methods of atomic spectrometry. It uses the radiation emitted by free excitated atoms or by elements in gaseous state for determining the element composition of samples in solutions as well as solid samples. The principle of the atomic emission spectrometry (AES) is to register photons that come into existence due to passes of valency electrons from higher energy levels to lower ones.

Tribotechnical diagnostics use the methods of AES to determine the degree of wear of friction pairs, the concentration of additives in a lubricant and the concentration of contaminants [5, 25, 26, 29]. By regular monitoring the concentration of abrasion metals in the oil, optimum preventive maintenance processes can be set and, hence, the reliability of the device can be increased.

Chosen abrasion metals and contaminants were evaluated on spectrometer Spectroil Q100 (Spectro Inc., USA).

Analysis on Particle Counters

The device that works on the principle of counting the particles in conjunction with the analysis of their shape is LaserNet Fines (LNF). The LNF detects particles larger than 5 μm using laser and advanced software for an image analysis. The outlines of images of the analysed particles in the 20–100 μm range are based on the algorithms of neural network methods.

The information about the captured particles is used for evaluating the current state of wear of the given system and can be observed directly on the screen of the LNF control computer.

The evaluation of the wear mode of the most exposed friction surfaces of the engine was realized by the particle analyzer LNF Q200 Laser Net Fines (Spectro Inc., USA)

2.3. Experimental results

The aim of the analysis of the engine oils from the buses was to consider whether the change in setting replacement intervals (in two buses the replacement interval was prolonged by 10 000 km, in one case it was shortened by 10 000 km) made on the basis of the results of elementary operating methods is optimal. Due to the limited size of the paper the analysis results are processed in detail only for the bus IrisBus CityBus 18 M. The results for other buses are shown in [4].

The first sample was taken when the mileage of the bus was 423 963 km. This articulated bus had been ridden on suburban routes. It had been used almost daily.

The spectral record in Fig. 1 demonstrates the penetrative decrease in antiwear and antioxidation additive based on ZnDDP (zone area 1050–950 cm^{-1} and 690cm^{-1}) after the mileage 23 828 km. This corresponds to a half replacement interval contrary to the interval given by the producer of the bus. The depletion of additives of this type can affect significantly the occurrence of increased wear of engine parts. The replacement of oil is recommended every time the decrease in anti-wear and antioxidant additives is about 80 %.

The quality of oils after a certain mileage was mainly marked by soot particles and aging products which is accompanied by oil darkening (the shift of the base line of the spectrum to higher values of absorbance in 2000 cm^{-1}). Some authors state [33] that higher temperature causes the formation of insoluble particles and more intensive oil carbonisation. .

Soot is formed by almost pure carbon and in some cases it is hard with sharp edges. That is why its higher concentration in engine oil causes increasing viscosity which later results in the increased wear of the engine. During pumping and circulation the polluted oil may occur (due to the flow of soot particles) and can cause abrasive wear on the metal surfaces. .

The limit concentration of soot in oil of 3 % wt. is used for older types of engines [34]. But modern engine oils replaced after longer intervals are made from hydrocracked based oil. These oils withstand higher load by soot without significant influence on the properties of the engine oil compared to a conventional solvent-refined base oils group I [34]. Of course this also very much depends on the used additive of the base oil and the dispersant effectiveness.

The fuel contaminates the oil charge (Fig. 1 zone with the peak of 816 cm^{-1}). Since the presence of fuel decreases oil viscosity and soot increases it, significant change in relative viscosity was not detected by a rotational viscometer.

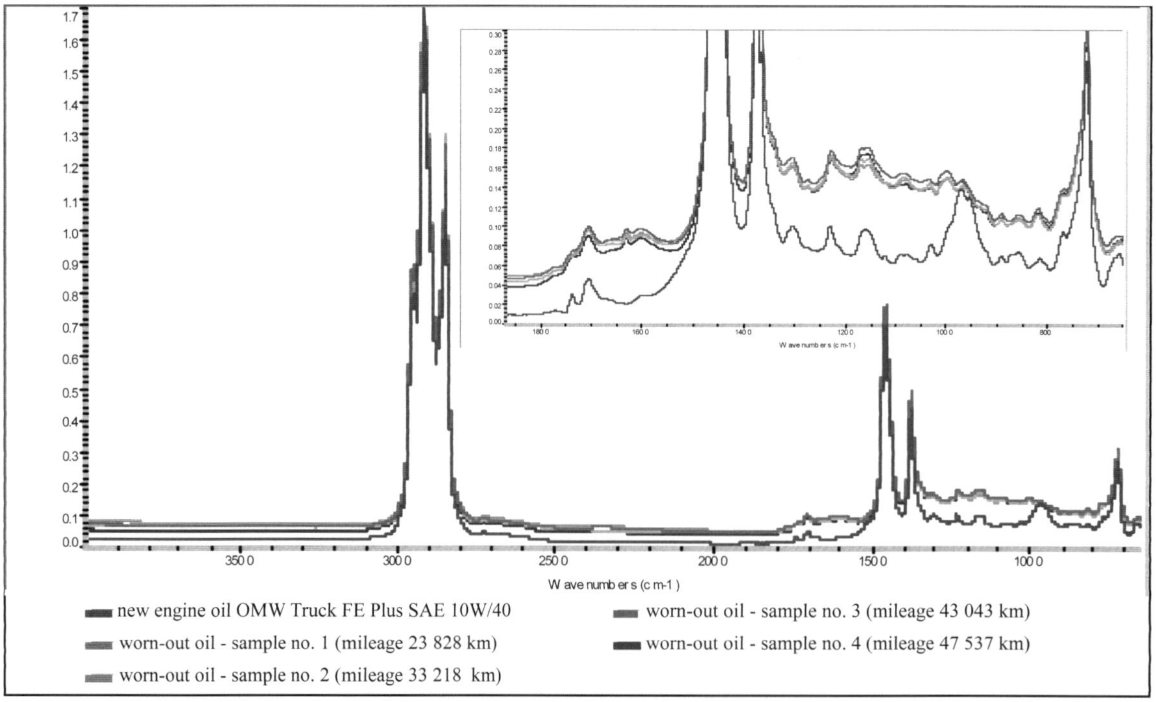

Fig. 1. The spectra of the operated oil charges in the bus after separate samplings.

The oil has a very good oxidation resistance, because the absorbance related to the oxidation of hydrocarbons did not increase in the range of 1710 cm^{-1}, i.e. the content of oxidation products containing the C=O bond in a carbonyl group did not increase.

The interval of 1610–1630 cm^{-1} which occurred in the samples is without operational risk. This area is associated with oil nitration which occurs due to the reaction of oil with nitrogen oxides from combustible gases. The increased content of nitrogen oxides may be caused by improper fuel/air mixture ratio in the combustion chamber. The range of nitrides in the FTIR spectrum indicates excessive wear of piston group, namely the leaks of piston rings.

The concentration of the elemental composition of wear metals and contaminants was monitored by AES with a rotating disk electrode. As shown in Fig. 2, in almost all oil charges the concentration of wear metals and contaminants changed when compared to the original values in the new oil.

It is apparent from Fig. 2 that the increased growth of the content of wear metals *Fe* and *Pb* was identified in all oil charges. At the third sample the concentration of *Fe*=50 mg.kg^{-1} reached the threshold of increased wear. The boundaries of increased wear were also reached by the concentration of *Pb*=15 mg.kg^{-1} for the sample no. 4. The concentration of *Cu*, which may indicate wear of bearings, did not acquire the values higher than 30 mg.kg^{-1} in any analysed oil sample. This limit is given as the lower limit on increased wear. The concentration of Al (the source may be bearing or cylinder head wear, for example) did not reach the value of 25 mg.kg^{-1} which is the limit on the increased wear. Fig. 2 shows that a sudden increase in the concentration of wear metals occurred at sample no. 2 (about in the mid of a replacement interval). This shows that the result is consistent with the conclusions of FTIR spectrometry - in this sample a marked decrease in antiabrasion additive was found.

Dust silicate particles did not get into the oil charge. Increase in *Na* could signalize the penetration of cooling liquid (coolant containing various sodium salts as an inhibitor of corrosion) into the oil charge. But this was excluded by performing FTIR spectrometry and simple operational test detecting the presence of water. Given that the sampling no. 2 and 3 was conducted during a winter season, it can be concluded that Na comes from the road salt which was used for treating winter roads.

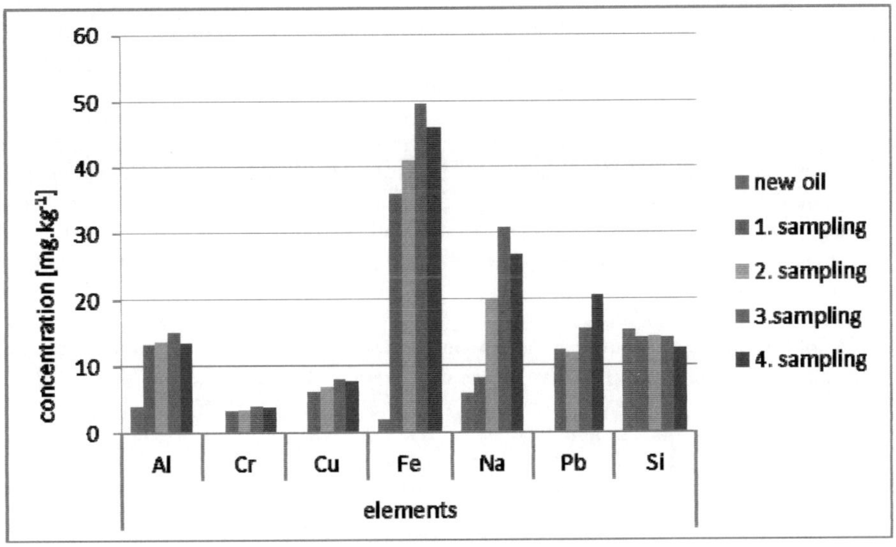

Fig. 2. The concentration of wear metals and contaminants in the oil of the operated bus.

When performing the analysis of the particle counter LNF it was found that although additional oil of 5.4l was put into the engine, this action did not provide significant increase in oil lubrication quality. Some parts of the engine showed signs of fatigue processes and increased wear processes which portend a possible accident (Table 2).

Table 2. Results of the particle counter analysis LNF Q200.

	THE NUMBER OF PERTICLES				
THE SAMPLE NUMBER	cutting wear	damage wear	fatigue wear	non-metalic particle	unclassified particle
1	6,2	15,4	52,3	18,5	0
2	1,5	6,2	33,9	12,3	0
3	1,6	0	25,6	6,4	1,5
4	69,4	35,5	673,3	313,3	3,2

Adhesive wear particles which range from 15 to 50 μ m (the particles of cutting and damage wear - Table 2, 3) were produced significantly since the manufacturer of the bus recommended the extension a replacement interval.

Table 3. The results of classification of particles according to size.

THE SAMPLE NUMBER	THE NUMBER OF PARTICLES ACCORDING TO SIZE				
	5–15 [μm]	15–25 [μm]	25–50 [μm]	50 and more [μm]	The total number of particles
1	8 894	168	48	3	9 113
2	8 637	68	37	6	8 748
3	16 686	77	6	3	16 772
4	26 386	1 718	589	42	28 735

There was a damage to the engine of the bus IrisBus CityBus 18 M; right in this case the replacement interval was prolonged by about 7500 km. Step increase in large particles (mainly those of the size ranging from 15 to 50 μm) might be considered to be a portent of unfavourable friction conditions. These particles were identified as particles connected with fatigue (fatigue wear often occurs at valve tappets [33]) and abrasive processes depending on their shape characteristics. Damage particles were also identified (Table 3).

3. Conclusion

The practical application of the TTD instrumental methods to oil samples taken from the buses proved that the generally recommended boundary values of abrasion metal concentration are not always sufficient when it comes to judging the course of wear of the engine since each engine is specific and is operated in different conditions. Another reason is that the AES is not sensitive enough to r detect the particles larger than 5 μ m. In order to judge the course of wear of the engine, it is necessary to have the data about the size-related composition and distribution of abrasion particles available; this can be achieved by analysing the particles using a laser analyser.

However, when interpreting the results of particle analysis it is also necessary to take into consideration the time intervals of sampling, re-filling the oil during operation and the efficiency of its filtering. The presence of oil ageing products (sludge and sediments) and the products of imperfect combustion of fuel and mechanical impurities (especially the abrasion from the wear of metal materials and dust particles from outside) is undesirable because it increases the wear and failure rate of the engine. Most often, oil analysis program combines the continuous attention to oil and disruptions groups of vehicles. The required contribution is indeed useful, but should not be considered significant in comparison with other objectives. Maintenance strategy based on the state defines the philosophy of a proactive maintenance which should be the real basis for the oil analysis program.

References

[1] Bekana D, Antoniev A, Yach M, et. al. Monitoring of Agricultural Machines with Used Engine Oil Analysis. *Acta Univ. Agric. Silvic. Mendelianae Brun* 2015; 63.1: 15-22.

[2] Kumbár V, et al. Oil additive and its effect. *Acta Univ. Agric. Silvic. Mendelianae Brun* 2013; 61.3: 691-697.

[3] Mihalčová J. Tribotechnical diagnosis in aircraft engine practice. *Applied Mechanics and Materials* 2013; 308: 57-62.

[4] Sejkorová M. *Analysis of Engine Oils for Transport Vehicles Using Instrumental methods*. Supervisor Dissertation work doc. RNDr. Jaroslava Machalíková, CSc. University of Pardubice, Jan Perner Transport Faculty; 2014.

[5] Sejkorová M. Tribotechnical diagnostics as a tool for effective management of maintenance. [in Czech: Tribotechnická diagnostika jako efektivní nástroj řízení údržby]. Perner´s Contacts, 2015; 10 (3): 126-136. [online]. Accessible at: http://pernerscontacts.upce.cz/40_2015/Sejkorova.pdf. [Accessed 2015, March 11].

[6] Hnilicová M, Kučera M, Pavlů J. Analysis of Hydraulic Oil in Handling Lines Baljer & Zembrod using the Methods of Tribotechnical Diagnostics. *Key Engineering Materials* 2016; 669: 451-458.

[7] Helebrant F. *Technická diagnostika a spolehlivost. Provoz a údržba strojů*, VŠB-TU, Ostrava; 2008.

[8] Latif U, Dickert FL. Conductometric sensors for monitoring degradation of automotive engine oil. *Sensors* 2011; 11.9: 8611-8625.

[9] Jakoby B, et al. An automotive engine oil viscosity sensor. *Sensors Journal*, IEEE 2003, 3.5: 562-568.

[10] Wang Simon S. Engine oil condition sensor: method for establishing correlation with total acid number. *Sensors and Actuators B: Chemical* 2002; 86.2: 122-126.

[11] Wang Simon S. Road tests of oil condition sensor and sensing technique. *Sensors and Actuators B: Chemical* 2001; 73.2: 106-111.

[12] Agoston A, Ötsch C, Jakoby B. Viscosity sensors for engine oil condition monitoring—Application and interpretation of results. *Sensors and Actuators A: Physical* 2005; 121.2: 327-332.

[13] Capone S, et al. Application of a gas sensors array to the detection of fuel as contamination defect in engine oil. In: *Sensors*, 2008 IEEE; 2008. p. 442-445
[14] Zhan Y, et al. Automatic assembly control system of bar-type engine oil cooler. In: *Mechatronics and Automation* (ICMA), 2011 International Conference on. IEEE; 2011. p. 2297-2302.
[15] Král J, Konečný B, Madač K, et al. Degradation and chemical change of longlife oils following intensive use in automobile engines. *Measurement* 2014; 50: 34-42.
[16] Machalíková J, Sejkorová M, Livorová M. Aplikace instrumentálních metod v analýze provozních hmot pro dopravní prostředky. *Scientific papers of the University of Pardubice. Series B, The Jan Perner Transport Faculty* 2012; 18: 33-54.
[17] Machalíková J, et al. Advanced instrumental analytical methods and their application for sustainable transport. *WSEAS Transactions on Environment and Development* 2006; 2.11: 1390-1395.
[18] Kastelik M, Żółtowski B. FT-IR Method Used for Evaluation of Motor Oil Condition and Quality. *Journal of POLISH SIMAC*. [online]. Accessible at: http://www.polishcimeeac.pl/Papers1/2010/025.pdf. [Accessed 2015, March 11].
[19] Tomášková M, Chýlková J, Jehlička V, et al. Simultaneous determination of BHT and BHA in mineral and synthetic oils using linear scan voltammetry with a gold disc electrode. *Fuel* 2014; 123: 107-112.
[20] Glos J, Svoboda M. Application infrared spectroscopy for monitoring the quality parameters of engine oils. In: *Transport Means - Proceedings of the International Conference* 2015; 103-106.
[21] Sejkorová M. Determination of Total Alkalinity of Motor Oil by FTIR Spectroscopy. *Chemické listy* 2013, 107: 643-647.
[22] Van de Voort FR, Sedman J, Cocciardi RA, et. al. FTIR condition monitoring of in-service lubricants: ongoing developments and future perspectives. *Tribology Transactions* 2006; 49.3: 410-418.
[23] Al-Ghouti MA, Al-Degs YS, Amer M. Application of chemometrics and FTIR for determination of viscosity index and base number of motor oils. *Talanta* 2010; 81.3: 1096-1101.
[24]Glos J, Sejkorová M. Monitoring an Engine Condition based on Tribological Diagnostics in MilitaryVehicles. *Machines, Technologies, Materials* 2012; 6: 7-10.
[25]Glos J. Modern Methods of Tribological Diagnostics. *Transport means - Proceedings of the International Conference* 2011; 26-29.
[26] Mihalčová J. Using Atomic Spectrometry and Volumetry Method for Determination of Bearing Corrosion in Tribotechnical Diagnostics of Engines. *Applied Mechanics and Materials* 2014; 616: 110-117.
[27] Mihalčová J, Rimár M. Control of Activity of Aircraft Engines by Analysis of Wear Debris Particles in Terms of Shape and Size. *International Journal of Engineering Research in Africa* 2015; 18: 136-143.
[28] Kučera M, Aleš Z, Pavlů J, et al. Applying of Automatic Laser Particle Counter as Technique to Morphology Assessment and Distribution of Wear Particles during Lifetime of Transmission Oils. *Key Engineering Materials* 2016; 669: 417-425.
[29] Kumbár V, Glos J, Votava J. Monitoring of Chemical Elements During Lifetime of Engine Oil. *Acta Univ. Agric. Silvic. Mendelianae Brun.* 2014; 62.1: 155-159.
[30]Machalíková J, Schmidová E, Sejkorová M, Černý Š. Morphology of Wear Particles from Motor Oils and Oil Filters. *Polytransport systems: Proceedings of the 5th Russian Scientific and Technical Conference Polytransport Systems*. Krasnoyarsk : Siberian Federal University; 2007, p. 21-30.
[31] Lucas M, Anderson D. Metody laboratorních rozborů olejů. Příručka mazání a tření. SPECTRO Inc., Littleton, Massachusetts, U.S.A.; 1997.
[32] Guo B, Lyons W, Ghalambor A.: *Petroleum production engineering*. Elsevier Science and Technology Books; 2007.
[33] Straka B. *Motorové oleje a tribotechnická diagnostika naftových motorů*, Praha: Nadas; 1986.
[34] Černý J. *Vlastnosti motorových olejů – nečistoty a saze v motorovém oleji*. [online]. Accessible at: https://www.oleje.cz/clanek/Vlastnosti-motorovych-oleju---Necistoty-a-saze-v-motorovem-oleji. [Accessed 2015, March 11].

Forum on Innovative Technologies and Management for Sustainability (ITMS'2016)

ktu 1922 panevėžys faculty of technologies and business

The 11th International Conference Intelligent Technologies in Logistics and Mechatronics Systems (ITELMS'2016), 28-29 April 2016, Panevėžys, Lithuania

Research of dynamic stability of articulated UGV

Jaskółowski M.B.[a]*, Konopka S.[a], Łopatka M.J.[a]

[a]*Military University of Technology, Gen. Sylwestra Kaliskiego Str. 2, 00-908, Warsaw, Poland*

Abstract

This paper provides a comparison of three structures articulated unmanned ground vehicles. The compared parameters were the maximum value of speed, and lateral acceleration while driving on prepared trajectories. The comparison was made on the results of simulation research of dynamic stability. The analyzed structures different from each other in: numbers of wheels, location of oscillation and turn joint, distribution of basic components and cargo space. Moreover, in this paper as compared of influence location of cargo on dynamic stability. Simulation tests were performed using the MSC. Adams software for modelling multibody systems. The paper also describes the simplifying assumptions and parameters used in the construction of the numerical models for the created platforms structures. In addition, the paper includes the methodology adopted to carry out simulation tests, together with the criterion for dynamic stability assessment. On the basis of simulation research specified which of the created structures of unmanned ground vehicle characterized by the most advantageous dynamic stability.

© 2016 M. B. Jaskółowski, S. Konopka, M. J. Łopatka.
Peer-review under responsibility of the Kaunas University of Technology, Panevėžys Faculty of Technologies and Business

Keywords: articulated unmanned ground vechicle; MSC Adams; dynamic stability; roll over tests.

---

1. Introduction

In the last years increasing use of unmanned ground vehicle (platform). The increase in use caused by better operating parameters compared with other construction solutions such as skid steer system. Details on research of basic systems in platforms with skid steer system are described in [5, 6, 7]. The advantages of platforms with articulated steering system are simple construction, high ability to copy terrain, high maneuverability, low turning resistance and low power consumption. However, a significant drawback of these platforms is their stability. Comparison of stability platform with articulated steering system and skid steer system are described in the [3, 4]. Moreover, there are lots of publications [1, 2] a defining influence of parameters such as the position of turn joint and oscillation joint or cargo placement on the stability of platforms with articulated steering system. However, these studies refer only to the static stability. This means that these studies did not indicate the impact of these parameters on the stability of the platform when moving. A similar problem also concerns articulated building machines such as loaders and dumpers. For these machines also mainly performed are static tests more fully described in [11] Moreover, in [11] it was attempt to develop a method for research of dynamic stability these type machine. Nevertheless, due to significant differences in both the structure, overall dimensions as well as operating conditions causing that they cannot directly implement these methods for research of dynamic stability articulated platforms. Is required that make attempt to identify the research method of dynamic stability

* Corresponding author. Tel.: 48 261 83 74 16.
E-mail address: miroslaw.jaskolowski@wat.edu.pl

© 2016 M. B. Jaskółowski, S. Konopka, M. J. Łopatka.
Peer-review under responsibility of the Kaunas University of Technology, Panevėžys Faculty of Technologies and Business

articulated platforms. The developed method should allow to define influence of parameters such as mass distribution or position of the basic elements and cargos on the dynamic stability.

2. Structures of articulated platforms

In order to determine the effect of referred parameters on dynamic stability created three structures platforms articulated. It was assumed that each of created structure must consist of the same basic elements and cargo. As basic elements adopted: power unit, drive system, fuel tank and hydraulic oil, tires and wheels, and elements of the steering system. Moreover, cargo takes the form of equipment for eight persons and his maximal weight is equal 240 kg. Furthermore, been assumed that the weight of the cargo can be split into parts with a mass of 20 kg with dimensions 450 mm x 360 mm x 160 mm.

First from created structures reminds own construction solution applied in the scrapers. It equipped with a joint turn located directly on the front axle at a height of 600 mm above the ground, which provides a maximum turn angle of the frame equal to 90°. In addition to this front and rear section of platform are interconnected by an oscillation joint. This allows to unlimited oscillation section of platform to each other. Moreover, this structure is equipped with a drive system consisting of the four wheels independently driven. Except that, in front section is placed drive unit while in rear section is located cargo space. Adopted the distribution of the basic components results that mass of the front section is equal 220 kg. The combined center of gravity front section is located at a height of 500 mm above the ground at a distance of 80 mm behind the front axle. The mass of the rear section is equal 100 kg and its combined center of gravity is located at a height of 470 mm above the ground at a distance of 240 mm before of rear axle. however, the center of gravity cargo is located at a height of about 510 mm above the ground at a distance of 370 mm before rear axle (Fig. 1). This means that the total weight of the platform without load is equal to about 320 kg, and with a cargo of 560 kg.

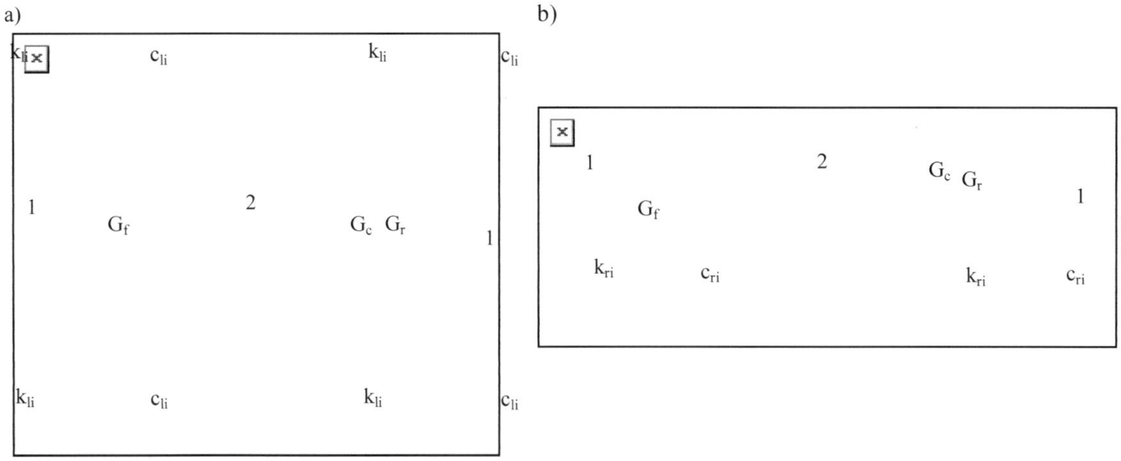

Fig. 1. Scheme of structures four-wheeled platform a) view from top, b) view from side: G_c –center of gravity front section of platform, G_r – center of gravity rear section of platform, G_c – center of gravity cargo, k_{ri} – radial stiffness coefficient of tires, c_{ri} – radial damping coefficient of tires, k_{li} – lateral stiffness coefficient of tires, c_{li} – lateral damping coefficient of tires, 1 – locations of basic components, 2 – location of space cargo.

The second structure is six wheels. Compared to the first structure in the position of the rear axle inserted bogie. Furthermore, it has changed the distribution of mass. Adopted that the drive unit is placed in rear section and a cargo space was split between front and rear section. Such a distribution of basic elements causes that the mass of the front section is equal 130 kg. the Combined center of gravity front section is 150 mm behind the front axle at a height of 340 mm above ground. However, mass of rear section is equal 230kg and his combined center of gravity is 60 mm before the rotation axis of bogie at a height of 450 mm above ground. In addition, the mass of the cargo attributable to a front section is equal 80 kg and his combined center of gravity is located in distance of 130 mm before of front axle at a height of 550 mm above ground. Whereas weight of cargo attributable to the rear section is equal 160 kg and its center of gravity is positioned at a distance of 420 mm behind the rotation axis of bogie at a height of 760 mm above the ground (Fig. 2). Mass this structure of platform is 360 kg without cargo and 600 kg with cargo.

a)

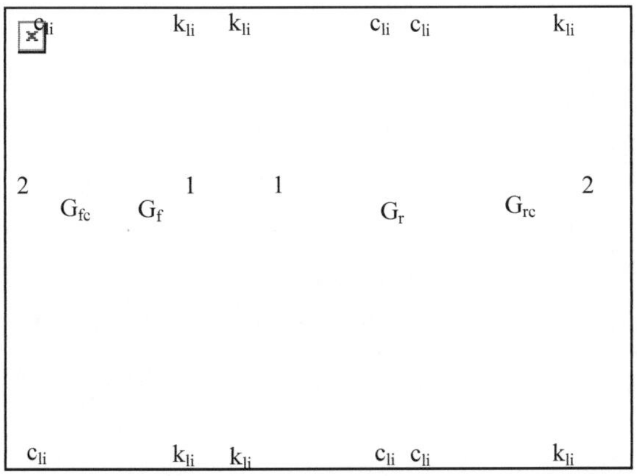

b)

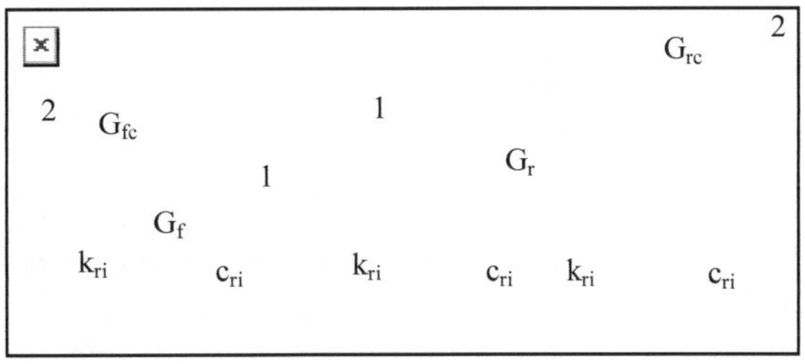

Fig. 2. Scheme of structures six-wheeled platform a) view from top, b) view from side: G_c –center of gravity front section of platform, G_r – center of gravity rear section of platform, G_c – center of gravity cargo, G_{fc} – center of gravity cargo in front section, G_{rc} – center of gravity cargo in rear section k_{ri} – radial stiffness coefficient of tires, c_{ri} – radial damping coefficient of tires, k_{li} – lateral stiffness coefficient of tires n, c_{li} – lateral damping coefficient of tires, 1 – locations of basic components, 2 – locations of space cargo

The last third structure of platform is significantly different from the others. First substantial change is increasing the number of wheels in the front section. In addition, couples of wheels in side connected in boogie. Another important change is the position of the joint turn with integrated oscillation joint. The joint turn is located

700 mm behind the axle rotation of front bogie at a height of 400 mm above the ground. Furthermore, was lowered to 45 ° a maximum value of turn angle for this structure. Moreover, an oscillation joint is integrated with a turn joint and a front section. The third significant change is the weight distribution. These mean that position of base elements and their local centers of gravity also will change. It has been assumed so that in the front section is located, the drive unit and the hydraulic oil tank, while in the back section is positioned fuel tank. In addition, it is assumed that the cargo space is split between the front and rear section of platform (Fig. 3).

In created structure the weight of front section is equal 260 kg. Moreover, the combined center of gravity front section is located at a height of 680 mm above the ground at a distance of 290 mm behind the rotation axis of front bogie. In additional, weight of rear section of created structure is equal 140 kg. The combined center of gravity rear section is 40 mm behind the rotation axis of rear bogie and at a height of 470 mm above the ground. In created structure the weight of cargo on a front section is 60 kg. Its center of gravity is located at a distance of 490 mm in front of the rotation axis of front bogie at a height of 600 mm above the ground.

a)

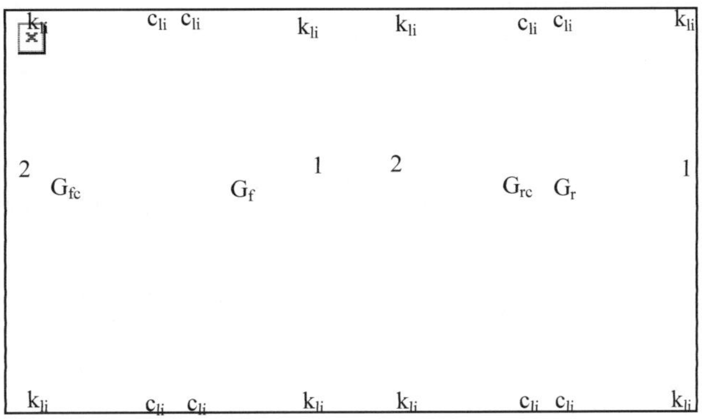

b)

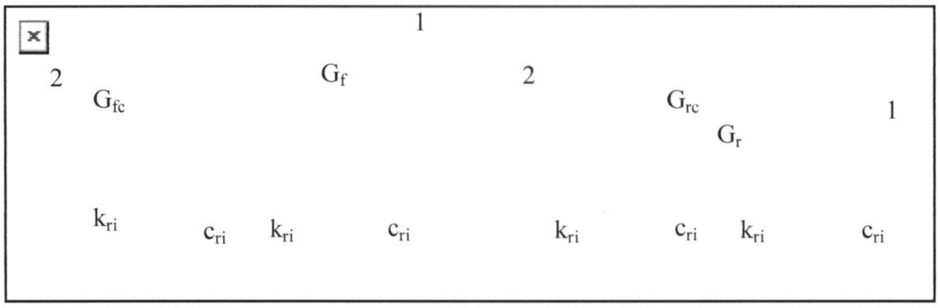

Fig. 3. Scheme of structures eight-wheeled platform a) view from top, b) view from side: G_c –center of gravity front section of platform, G_r – center of gravity rear section of platform, G_c – center of gravity cargo, G_{fc} – center of gravity cargo in front section, G_{rc} – center of gravity cargo in reart section k_{ri} – radial stiffness coefficient of tires, c_{ri} – radial damping coefficient of tires, k_{li} – lateral stiffness coefficient of tires n, c_{li} – lateral damping coefficient of tires, 1 – locations of basic components, 2 – locations of space cargo.

Furthermore, weight of the cargo on the rear section is equal to 180 kg. the center of gravity cargo on the rear section is located at a height of 600 mm at a distance of 130 mm before the rotation axis of rear bogie. Weight of created platform structure is 400 kg without cargo and 600 kg with cargo.

3. Simulation models created structures of articulated platforms

In the order perform models of created structures platforms articulated was used software for modelling multibody systems MSC. Adams. Performed models of structures started off create a model of pneumatic wheels. Model pneumatic wheels are made on the basis of the existing literature [8, 9]. Created model is comprised of a finite number of spheres with an adopted diameter. The spheres are connected with the wheel rim by using stiffness-damping elements in the radial and transverse. The coefficients of the radial and transverse elasticity were determined on the basis of data given by the tire manufacturer, and from the relation described in [10]:

$$k_r = \frac{P_o}{f_n} \tag{1}$$

where: k_r - radial stiffness coefficient of tires, P_o - nominal load of tire, f_n - nominal deflection of the tire expressed in equation:

$$f_n = P\frac{1}{\pi \cdot p_o \cdot \sqrt{2 \cdot r \cdot D}} \tag{2}$$

where: P - the maximum permissible load on the tire, p_o - nominal pressure in the tire, r - radius of curvature of the tread, D - diameter of the outer tire.

In the next step based on [1], estimated value of the lateral stiffness coefficient:

$$k_l = \frac{k}{r_D} \tag{3}$$

Where: k_l - lateral stiffness coefficient of tires, k - coefficient of resistance tires to drift side equal to: ⍰⍰ = 5.7·⍰⍰,
rD - radius of deformation tire.

Then based on [10] was determined value of radial and lateral damping coefficient tire according to the formula:

$$c_{r,l} = 2 \cdot k_{p,l} \cdot T \cdot \xi \tag{4}$$

where: $c_{r,l}$ - radial and lateral damping coefficient of tires, ξ - dimensionless damping coefficient, T - time constant defined relationship:

$$c_{r,l} = 2 \cdot k_{p,l} \cdot T \cdot \xi \tag{5}$$

where: cp - radial and lateral damping coefficient of tires, ξ - dimensionless damping coefficient, T - time constant defined relationship:

$$T = \sqrt{\frac{m}{k_r}} \quad (6)$$

where: m - load on single wheel of the platform structure.

In addition, to the model it was assumed that the weight of the basic elements and cargo would be treated as concentrated points, whose position is dependent on the structure of the platform. Moreover, was assumed that the combined centres of gravity basic elements and cargo are rigidly connected to the main frame of the model. Furthermore, was assumed that the turn joint and oscillation joint are position is dependent on the structure of the platform. The ground on which the model moves is non-deformable. Created in MSC. Adams model of eight - wheels platform is presented on figure 4.

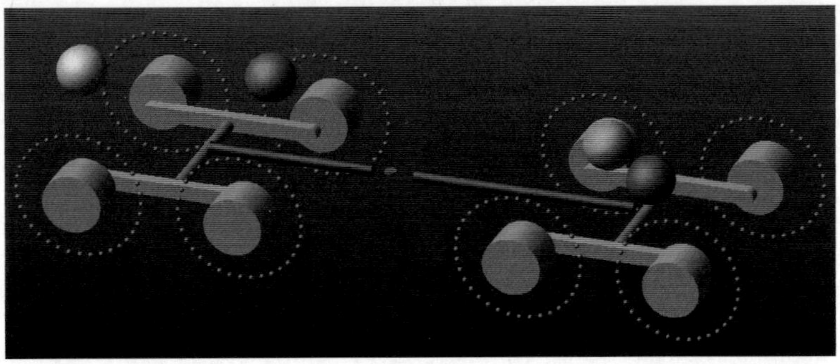

Fig. 4. Model of eight-wheels structures of articulated platform created in MSC. Adams.

4. Methodology of research dynamic stability

Development of research methodology of dynamic stability began with an analysis of existing tests. For the analysis adopted tests of dynamic stability of the cars and trucks, construction machines with articulated steering system and military vehicles [11, 12, 13, 14, 15]. Following this analysis assumes that the most effective method for determining the dynamic stability is a test of constant velocity. This test are subjected cars and trucks, some construction machines and military vehicles. In this test measured is value of lateral acceleration. According to the standards of a car should be stable with the lateral acceleration equal to 0.2g. Whereas for trucks the value of this parameter cannot be less than 0.4g. The discussed test can be performed in two ways. In the first of test the examine object accelerates to a certain speed and then gradually reduces turning radius until it reaches the limit values of lateral acceleration. While in the second test the examined object moving on a trajectory in the shape of a circle with a diameter of 30m. When driving gradually increased the speed until the limit values of lateral acceleration. However, a significant disadvantage of this test is the possibility of falling of the object to slide before reaching required value of lateral acceleration. In view of significant differences in the construction, character of work and achieve speeds between platforms and the cars, decided to modernize the methodology of the test of constant speed. Modernization started by identifying the type of test. To the research adopted a second type of test. However, in this test was changed trajectory shape on which examined object are moving. In order to accurately determine the influence of selected parameters on dynamic stability adopted two trajectories. Adopted trajectories different from each minimum value of the turning radius. In the first trajectory the minimum value of turning radius is equal 5 m which corresponds to the radius turning in cars of middle class. However, due to requirements of platforms for high maneuverability assumed that in the second trajectory value of turning radius is equal 2.5 m which is about half less than in the first trajectory. In order to ensure of examined object, the gradual acceleration it assumed that the first part of the trajectory must be a straight line with a length of not less than 50m. Moreover, in order to reduce influence of a violent changes in steering angle in both trajectories adopted segments in the shape of a spiral of decreasing radius. The final stage of both trajectory is a circle with an adopted value of turning radius. An exemplary trajectory is shown in Figure 5.

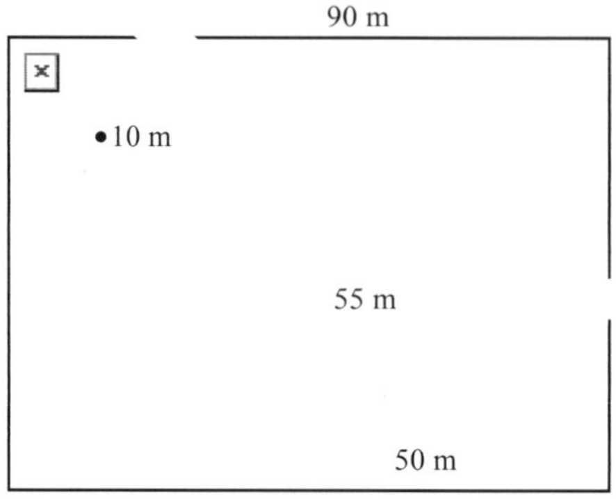

Fig. 5. Basic dimensions of first trajectory.

Moreover, due to the unavailability of information on required values of lateral acceleration relating to the platform articulated, developed own guidelines. Assume that during the study will be determined maximum value of speed and lateral acceleration at which the examined object is not lost stability. Loss of stability is related to the against removing wheels from ground examined object. This is equivalent to the decrease the value of reactions 0kN on wheels. This means that it is necessary to measure the response of the individual wheels. Additionally, in order to determine the effect of the position of cargo on dynamic stability research will be conducted for the platform with cargo and without cargo.

In summary prepared research method of dynamic stability is based on movement of the created structures of articulated platforms the defined trajectory at a constant speed. In the case of crossing platform for the entire trajectory the research must be repeated by increasing the value of speed by 1 km/h. However, in the event of rollover platform during the research assumed that the maximum value of speed should be lower about 1 km/h than the currently accepted. For maximum speed you should read the obtained values of lateral acceleration and reaction on wheels.

5. Simulations research of dynamic stability articulated platforms

Simulation research of dynamic stability platform articulated conducted according to the established methodology. However, the study also adopted the following guidelines. In order to eliminate the possibility of wheel slip under study assumed that the coefficient of traction to the ground is equal to 1. Moreover, it was assumed that drive is transmitted simultaneously on all wheels which is consistent with character of the work of the hydraulic drive system. Acceleration time has been adjusted so that the test platform has reached a predetermined speed before starting the process turning. Sample waveforms of the lateral acceleration and reaction to wheels shown in figures 7 and 8.

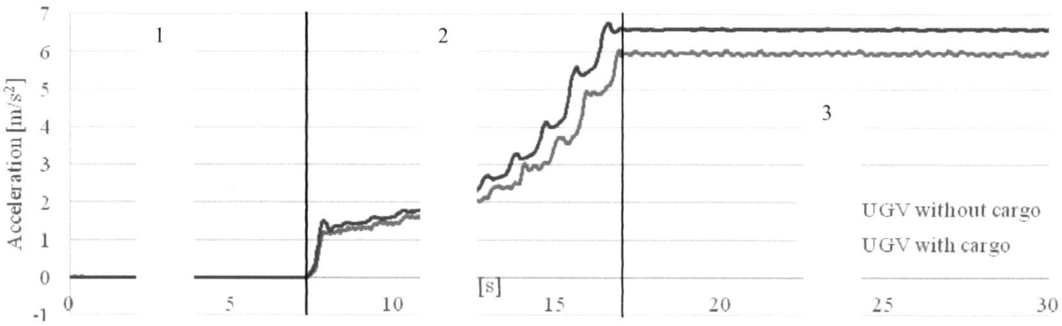

Fig. 6. Lateral acceleration in rear section of four wheeled articulated platforms while dri the first trajectory, 1 - stage of moving straight ahead – acceleration, 2 - stage of increasing steering angle - consi iving speed, 3 - stage of maintaining constant steering angle - constant driving speed.

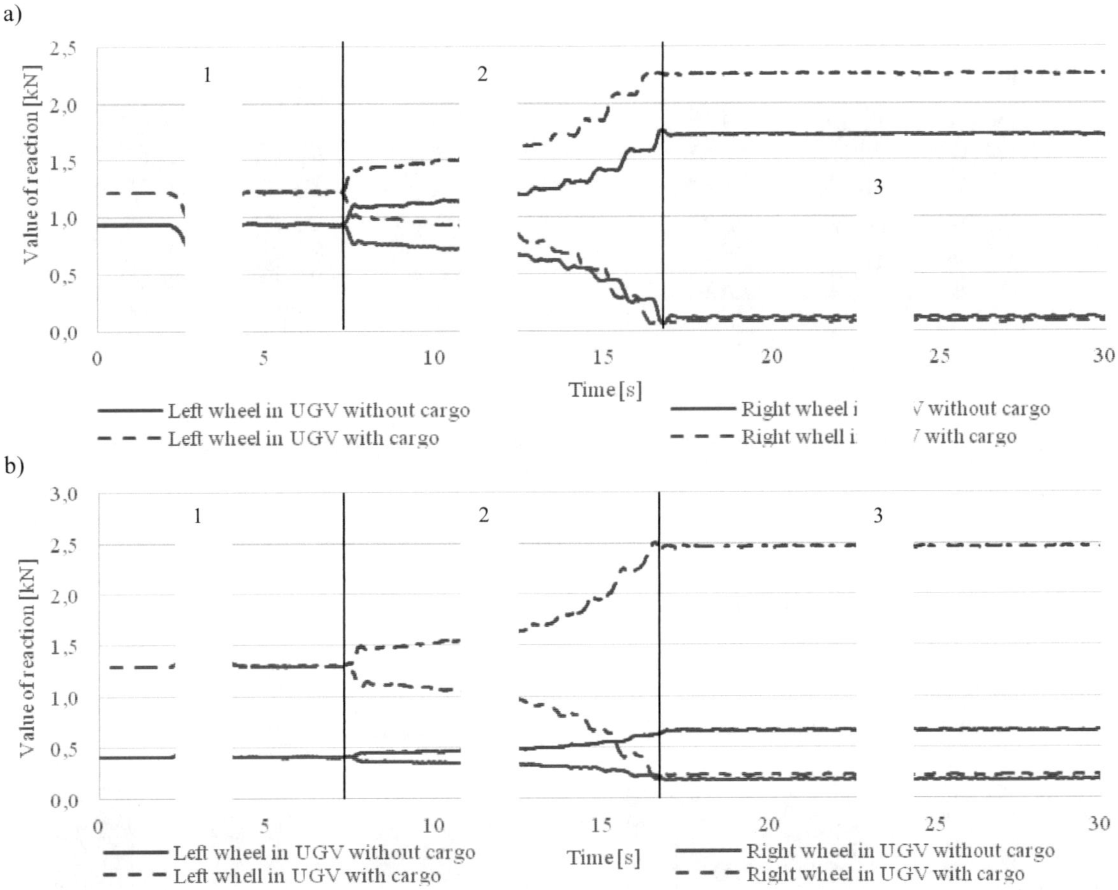

Fig. 7. Value of reaction on wheels of four wheeled articulated UGV while driving on the first trajectory, a) reaction of front section, b) reaction of rear section, 1 - stage of moving straight ahead – acceleration, 2 - stage of increasing steering angle - constant driving speed, 3 - stage of maintaining constant steering angle - constant driving speed.

On the presented diagrams marked the three stages of the performed research. In the first stage of examined object moves along a straight line gradually increasing its speed. This can be seen on the graphs presented as the values of the measured variables are subject to minor changes. The occurrence of a change of reaction to wheels it exerted by start of acceleration process. In the second stage, after reached a required value of speed platform, is gradually increase of angle of turn. In the chart lateral acceleration is shown by the gradual increase in the value. While the graphs of reaction on wheels this is shown through a gradual increase the load of outer wheel and reducing load of the inner wheel. The last third step of maintaining a constant angle of turn is shown by stabilizing the measured variables. On the received waveform depicted in the charts also shows influence of cargo on values measured variables. The value of lateral acceleration is greater than the value obtained for the platform unloaded in second and third stage of research. Also shows a differential in value of reaction on the wheels in all stages. Moreover, it can be seen that the value of the reaction the inner wheel is reduced to nearly the same level as the platform without cargo. This value of the reaction on wheel say about the fact that a further increase in speed can result in detachment of the wheel from the ground which is equivalent to the loss of stability. n the figures 8 and 9 show the obtained values of the maximum speed and the lateral acceleration created structures.

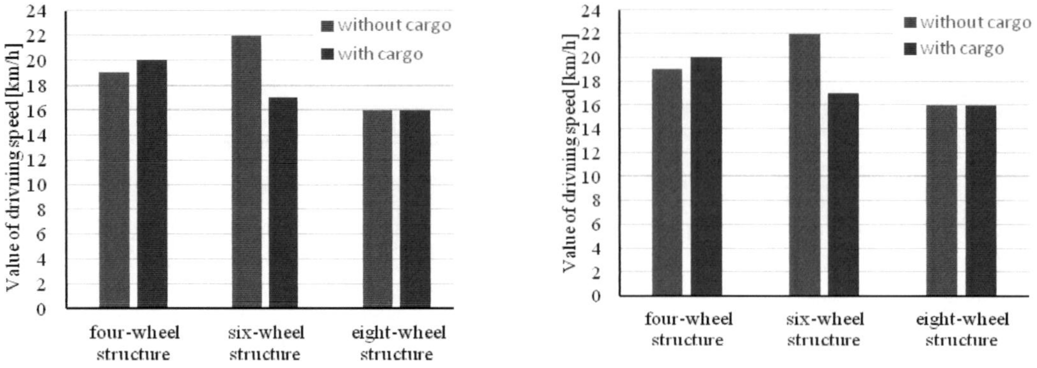

Fig. 8. Maximum value of speed created structures of platforms a) driving on first trajectory, b) driving on second trajectory.

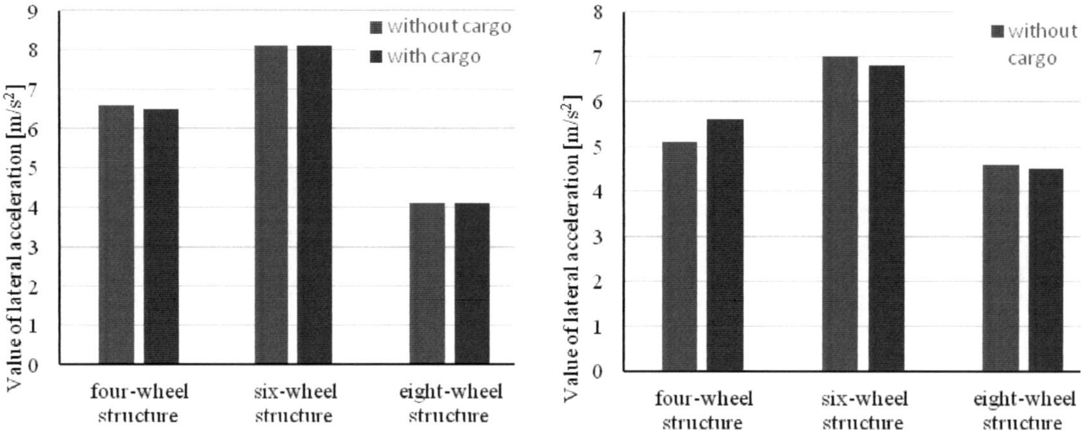

Fig. 8. Maximum value of lateral acceleration created structures of platforms a) driving on first trajectory, b) driving on second trajectory.

When comparing the values obtained for each trajectory it is evident that the values obtained from research performed on the second trajectory are much lower. This is due to higher value of turn angle in the second trajectory. Furthermore, the results show that the location of cargo has a significant influence on the dynamic stability. However, the impact is different for each of the structures. In the four- wheeled structure adding cargo increases the measured variables in the structure of their reduction six- wheeled. However, in the structure eight-wheeled not affect the resulting value. This means that the location of the load is one of the decisive parameters determining the dynamic stability. From the obtained value of the lateral acceleration it is apparent that each of the structures reaches a value higher than 0.4 g. Assuming that 10% slope of the land is tantamount to forcing the lateral reaching 0.1g. It was found that the created structures should be able to move around the 30% slopes of the stability reserve equal to at least 0.1 g. Furthermore, the results show that the highest vehicle speed and the lateral acceleration is characterized by a six-wheeled structure.

Conclusion

Based on results made simulation research it was found that the weight distribution and the structure types of articulated platforms has a significant influence on the dynamic stability. This is particularly noticeable through a variable value of maximum speed and lateral acceleration for each of created structures. The differences between the values obtained may also derive from different levels of maneuverability determined by the position of steering system components. Each of created structures was characterized by a different value of turn angle when moving along the same trajectory. Through that appears is the relationship that the higher value turning angle the lower value speed and lateral acceleration. Moreover, the made research shows that the location of cargo also influences the dynamic stability. Location of cargo may improve, worsen or have no effect on the value of parameters analyzed.

In summary described research method enables to define influence of parameters such as the position of the steering system components, weight distribution and location of cargo on the dynamic stability of the articulated platform. Moreover, from made simulation research the most favorable dynamic stability characterized by a four-wheeled structure. This structure is characterized by the highest values of speed and lateral acceleration when moving along trajectories. It should be noted that the described method does not take into account the effect of the slope of the terrain which significantly can affect the results. This means that further research should be developed a method of determining dynamic stability in areas of different slope.

References

[1] Dąbrowska A, Jaskółowski M, Konopka S, Przybysz M. Stability research of lightweight six wheeled ground vechicles. *ITELMS'2015 10th International Conference Intelligent Technologies in Logistics and Mechatronics Systems*. Panevezys. Lithuania; 2015.

[2] Dąbrowska A, Jaskółowski M, Krogul P, Rubiec A. Stability evaluation of lightweight track ground vehicles. *ITELMS'2015 10th International Conference Intelligent Technologies in Logistics and Mechatronics Systems*. Panevezys. Lithuania; 2015.

[3] Dąbrowska A, Jaskółowski M, Krogul P, Spadło K. Mobility evaluation of a lightweight four-wheel unmanned ground vehicle with articulated steering system. *10th International Conference Intelligent Technologies in Logistics and Mechatronics Systems*. Panevezys. Lithuania; 2015.

[4] Dąbrowska A, Konopka S, Przybysz M, Rubiec A. Ability to negotiate terrain obstacles by lightweight six wheeled unmanned ground vehicles. *10th International Conference Intelligent Technologies in Logistics and Mechatronics Systems*. Panevezys; Lithuania. 2015.

[5] Łopatka MJ, Muszyński T, Rubiec A. Simulation identification of Fire Rescue Robot suspension loads. *18th International Conference on Methods and Models in Automation and Robotics (MMAR)*. Miedzyzdroje. Poland; 2013.

[6] Łopatka MJ, Muszyński T, Rubiec A. Loads analysis of skid steer robot drive system. *18th International Conference on Methods and Models in Automation and Robotics (MMAR)*. Miedzyzdroje. Poland; 2013.

[7] Bartnicki A, Sprawka P, Rubiec A. Remote control system for rescue robot. *Symposium on Mechatronics Systems, Mechanics and Materials*. Jastrzebia Gora. Poland; 2013.

[8]Nakashima H, Yamazaki M. 3D FEM procedure for static sinkage problems of tire on soil surface *Japanese Society of Agricultural Machinery*. Kobe, Japan; 1995.

[9]Wong J. Y. *Theory of ground Vehicles*. John Wiley & Sons. Inc, New York; 1993.

[10] Анкуаинов Д. Т., Малиновский Е. Ю., Гайцгори М. М.: *Динамика самоходных машин с шарнирной рамой*. Москва; 1974.

[11] ISO 20474: 2008: *Earth-moving machinery –Safety, Part 1-14.*

[12] ISO 3888: *Part 1 – Test procedure for a severe lane change.*

[13]ISO 7975: Braking in a turn - open loop test procedure.

[14] ISO 8379: Rough terrain truck – stability test.

[15] The Engineering Society for Advancing Mobility Land Sea Air and Space: *Surface Vehicle Standards*, SAE Handbook; 2003.

Forum on Innovative Technologies and Management for Sustainability (ITMS'2016)

panevėžys faculty of technologies and business

The 11th International Conference Intelligent Technologies in Logistics and Mechatronics Systems (ITELMS'2016), 28-29 April 2016, Panevėžys, Lithuania

Issues with controling hydrostaticly driven manipulator using an intuitive human - machine interface

Krogul P.[a]*, Typiak R.[a]

[a]Military University of Technology, Kaliskiego 2 street, Warsaw 00-908, Poland

Abstract

Continuous aspiration to improve the intuitive control of manipulators with hydrostatic drive systems has become the basis for researches of many authors. In most cases this type of control is being accomplished thanks to the usage of specially shaped lever mechanism which are capable of copying and tracking the operator's hand movements. The following article present one such system based on recreating the scaled version of the controlled manipulator. During the development of the intuitive control system, there has been a number of issues which needed to be resolved in order to achieve a desired control accuracy. The following paper present selected issues and steps taken in order to address them.

© 2016 P. Krogul, R.Typiak.
Peer-review under responsibility of the Kaunas University of Technology, Panevėžys Faculty of Technologies and Business

Keywords: control; manipulator; hydraulic drive system.

1. Introduction

The diversity of currently emerging structures of manipulators are a result of ever newer challenges posed by man. Apart from the industrial manipulators performing strictly programmed tasks, the remaining structures are controlled directly by the operator. Taking into account the complexity of current manipulator structures, the conventional manner of their control becomes very unintuitive, especially for inexperienced operators. Therefore, there is a lot of stress being put on improving control efficiency by utilizing dedicated controllers making use of the natural psycho-physical abilities of the human body. Most of the currently available solutions are based on lever mechanisms which aim to project the operator's hand position onto controlled machine's sections [1,2,3,4,5,6]. With this approach, the operator's task is reduced to positioning and orientating the tip of the manipulator which greatly relieves him mentally and physically and also improves the accuracy and speed of the conducted task. Such solutions are being introduced in backhoe excavators, medical robots, casting manipulators, mobile robots and anthropomorphic manipulators. One of the fastest growing areas of application for these control systems are the hydraulically driven manipulators [7,8]. Their significant load capacity which can exceed 100 kg and their reach of more than 4 m make them hard to control using traditional means. Additionally, they are most often teleoperated as part of high mobility unmanned ground platforms [9,10,11,12,13,14,15,16]. Within those systems, the hydraulically driven manipulators may serve additional purposes. For example, because of an unknown terrain structure [17,18,19], they sometimes may be used as means of improving platform's stability [20,21]. In such a scenario the intuitive control system's task is to translate slight operator's hand movements onto significant translations of the manipulator's actuators. To date there has been a number of scientific papers describing different kinematic structures of intuitive controllers in

* Corresponding author. Tel.: +48-261-83-74-16; Rafał Typiak. Tel. +48-261-83-73-06.
E-mail address: piotr.krogul@wat.edu.pl; rafal.typiak@wat.edu.pl

© 2016 P. Krogul, R.Typiak.
Peer-review under responsibility of the Kaunas University of Technology, Panevėžys Faculty of Technologies and Business

terms of their ergonomics and hand copying capabilities. Unfortunately, there is no data regarding the accuracy they provide when interacting with manipulation, real world objects and the issues needed to be addressed in order to increase it. The following paper identifies and addresses several of the issues face when trying to develop an intuitive control system for a hydrostatically driven manipulator.

2. Selected issues

In order to achieve a properly functioning intuitive control system, it's crucial to accurately follow the control signals by the controlled machine. In order to generate said signals, a series of linkage mechanisms need to be carefully configured in order to achieve the highest position accuracy within the special ranges achieved by the operator's hand. Most often they take form of scaled down kinematics of the controlled manipulator. In such a scenario, the changes in the measured angles between the various controller's sections are converted into control signals for the hydraulic cylinders. The omission of a nonlinear relationship between the movement of the actuators and the change in the rotation angles occurring in this case, leads to a decrease of movement accuracy of the working tool manipulator. Another problem which must be tackled is the time delay between the control signal generation and controlled manipulator's response. Controlling the movements of these types of manipulators involves the movement of large masses. This in turn results in the formation of significant dynamic loads in the initial and final phases of movement. Therefore, high copying speeds of the operator's movements are not desirable and careful tuning of this parameter should be made based on a compromise between the speed and movement stability. Frequently, during the implementation of intuitive control systems it's desired to eliminate the displacement error. This, however may lead to constant movements of the hydraulic cylinders around the target value. This phenomena may also be a result of tremors of the human hand, which indirectly generates a control signal, or the accuracy of displacement sensors. All of said error sources will cause actuator's displacement fluctuations because of a zero deadband. The amplitude of these fluctuations is further increased because of the fact, that slight movements of the linkage mechanism generate significant movements commands for hydraulic cylinders, which in turn limit the control capabilities of the manipulator.

3. Tested manipulator

The study was conducted using a manipulator from the EOD / IED support robot [22]. It has a serial-parallel kinematic structure with 6 degrees of freedom and a hydrostatic drive system (Figure 1). The manipulator consists of: a turntable, boom, the long and the short arm and a rotating gripper. The ability to simultaneously control multiple actuators was carried out using load sensing principle. In the examined solution, the control of hydraulic oil flow volume and direction, is being achieved using a Danfoss PVG32 hydraulic distributor with proportional PVED-CC electromagnetic valves using CAN bus for communication. In order to measure the displacement of individual pistons inside hydraulic cylinders, analog distance measuring sensors had been used. Their accuracy is of 0.1% of the 1250 mm measuring range.

a) b)

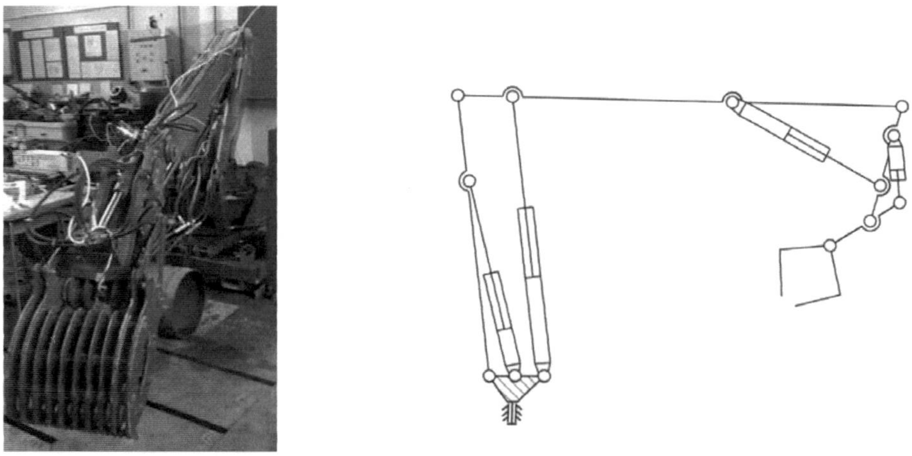

Fig. 1. (a) tested manipulator; (b) kinematic structure of the tested manipulator.

4. Controller's structure

The controller's structure is lever based, which allows to copy and track each individual manipulator section. It has been designed and developed using Catia V5 development software. The structure of the mechanism reflects the scaled kinematics of the controlled manipulator (fig. 3). The controller's size has been scaled down tenfold in relation to the controlled object. This value has been chosen for ergonomic reasons, allowing the controller to remain inside the operational area of a human hand (fig. 4). During the development of the controller's structure a special handle has been created which not only allows the operator to easily drive the work tools into a desired position but additionally, he's give the ability to set the desired kinematic configuration using that same hand, if needed. The entire structure of the model has been implemented using 3D printing with an ABS material. This approach has made it possible to reduce the costs associated with the construction of this mechanism. Rotation sensors user for the controller consisted of single turn 14 bit absolute encoders using CAN bus and a single analog length sensor with a measurement range of 1250 mm, used to register the position of the long arm.

a) b)

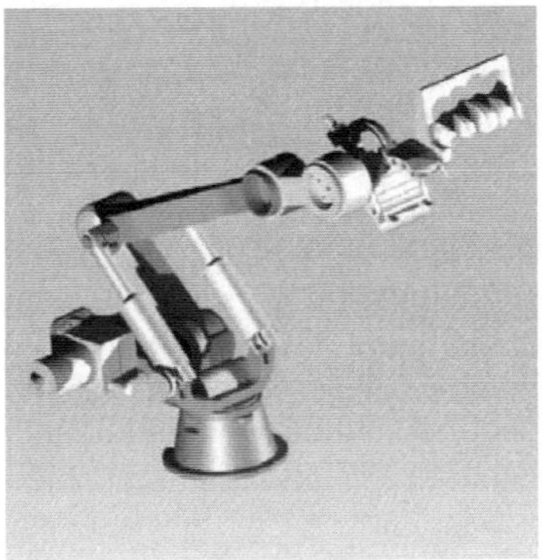

Fig. 3. (a) controller's mechanism's 3D model; (b) finial controller's mechanism.

a) b)

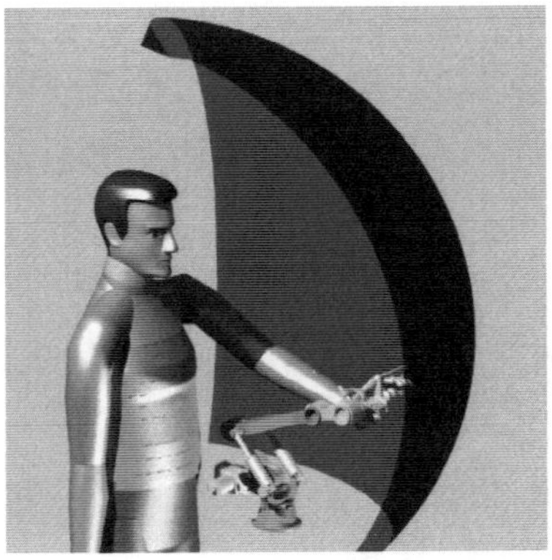

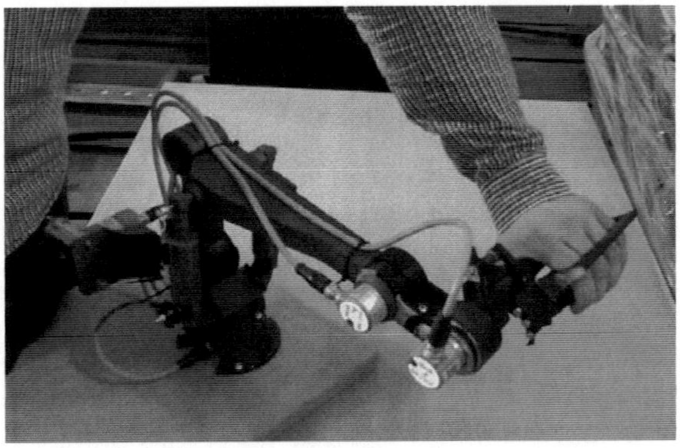

Fig. 4. (a) 3D model of the controller's manipulation area; (b) controller's manipulation in progress.

5. Control system

The developed intuitive control system consists of two main components: the hardware platform and the software control algorithms. It's crucial to provide enough computing power and appropriate communication interfaces for the software side to be able to run smoothly while maintaining minimal costs so that using the developed solution may be economically justifiable. From the software side however, it's important to make use of specific functions and resources made available by selected hardware. Thus, creating a control system capable of computing "real time" control signals for intuitive control of a hydraulically driven manipulator has to take both elements into account. The developed solutions runs in an industrial PLC which accepts a wide range of power voltages (9-36V), utilizes two independent CAN bused for communication and has an IP54 casing. All of the mentioned characteristics make it an appropriate choice for industrial applications. Additionally, it's equipped with 16MB of flash memory which was used for event logging purposes and storing position maps for the manipulator's sections. The control system consists of smaller sub system for managing CAN communication, reading input data both locally and from the network, conversion of acquired data into human-friendly format, controlling the deviation regulators and encapsulation of control data into a J1939 protocol for hydraulic control of PVG valves. The available 16bit ADC and configurable measuring ranges allow to register local input data with the same accuracy as the sensors provide, thus taking full advantage of the hardware configuration of the intuitive control system. Stress test have shown that the developed control system is capable of calculating speed data at the time of 2ms. Additional delay from CAN based sensor communication increases it to 7ms.

6. Results

To assess the impact each of the defined issues has on the accuracy of controlling a hydrostatically driven manipulator, a series of tests has been carried out. First one tackled the error value resulting from a simplification of the control systems software algorithms, which were linearizing the relation between angles of rotation of manipulator's sections and their corresponding hydraulic actuators translations. Research has shown that, in the example of the tested manipulator, the maximum difference between the required angle and that set by the control system was approx. 8^0 for the boom section (fig. 5). This deviation alone was able to offset the horizontal component of the work tool's position by approx. 200 mm. Additionally; this error has not created major deviations for the vertical component of the work tools position. When analyzing the linearization impact on the position of the short arm it has been calculated, that the difference of angles desired and achieved was approx. 5^0 (fig. 6), which was generating the horizontal deviation of 30 mm and vertical deviation of 50 mm. The last section tackled was that of the work tool's rotation. In this case, di difference was approx. 9^0 (fig. 7), which was not only marginally influencing the work tool's position in an absolute coordinate system, when compared to other sections. During testing, there had been voltage fluctuation registered for the analog position sensors. While these were not influencing the calculation of the actual manipulator's configuration, they were problematic for the controller, as were generating tenfold fluctuations of the control system (the initial phase of movement registered in fig. 8).

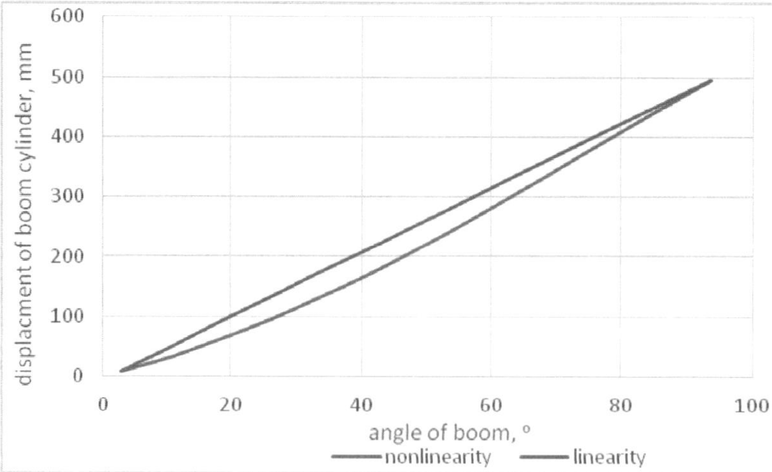

Fig. 5. The characteristic of the boom cylinder's piston displacement in relation to its desired angle.

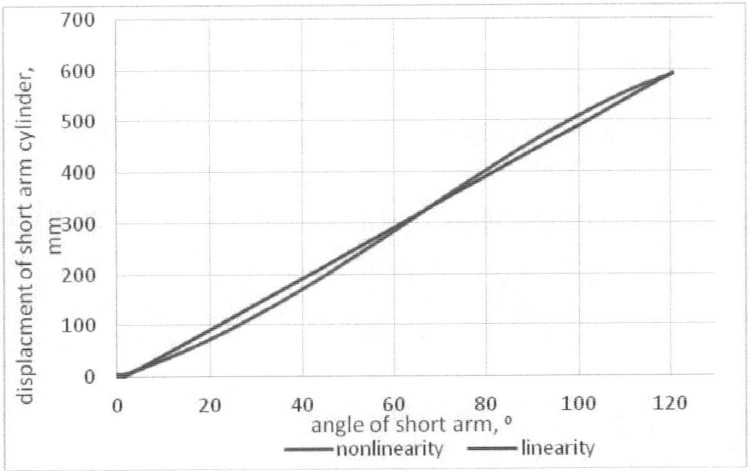

Fig. 6. The characteristic of the short arm cylinder's piston displacement in relation to its desired angle.

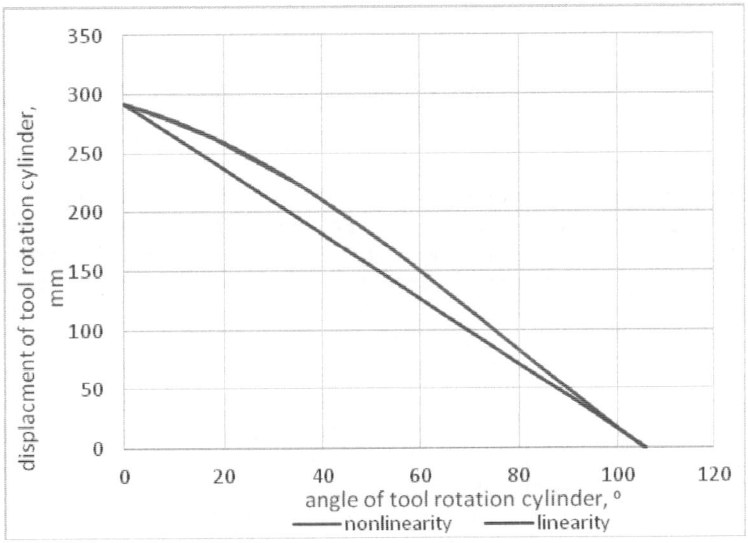

Fig. 7. The characteristic of the toll rotation cylinder's piston displacement in relation to its desired angle.

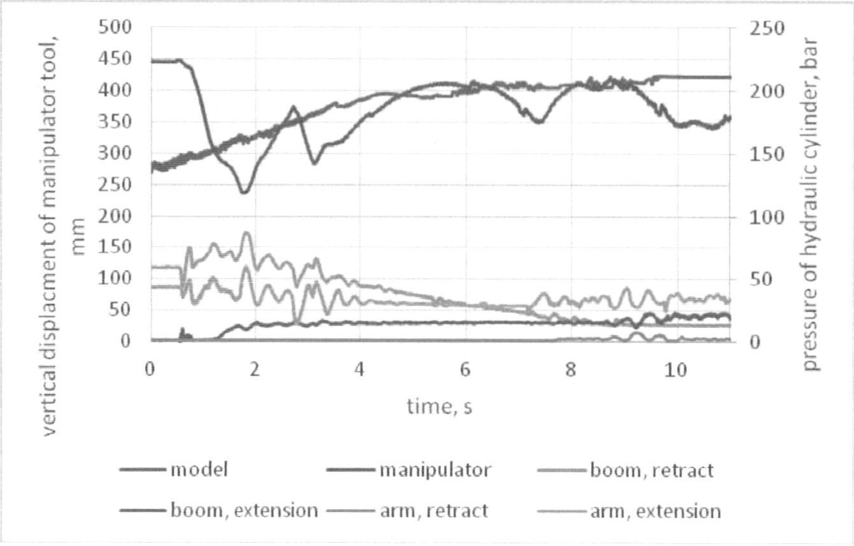

Fig. 8. Time function of vertical displacement of manipulator's tool position and pressure of hydraulic cylinders.

Further studies were aimed at tackling the issue of time delay between the operator's hand movement and the manipulator's section movement. Figure 9 illustrates the transition of the long arm actuator's piston in response to a corresponding movement of the lever mechanism. Data shows that the delay in manipulator's movement is approx. 1 s. The designated value of the delay is mainly the result of a subjective assessment of the motion's stability as well as the fluctuations of the measured values. They were the main reason for the introduction of deadzones for each of the section. On one hand, this procedure has helped to mitigate or eliminate completely (depending on the deadzone's value) the desired position's oscillations. On the other hand, it resulted in an increase in time delay which was affecting the initial phase of movement, leading to an increased time delay.

After conducting preliminary research and analysis which helped to choose the most appropriate deadzone values, final tests had been conducted. Their initial stage focused on registering a selected movement of the controller which was later on recreated numerous times by the control system with the manipulator's response and actuator's pressure values being measured every time (fig. 8). The recorded movement involved controlling the boom and the long arm with the rest of the sections being locked at a set position (fig. 10). Research has shown that the work tool's position during movement is oscillating around the designated position with an acceptable amplitude. The difference in position values during the initial movement phase are the result of the boom's deadzone values. However, after the manipulator starts moving, the maximum vertical deviations from the desired path are approx. 80 mm with the designated position being missed by an average of approx. 60 mm. The registered oscillations during movement were a result of different reaction times of the hydraulic cylinders due to the load sensing method of control of the PVG32.

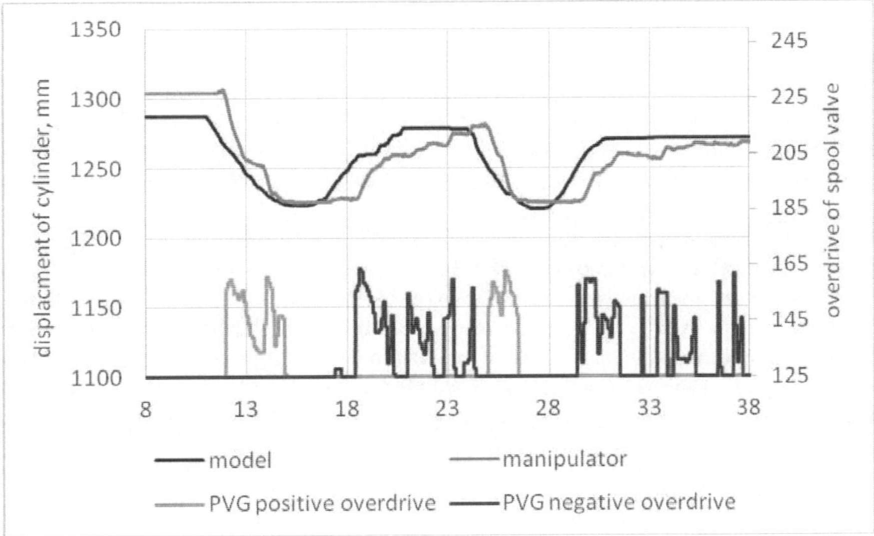

Fig. 9. Time function of the long arm cylinder's piston displacement and the overdrive of its corresponding spool valve.

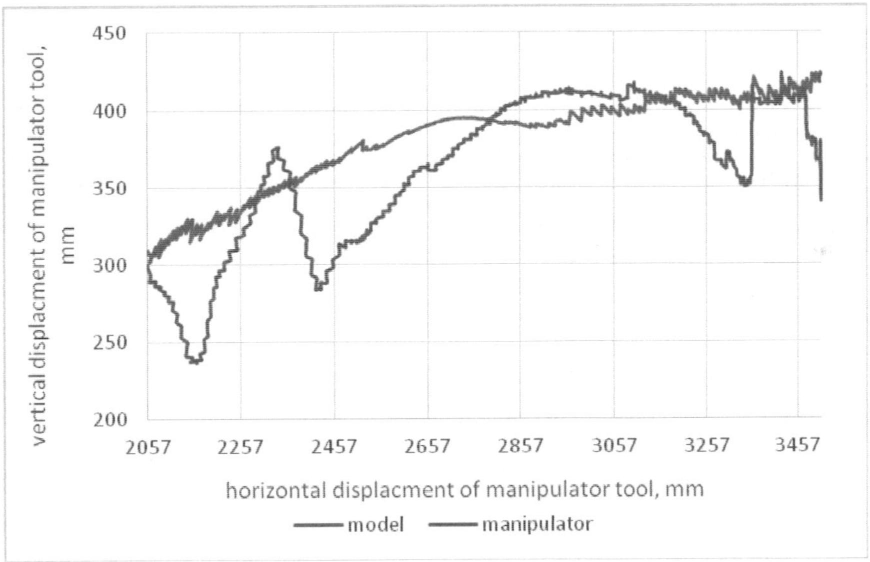

Fig. 10. The characteristic of the manipulator's tool vertical displacement in relation to its horizontal displacement.

7. Conclusions

The study presented in this paper showcases an intuitive control system for hydrostatically driven manipulators. On its basis, a selected set of issues has been described which influence the manipulator's effector control accuracy. First one has shown, that using linearization od relations between angles and actuator translation may lead to significant deviations of the outputted effector's position. In the example presented here, these errors were reaching levels of approx. 200 mm in offshoot from the desired vertical position. Another aspect which needs to be taken into account is the time delay value of the entire control and drive system. It has been assumed that a stable and thus safe effector's trajectory is more important than the system's reaction time. That is why, on the basis of research carried out it has been proposed that the acceptable level of delay between the controller's and the manipulator's movements may be up to 1 s. This value incorporates both the inertia of the hydraulic system as well as the time it takes to calculate control command for the PVG32. When developing a controller for a hydrostatically driven manipulator, especially that of an industrial kind, one should take into account the possibility of sensor data fluctuations which may influence the output signal quite a bit. Introducing deadzones for sensor data may mitigate this problem but it increases the time delay of the entire system and in extreme cases may lead to control accuracy degradation. The presented control method allows reaching an acceptable level of control accuracy of a hydrostatically driven manipulator and time delay of the control system. However, further work on introducing additional subsystems and sub routines on the control system's software side will improve both of these parameters. As of the time of writing, the achieved error in vertical displacement of a 4m long, hydrostatically driven manipulator was approx. 80 mm.

Further work on the presented intuitive control system include modifications to the controller's structure to increase its stiffness and incorporate damping nodes. This will help to mitigate fluctuations of registered data and make movements easier to repeat. It will also allow to filter out operator's hand shaking during the tasks of minting the desired position. Additionally, a series of software tweaks will be introduced to increase system's response time and achieve higher travel speed while maintaining the same level of control and safety.

Acknowledgements

We would like to thank of National Centre for Research and Development for founding grant number LIDER/007/424/L-4/12/NCBR/2013 within have been conducted investigation described in this paper.

References

[1] Kim D,Oh KW, Hong D, Kim YK, Hong S. Motion control of excavator with tele-operated system. *26th International Symposium on Automation and Robotics in Construction*; 2009. p. 341-347

[2] Kim D, Kim J, Lee K, Park C, Song J, Kang D. Excavator tele-operation system using a human arm. *Automation in Construction 18*; 2009. p. 173-182

[3] Hu H, Li J, Xie Z, Wang B, Liu H, Hirzinger G. A Robot Arm/Hand Teleoperation System with Telepresence and Shared Control. *International Conference on Advanced Intelligent Mechatronics*. Monerey, USA; July 2005. p. 24-28

[4] Osafo-Yeboah B, Jiang S, Delpish R, Jiang Z, Ntuen C. Empirical study to investigate the range of force feedback necessary for best operator performance in a haptic controlled excavator interface. *International Journal of Industrial Ergonomics*; 2013. Vol. 43, p. 197-202

[5] Mullins J, Fielding M, Nohavandi S. OzBot and haptics: remote surveillance to physical presence. *Unmanned Systems Technology XI, SPIE;* April 2009. vol. 7332. p. 73321R-73321R-9

[6] Oh KW, Kim D, Kim NH, Hong D. The virtual environment for force-feedback experiment of excavator using a novel designed haptic device. *28th International Association for Automation and Robotics in Construction*. South Korea. Seul; 2011. p. 51-56

[7] Łopatka MJ, Muszyński T, Rubiec A. Simulation identification of Fire Rescue Robot suspension loads. *Proceedings of 18th International Conference on Method and Models in Automation and Robotics Międzyzdroje 2013*. IEEE. New York; 2013. p. 408-413

[8] Bartnicki A, Łopatka MJ, Muszyński T. Stiffness evaluation of Fire Rescue Robot suspension with hydropneumatic components. *MECHATRONIC SYSTEMS, MECHANICS AND MATERIALS II. Solid State Phenomena*. Trans Tech Publications Ltd. Zurich; 2014. vol. 210. p. 301-308

[9] Dabrowska A, Konopka S, Przybysz M. Ability to negotiate terrain obstacles by lightweight six-wheeled unmanned ground vehicles. *Proceedings of 10th International Conference on Intelligent Technologies in Logistics and Mechatronics Systems ITELMS 2015*. Kaunas University Technology Press; 2015. p. 102 – 109

[10] .Dabrowska A, Jaskolowski M, Konopka S. Stability research of lightweight six-wheeled ground vehicles. *Proceedings of 10th International Conference on Intelligent Technologies in Logistics and Mechatronics Systems ITELMS 2015*. Kaunas University Technology Press; 2015. p. 83 – 89

[11] Konopka S, Sprawka P, Muszyński T, Spadło K. Investigating Resistance of Turn Six-Wheel Skid-Steer Vehicle. *Proceedings of 8th International Conference on Intelligent Technologies in Logistics and Mechatronics Systems ITELMS 2013*. Kaunas University Technology Press; 2013. p. 107 – 115

[12]. Łopatka MJ, Muszyński T, Rubiec A, Sprawka P. Preliminary Simulations of High Mobility IED Resistance Suspension with Casting Arms. *MECHATRONIC SYSTEMS, MECHANICS AND MATERIALS II. Solid State Phenomena*. Trans Tech Publications Ltd. Zurich; 2014. vol. 210. p. 115-121

[13] Konopka S, Łopatka MJ, Przybysz M. Kinematic Discrepancy of Hydrostatic Drive of Unmanned Ground Vehicle. *Proceedings of 8th International Conference on Intelligent Technologies in Logistics and Mechatronics Systems ITELMS 2013*, Kaunas University Technology Press. Kaunas; 2013. p. 116 – 125

[14] Dabrowska A, Przybysz M, Rubiec A. Hydropneumatic suspension efficiency in terms of the teleoperated unmanned ground vehicle tests. *Proceedings of 10th International Conference on Intelligent Technologies in Logistics and Mechatronics Systems ITELMS 2015*, Kaunas University Technology Press. Kaunas; 2015 pp. 110 – 116

[15] Bartnicki A, Łopatka MJ, Muszyński T. Sprawka P. Track interaction study of the ground in terms of the use of electric driving systems in the robot. *Proceedings of 18th International Conference on Method and Models in Automation and Robotics Międzyzdroje 2013*, IEEE. New York; 2013. p. 414-419

[16] Łopatka MJ, Muszyński T,Rubiec A. Loads analysis of skid steer robot drive system. *Proceedings of 18th International Conference on Method and Models in Automation and Robotics Międzyzdroje 2013*. IEEE. New York; 2013. p. 362-367

[17] Typiak A, Gnatowski M. Map Building System for Unmanned Ground Vehicle. *MECHATRONIC SYSTEMS, MECHANICS AND MATERIALS, Solid State Phenomena*. Trans Tech Publications Ltd. Zurich; 2012. Vol. 180. p. 131-136

[18] Typiak A. Ladar - Aided Navigation System for Unmanned Ground Vehicles. *Proceedings of 8th International Conference on Intelligent Technologies in Logistics and Mechatronics Systems ITELMS 2013*. Kaunas University Technology Press. Kaunas; 2013. p. 254 - 261

[19] Typiak A, Gnatowski M, Szklarski J. Create Terrain Maps for Unmanned Ground Platform. *MECHATRONIC SYSTEMS, MECHANICS AND MATERIALS II, Solid State Phenomena*. Trans Tech Publications Ltd. Zurich; 2014. Vol. 210. p. 32-39

[20] Łopatka MJ. Influence of Power Steering System on Directional Stability of Articulated Wheel Loader. *Proceedings of 8th International Conference on Intelligent Technologies in Logistics and Mechatronics Systems ITELMS 2013*. Kaunas University Technology Press. Kaunas; 2013. p. 148 – 155

[21] Bartnicki A, Dabrowska A, Łopatka MJ, Muszyński T. Experimental research on directional stability of articulated tractors. *MECHATRONIC SYSTEMS, MECHANICS AND MATERIALS II. Solid State Phenomena*. Trans Tech Publications Ltd. Zurich; 2014. vol. 210. p. 77-86

[22] Inżynieryjny robot wsparcia EOD/IED. Wojskowa Akademia Techniczna *Research Project nr. 0012/R/T00/2008/05*. Military University of Technology. Warsaw; 2011.

Forum on Innovative Technologies and Management for Sustainability (ITMS'2016)

panevėžys faculty of technologies and business

The 11th International Conference Intelligent Technologies in Logistics and Mechatronics Systems (ITELMS'2016), 28-29 April 2016, Panevėžys, Lithuania

Analysis of concrete bridge pavements' durability by calculation of water uptake by immersion

Navickienė T.[a]*, Sušinskas S.[a]

[a]Kaunas University of Technology, Nemuno s. 33, Panevėžys, Lithuania

Abstract

During exploitation period concrete is affected by many conditions. They have negative effect on material's durability. Gradual decay of constructions and materials depends on moisture affection and temperature variation. This paper analyzes how concrete's structure affects it's durability.

© 2016 T. Navickienė, S. Sušinskas.
Peer-review under responsibility of the Kaunas University of Technology, Panevėžys Faculty of Technologies and Business

Keywords: concrete; porosity; capillary moisture; freeze resistance.

1. Introduction

Here Experience of building construction and exploitation, field and laboratory tests shows that gradual decay of constructions and materials depends on moisture affection and temperature variation [1] Because of capillaries, cracks and pores that appear because of improper consolidation, fluids leak through concrete. Because of freeze-thaw cycles that take place during exploitation period, conduction channels may appear. During these cycles fluids in concrete become solids, change their volume and appear forces, damaging concrete from the inside. Concrete's durability depends on how much aggressive material's can penetrate into it and how much damage they can do. Permeability depends on porosity and pore connections. The more open pores are, the more vulnerable to environmental conditions material is.

Concrete bridge pavements, if they are in cold regions, are affected not only by freeze, but also by sea or ocean tides, which water contains chlorides, sulphates and other aggressive materials. On winter, in order to have functional and safe pavement, they are sprinkled with de-icing salts which also have negative effects.

2. Investigation of concrete structure

Concrete's structural characteristics can be seen from its structural indicators. The main structural characteristic is porosity. Concrete's structural characteristics are: effective porosity, total porosity, porous space reserve, relative pore and capillary wall thickness, structural singleness distortion rate. Effective porosity describes the pore cavity, which during the test is filled with water. Total porosity describes the pore and capillary cavity in specimen. Porous space reserve is the space, which is not filled with water in the beginning, but during the test slowly fills in. Compositions of concrete cubes [3] tested in laboratory, are presented in Table.1. [1]

* Corresponding author. E-mail address: navickiene.toma@gmail.com

© 2016 T. Navickienė, S. Sušinskas.

Peer-review under responsibility of the Kaunas University of Technology, Panevėžys Faculty of Technologies and Business

Table 1. Composition of investigated concrete cubes

Materials	Units	Content in $1m^3$ of concrete mixture		
		C25/30 (D1 series)	C35/45 (D2 series)	C/35/45 (D3series)
Content of cement	kg	366	457	457
Content of water	l	205	160	160
Content of dolomite aggregate	kg	1171	1157	1157
Content of sand	kg	601	656,14	656,14
V/C	-	0,56	0,35	0,36
Flormix plasticizer	ml	-	-	90
Mobility	mm	7 (S2)	-	9 (S2)
Vebe rate	s	-	32 (V0)	-

Concrete cubes were hardened in accordance with the requirements, specified in LST EN 12390-2:2003. 28 days cubes were kept in water in constant temperature.

For measuring water penetration into concrete cubes a following system was constructed. There was a vessel, where dry and weighed before cubes were placed in. After that, the vessel was closed, air from it was taken out and vacuum created. Cubes were left in vacuum for an hour and then water was made to flow into the vessel, until it reached concrete surface from the below or until cubes were all in the water, that depended on the test, that was made. After some time vessel was opened and cubes weighed in the moist state. Also there were made similar tests, just not in the vacuum, but in the natural conditions.

After tests concrete's structural indicators were calculated according to the following formulas. [2]

$$W_E = \frac{m_0}{V} = \frac{m_1 - m_0}{m_0} \cdot 100\%, \quad (1)$$

$$W_R = \frac{m_0}{V} \cdot \frac{m_4 - m_0}{m_0} \cdot 100\%, \quad (2)$$

$$R = \left(1 - \frac{W_E}{W_R}\right) \cdot 100\%, \quad (3)$$

$$D = \frac{100 - W_R}{W_R} \cdot 100\%, \quad (4)$$

$$N = \frac{h_{max} - h_{min}}{h_{min}}, \quad (5)$$

where: W_E – effective porosity, %; m_0 – weight of dry specimen, g; m_1 – weight of moist specimen, soaked in natural conditions, g; V – volume of specimen, determined after test of absorption in vacuum, cm^3; W_R – total porosity %; m_4 – weight of moist specimen, soaked in vacuum, g; R – porous space reserve, %; D – relative pore and capillary wall thickness, %; h_{max} – maximum value of moisture capillary rise, mm; h_{min} – minimum value of moisture capillary rise, mm; N – degree of structural heterogeneity, units.

Calculated values of structural indicators are shown in Fig.1. [1]

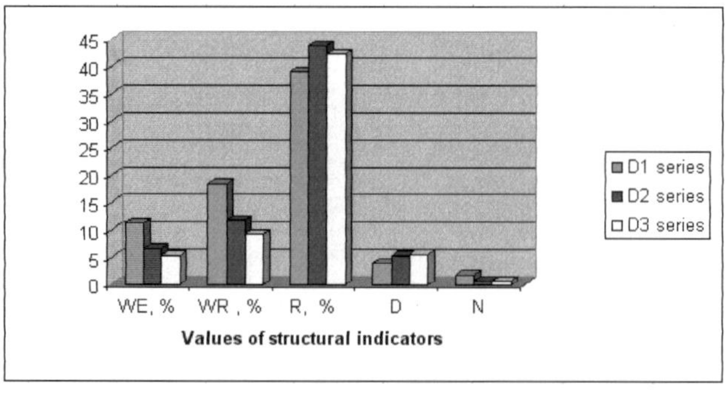

Fig.1. Values of structural indicators of investigated concrete cubes

Test results showed that structural specimen indicators depend on concrete's composition. The more composition contains large aggregates and water, and the less small aggregates and cement, the bigger concrete's porosity and absorption is.

Since tests showed that water absorption by immersion is not trustworthy parameter to evaluate concrete's durability issues such as carbonization and chloride migration, and the immersion test shows total porosity, but not permeability, which is more important in these cases [5], only concrete's resistance to freeze-thaw cycles will be reviewed.

3. Forecasted freeze resistance

In order to evaluate forecasted freeze resistance, capillary mass speed is calculated.

$$g_1 = \left(\frac{m_2 - m_0}{S_1}\right) \qquad (6)$$

where: g_1 – capillary mass speed in normal conditions, g/cm^2·h; m_2 – mass of the specimen, after 30 minutes capillary immersion in normal conditions parallel to freezing direction, g; S_1 – beneficial specimen' surface area parallel to freezing direction, cm^2.

$$G_1 = \left(\frac{m_3 - m_0}{S_2}\right) \qquad (7)$$

where: G_1 – capillary mass speed in vacuum, g/cm^2·h; m_3 – mass of the specimen, after 30 minutes capillary immersion in vacuum parallel to freezing direction g; S_2 – beneficial specimen' surface area parallel to freezing direction, cm^2.

$$G_2 = \left(\frac{m_5 - m_0}{S_3}\right) \qquad (8)$$

where: G_2 – capillary mass speed in vacuum perpendicular to freezing direction, g/cm^2 ·h; m_5 – mass of the specimen, after 30 minutes capillary immersion in vacuum perpendicular to freezing direction, g; S_3 – beneficial specimen' surface area perpendicular to freezing direction, cm^2.

The smaller capillary mass speeds are, the less specimens soaks in with water. That means that operational freeze resistance should be bigger. [6]

$$F_{RE} = 0{,}231 \cdot \frac{R^{1{,}068} \cdot D^{1{,}348} \cdot G_1^{0{,}278} \cdot G_2^{0{,}668}}{N^{0{,}288} \cdot g_1^{1{,}084}} \qquad (9)$$

where: F_{RE} – forecasted operational freeze resistance, cycle quantity.

Since in this work capillary mass speed were not tested and calculated, in following calculations would be agreed that they are equal to one.

$$G_1 = G_2 = g_1 = 1 \qquad (10)$$

Calculated forecasted operational freeze resistance values (F_{RE}) are shown in Fig.2:

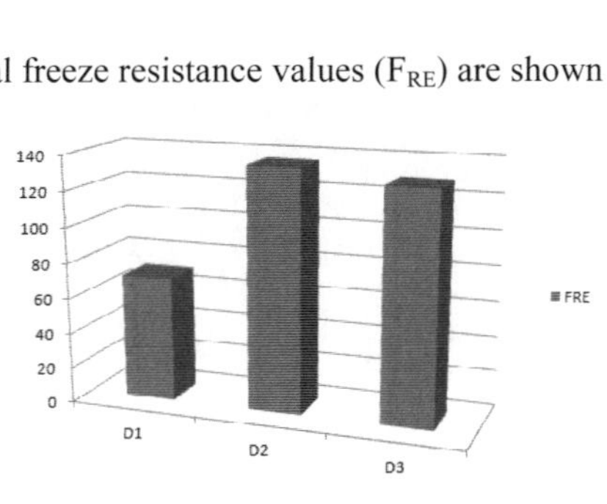

Fig.2. average forecasted operational freeze resistance values

From diagram and calculations it is seen that, while not evaluating capillary mass speed, concrete's resistance to freeze depends directly on its capillary and pore wall thickness and porous space reserve. The bigger these indicators are, the more resistant to freeze concrete is.

Other studies have proved that plasticizers have positive effects on concrete's durability. [4] Although in this paper calculated average forecasted operational freeze resistance for D3 specimens is shorter than for analogical D2 specimens with plasticizers, as long as capillary mass speeds were not calculated and only their comparative values were used, it would be inaccurate to make conclusions about influence of plasticizers for concrete's durability.

Conclusions

1. Concrete resistance to freeze depends directly on its capillary and pore wall thickness and porous space reserve. When porous space reserve increased by app. 16% and capillary and pore wall thickness increased by app. 25%, quantity of freeze cycles that concrete resists to increased by app. 109%.
2. Usage of plasticizers affects moisture motion in concretes with the same composition, comparing with concrete without them.

References

[1] Kaupienė J, Pakėnaitė T, Navardauskaitė M. Moisture motion kinetics in concrete. In *Intelligent technologies in logistics and mechatronics systems (ITELMS'2010)*. Kaunas University of Technology; 2010. p. 111-113

[2] Nagrockienė D, Žurauskienė R, Mačiulaitis R, Kičaitė A, Ravnialičevas V. *Medžiagų mokslas. Metodikos nurodymai laboratoriniams darbams atlikti*. Vilnius: Technika; 2005. p. 51-56, 100

[3] Malakauskas M. *Gelžbetoninių gaminių technologija*. Vilnius: Mokslas; 1990. p. 39-43

[4] Abasova A, Nagrockienė D. Plastifikatoriaus poveikis sukietėjusio betono savybėms. *Science – future of Lithuania*; 2012. p. 289-293

[5] De Schutter G, Audenaert K. Evaluation of water absorption of concrete as a measure for resistance against carbonation and chloride migration. *Materials and constructions/Materiaux et Constructions* 2004; 37: p. 591-596.

[6] Kizinievič O. *Pramoninių atliekų panaudojimo statybinėje keramikoje tyrimai*: Daktaro disertacija; Vilnius; 2006.

Forum on Innovative Technologies and Management for Sustainability (ITMS'2016)

panevėžys faculty of technologies and business
1922

The 11th International Conference Intelligent Technologies in Logistics and Mechatronics Systems (ITELMS'2016), 28-29 April 2016, Panevėžys, Lithuania

Model of a steering axle of a military truck

Neumann V.[a]*

[a]*University of Defence, Kounicova 65, Brno 662 10, Czech Republic*

Abstract

Fire power, protection and mobility are the main capabilities of the military vehicles. Fire power and protection improvements during a process of a vehicle modernization are current trends in construction of the military vehicles. All this improvements bring a weight increasing and negatively influence vehicle mobility. Vehicle capability of movement influences the vehicle survivability in the battle field, too. From this point of view, an analysis of a vehicle mobility, load of the main vehicle undercarriage parts and influence on the crew should be a part of the vehicle modernization process. Simulating technologies are suitable manner for mentioned purpose. The paper is focus on the modelling and simulation of the steering axle of a military truck. The analysis is focus on the load of the main parts and mechanism behavior during the vehicle movement. A multibody software MSC Adams was used for the model development and simulations.

© 2016 V. Neuman.
Peer-review under responsibility of the Kaunas University of Technology, Panevėžys Faculty of Technologies and Business

Keywords: steering; analysis; military; simulations; MSC ADAMS.

1. Introduction

Fire power, protection and mobility are the main capabilities of the military vehicles. Weight reserve and modernization capability are the next requirements typical for military vehicles. Military expects life cycle of military vehicles about 20 – 30 years. It means that the vehicles have to enable vehicle updates and technology exchange. Protection improvements against mines and IEDs (improvise explosive device) during a process of a vehicle modernization are current trends in construction of the military vehicles. Utilization of the additive armor and slat armor are nowadays trends in vehicle protection. All improvements of the vehicle protection bring a vehicle weight increasing. Weight increasing negatively influences vehicle maneuverability, tactical and strategical mobility and a load of main parts of a vehicle (body, undercarriage, suspension, steering mechanism, axles, gearbox, wheels). Improvements of the vehicle protection must follow an analysis of a vehicle mobility change, because vehicle movement capability influences the it´s survivability in the battle field, too. Analysis of the vehicle maneuvering should be included in a development process of a new vehicle and in a modernization process, too.

Utilization of high performance computers with special simulating software support (simulating technologies) is current trend in construction of mechanical parts, mechanical groups, various mechanisms and vehicles, too. Simulating technologies can be effectively utilized in assessing, evaluating and comparison of basic vehicle capabilities, too. The main general advantage of simulation technologies are:

- Time reduction – design, development and testing processes,
- Optimization of vehicle manufacturing – précising of manufacturing technology and material saving,
- Analysis and testing of various variants of product,
- Expenses saving.

* Corresponding author. Tel.: +420-793-442-671; fax: +420-793-443-384.
E-mail address: vlastimil.neumann@unob.cz

From military point of view we can find next main advantages of this manner:

- Testing area and conditions are the same for all tested vehicles – analysis is not fixed on the particular place,
- Possibility of various operational vehicle conditions and it´s in-service behavior – movement of overloaded vehicle, drive with damaged wheel,
- Analysis of maneuvering of different vehicle modifications – movement with different weapon station, vehicle behavior with additive armour – beneficial during the vehicle modernization,
- Development of vehicle mobility, survivability and reliability capabilities.

FEM (Finite Element Method) was in the beginnings of the utilization of the CAD (Computer Aided Design) technologies. Huge computational time and hardware requirements were the main drawbacks of this tool. Utilization of the MBD (MultiBody Dynamics) software enabled to shorten the computational time and bring another more effective computational solution.

On the other side simulating technologies have some disadvantages, too. Credibility of this manner can be one of the first drawback. No one believes outputs from simulations without experimental verification, yet every one believe experimental data without it´s validation. Input data is the next main drawback of this tool. The core of this key point is insufficient data or lack of them. During the model creation are taking into account some simplification, which can influence the overall outcomes of the simulations.

Simulation technologies are highly utilized for several years. People who work with this tool are more experienced and softwares are more sophisticated than in the beginning. The mass utilization of simulating technologies and experience improvement gradually decrease mentioned the main drawbacks.

Demonstration of the utilization of simulating technologies (mathematic model development) for the analysis of the military truck steering axle is a purpose of this paper. A multibody dynamics software MSC. Adams was utilized for model of the steering axle development, for simulation and analysis.

As the world's most famous and widely used Multibody Dynamics (MBD) software, Adams improves engineering efficiency and reduces product development costs by enabling early system-level design validation. Engineers can evaluate and manage the complex interactions between disciplines including motion, structures, actuation, and controls to better optimize product designs for performance, safety, and comfort. Along with extensive analysis capabilities, Adams is optimized for large-scale problems, taking advantage of high performance computing environments. [1]

For dynamics vehicle development and testing MSC offers module ADAMS/Car. With Adams/Car engineering teams can build and test functional virtual prototypes of complete vehicles and vehicle subsystems. Working in the Adams vehicle environment, automotive engineering teams can exercise their vehicle designs under various road conditions, performing the same tests they normally run in a test lab or on a test track, but in a fraction of time. [2]

2. Mathematics model of the selected vehicle

I have selected Czech military off-road truck Tatra T-810 for application of simulating technology on the analysis of the steering mechanism. The vehicle Tatra T-810 you can see in the Figure 1.

Fig. 1. Tatra T-810 truck – left basic version [3], right armoured version [4].

2.1. Basic vehicle description

Acquisition the vehicles Tatra T-810 was approved by the Czech government in 2005. The first vehicles of Tatra T-810 were delivered to the Army in 2008. This vehicle meets the specific, nowadays requirements (constructional and operational) of the Army and the Czech legislature for a period of at least 20 years. Truck Tatra T-810 replaced the old legendary Czech Praga V3S truck. The Tatra T-810 is the medium off-road truck which meets requirements of high terrain passability, endurance and transportability (airway, seaway and railway). [5]

In the Figure 2 there is shown a front steering axle of the truck Tatra T-810.

Fig. 2. Front steering axle of the truck Tatra T-810 – left from the front side, right from the back side.

Basic tactical-technical data of the vehicle are presented in the Table 1.

Table 1. Basic tactical technical data of the Tarta T-810 [1, 2].

Quantity	Value
Engine	Renault DXi7 240 EC-01
	6 cylinders diesel, water cooled
	177 kW
Gearbox	ZF 6S 1000 T0
	12+2
Maximal speed	106 km/h
Minimal speed	3,2 km/h
Tank range	800 km
Maximal weight	13 t
Unladen weight	8,5 t
- basic version	8,5 t
- armoured modification	9,85 t
Vehicle length	7 365 mm
Vehicle height	3 300 mm
Vehicle width	2 550 mm
Clearance	430 mm
Fording	120 mm
Wheels	14,5 R20
Front suspension	wound springs
Rear suspension	leaf springs

2.2. Model description

Model development in Adams Car is divided into three stages: template builder, subsystem, assembly. Template is determined for the more experienced users. User in this stage can develop whole part or completely change the already created template. Template builder serves only for development of the templates, it does not enable running of simulations. Subsystem stage enables creation and editing of the vehicle systems, for example suspension system, steering system, chassis, wheels.... Assembly environment enables vehicle completion and running of simulations. Subsystem and assembly modes are the standard user interfaces which are determined for the basic user. User in this level works with the predefined templates of the basic vehicle parts, he can modify them and he can create individual subsystems. Each subsystem is defined by role, which determines its function in the vehicle (chassis, propulsion mechanism, track element, sprocket wheel...). Positions and dimension can be defined by the hardpoints (which can be edited in standard interface) or by construction frames (which ca be edited only in Template builder). Standard user can change position of the hardpoints, he can change unlock dimensions and he can change general physical characteristics of the body (weight, moments of inertia, central mass point position...).

Created assembly of the front steering axle consists of the three subsystems: axle with a suspension mechanism, steering mechanisms and tires with test rigs.

Picture of the axle subsystem you can see in the Fig. 3 - left. Axle subsystem consists of next main parts: Propelling shaft (1), tube (2), V rod (3), leading rods (4), king bolts (5), springs (6), dampers (7), steering linkage (8), steering arm (9), steering swivel arms (10), steering bar (11), hook joints, bumpstops and bushings (bushing are used in joints of the main axle hitches).

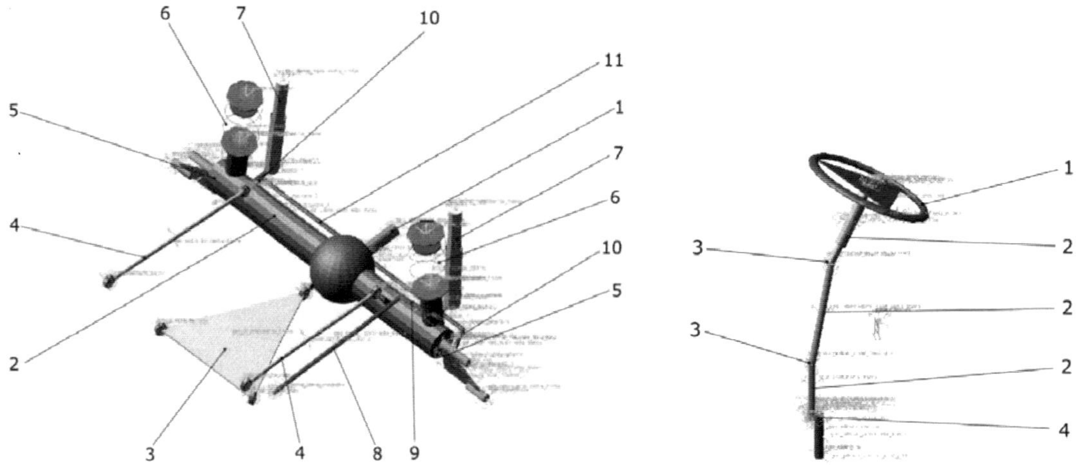

Fig. 3. Subsystems – left front axle subsystem, right steering subsystem.

The steering subsystem (Fig. 3 - right) consists of next main parts: steering wheel (1), shafts (2), hook joints (3), steering box (4) and joint. Gear ratio of the steering box was set up to 18,5.

Testing rids and tires are the next subsystem and they are one of the main part of the assembly, too. Characteristics of the tires, parameters of the other parts and joints and their position are editable – can be modified. Parameters of the utilized tires are presented in the Fig. 4.

```
[MDI_HEADER]
FILE_TYPE                  :'tir'
FILE_VERSION               :3.0
FILE_FORMAT                :'ASCII'
! : TIRE_VERSION :         PAC2002
! : COMMENT :              Tire                         365/80 R20
! : COMMENT :              Manufacturer
! : COMMENT :              Nom. section with   (m)      0.365
! : COMMENT :              Nom. aspect ratio   (-)      80
! : COMMENT :              Infl. pressure      (Pa)     800000
! : COMMENT :              Rim radius          (m)      0.216
! : COMMENT :              Measurement ID
! : COMMENT :              Test speed          (m/s)    16.7
! : COMMENT :              Road surface
! : COMMENT :              Road condition               Dry
! : FILE_FORMAT :          ASCII

$-------------------------------------
[UNITS]
LENGTH                     :'meter'
FORCE                      :'newton'
ANGLE                      :'radians'
MASS                       :'kg'
TIME                       :'second'
$-------------------------------------
[MODEL]
PROPERTY_FILE_FORMAT       :'PAC2002'
USE_MODE                   : 14
VXLOW                      : 1
LONGVL                     : 16.7
TYRESIDE                   : 'LEFT'
vehicle/test bench
$-------------------------------------
[DIMENSION]
UNLOADED_RADIUS            : 0.508
WIDTH                      : 0.365
tyre
ASPECT_RATIO               : 0.8
RIM_RADIUS                 : 0.216
RIM_WIDTH                  : 0.355
```

```
$-----------------------------------
[VERTICAL]
VERTICAL_STIFFNESS         : 1e+006
VERTICAL_DAMPING           : 15000
BREFF                      : 3.5

DREFF                      : 0.5
FREFF                      : -0.005

FNOMIN                     : 35000
$-----------------------------------
[LONG_SLIP_RANGE]
KPUMIN                     : -1.5
KPUMAX                     : 1.5
$-----------------------------------
[SLIP_ANGLE_RANGE]
ALPMIN                     : -1.5708
ALPMAX                     : 1.5708
$-----------------------------------
[INCLINATION_ANGLE_RANGE]
CAMMIN                     : -0.26181

CAMMAX                     : 0.26181

$-----------------------------------
[VERTICAL_FORCE_RANGE]
FZMIN                      : 1750

FZMAX                      : 78750
```

```
longitudinal force
PDX1                       : 0.77751
Fznom
PDX2                       : -0.24431
load
PDX3                       : -0.00015908
camber
PEX1                       : 0.46659
Fznom
PEX2                       : 0.393
load
PEX3                       : 0.076024
load squared
PEX4                       : 2.6509e-006
driving
PKX1                       : 14.848
Kfx/Fz at Fznom
PKX2                       : -9.8161
Kfx/Fz with load
PKX3                       : 0.15818
Kfx/Fz with load
PHX1                       : -0.00088873

PHX2                       : -0.00067818

PVX1                       : -5.5714e-007

PVX2                       : 6.2972e-006
load
RBX1                       : 11.13
Fx reduction
RBX2                       : -12.494
with kappa
RCX1                       : 0.97505
Fx reduction
REX1                       : -0.37196

REX2                       : 0.0017379
with load
RHX1                       : 0.0045181
Fx reduction
PTX1                       : 1.5
Fznom
PTX2                       : 1.4
load
PTX3                       : 1
exponent of load
```

```
[LATERAL_COEFFICIENTS]
PCY1                       : 1.5874
forces
PDY1                       : 0.73957
PDY2                       : -0.075004
load
PDY3                       : -8.0362
squared camber
PEY1                       : 0.37562

PEY2                       : -0.069325
load
PEY3                       : 0.29168
curvature Efy
PEY4                       : 11.559
camber
PKY1                       : -10.289
Kfy/Fznom
PKY2                       : 3.3343
maximum value
PKY3                       : -0.25732
camber
PHY1                       : 0.0056509

PHY2                       : -0.0020257

PHY3                       : -0.038716
camber
PVY1                       : 0.015216
Fznom
PVY2                       : -0.010365
load
PVY3                       : -0.31373
camber
PVY4                       : -0.055766
camber and load
RBY1                       : 13.271
reduction
RBY2                       : 5.2405
with alpha
RBY3                       : 1.1547e-005
reduction
RCY1                       : 1.01
reduction
REY1                       : 0.010513

REY2                       : 5.9816e-005
with load
```

Fig. 4. Parameters of the tires.

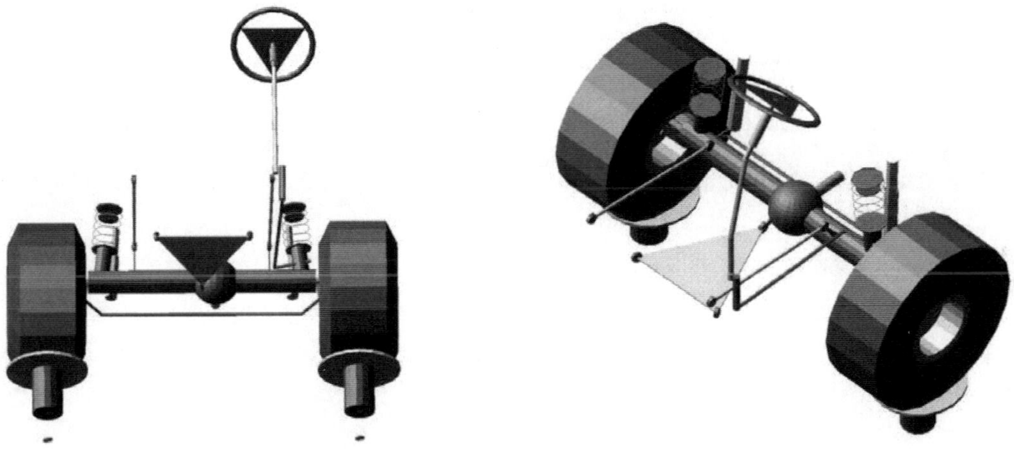

Fig. 5. Final model of the front steering axle.

3. Analysis

Created model and next simulations were used for comparison of the steering load of the standard and armoured vehicle. Both models were the same (tires, suspension mechanism, construction) except the initial static load of the axle. The parameters of the wound springs and dampers are presented in the Fig. 6.

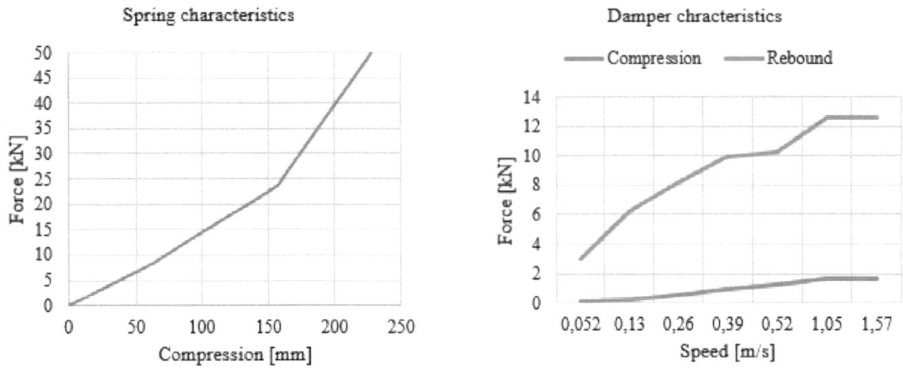

Fig. 6. Suspension characteristics.

3.1. Proceeded simulations

Maximal vertical movement of the wheels was set up to 75 mm, maximal rotation of the steering wheel was 7 RAD (in both directions). Simulation process was divided into basic 6 parts:

- A – left and right wheel are in steady position, steering wheel rotation is defined by curve (Fig. 7),
- B – left and right wheels move simultaneously according to curve (Fig. 7), steering wheel is turned left (7 RAD),
- C – left and right wheels move simultaneously according to curve (Fig. 7), steering wheel is turned right (-7 RAD),
- D – left and right wheels move oppositely according to curves (Fig. 7), steering wheel is turned left (7 RAD),
- E – left and right wheels move oppositely according to curves (Fig. 7), steering wheel is turned right (-7 RAD),
- F – left and right wheels move oppositely according to curves (Fig. 7), steering wheel rotation is defined by curve (Fig. 7).

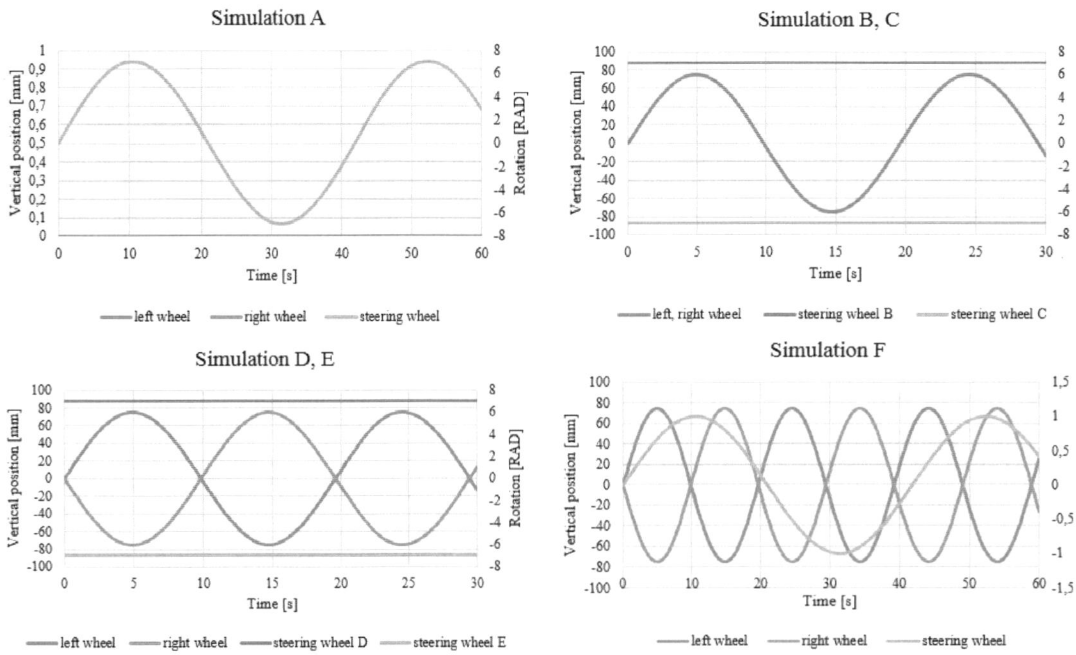

Fig. 7. Movements of the main parts during the simulations.

3.2. Measured outcomes

Torque of the steering wheel during the defined vertical wheels movement and during the defined steering wheel rotation was the recorded quantity. The other values (displacement, velocity, acceleration, force, torque…) of the main assembly parts were recorded, too. These outcomes can be used for the next analysis of the steering axle. There is presented examples of the axle behavior during the simulation in the Fig. 8.

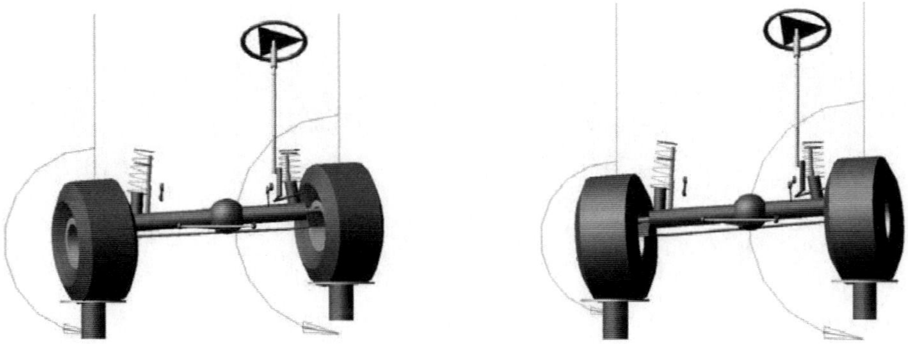

Fig. 8. Screenshots from the simulations.

There are presented the simulations outcomes in the graphical form in the next figures (Fig. 9, Fig. 10).

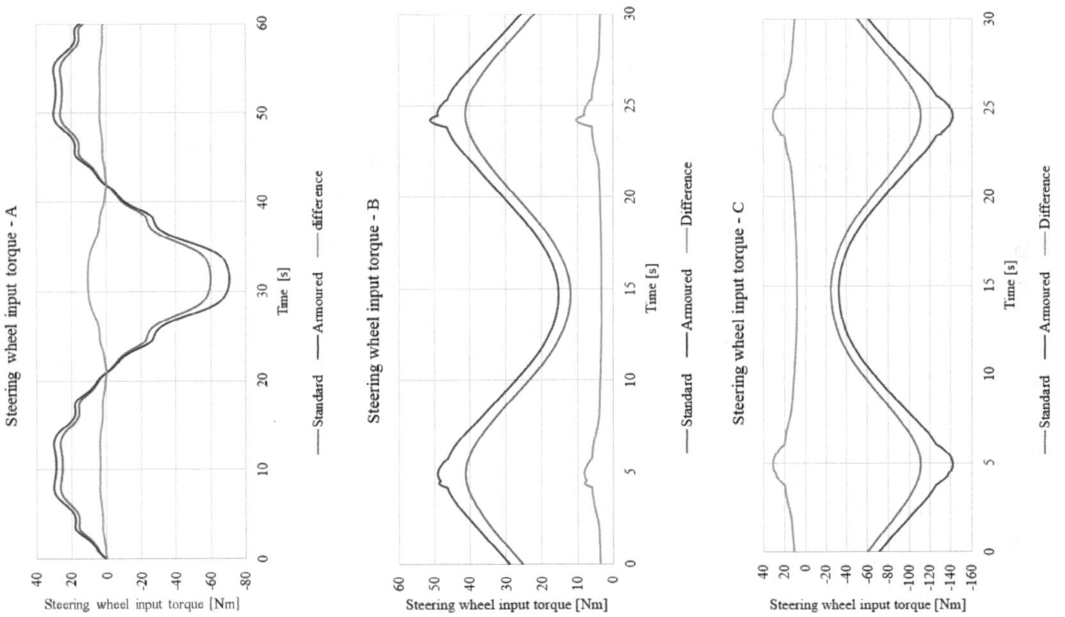

Fig. 9. Steering wheel course – simulations A - C.

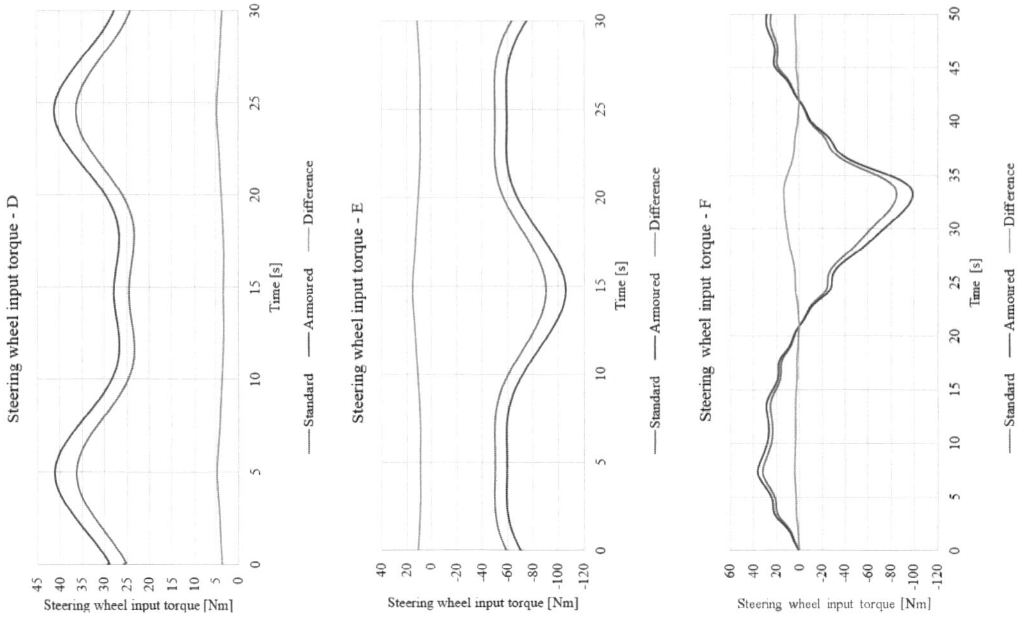

Fig. 10. Steering wheel course – simulations D - F.

4. Conclusion

Increasing of the steering wheel torque of the armoured version of the vehicle was expected and it was proven by the simulations. Value of the steering wheel torque influences the overall driver energy endurance. Maximal torque of the standard vehicle version (T_S), Maximal torque of the armoured vehicle version (T_A) and Maximal absolute difference of the standard and armoured vehicle torques (D_n) of the each simulation were:

- simulation A: T_S = 26,4 Nm, T_A = 30,3 Nm and D_n = 10,8 Nm;
- simulation B: T_S = 41,2 Nm, T_A = 51,3 Nm and D_n = 10,2 Nm;
- simulation C: T_S = -111,2 Nm, T_A = -142,4 Nm and D_n = 31,2 Nm;
- simulation D: T_S = 36,4 Nm, T_A = 41,1 Nm and D_n = 4,7 Nm;
- simulation E: T_S = -90,8 Nm, T_A = -106,0 Nm and D_n = 15,2 Nm;
- simulation F: T_S = 26,4 Nm, T_A = 36,5 Nm and D_n = 13,8 Nm.

We can express average magnitude of the difference (D_{na}) and the increasing of the maximal torque proportionally ($D_{n\%}$):

- simulation A: D_{na} = 3,7 Nm, $D_{n\%}$ = 14,6 %;
- simulation B: D_{na} = 4,5 Nm, $D_{n\%}$ = 24,5 %;
- simulation C: D_{na} = 14,5 Nm, $D_{n\%}$ = 28,0 %;
- simulation D: D_{na} = 3,9 Nm, $D_{n\%}$ = 13,0 %;
- simulation E: D_{na} = 10,6 Nm, $D_{n\%}$ = 16,7 %;
- simulation F: D_{na} = 3,9 Nm, $D_{n\%}$ = 38,2 %.

From the courses of the torque outcomes we can find that:

- Simulation with the steering wheel turned rightward gained the maximal values of the torque (-111,2 Nm and -90,8 Nm for standard version of vehicle; -142,4 Nm and -106 Nm for armoured version of the vehicle).
- Turning direction of the steering wheel influences the magnitude of the steering wheel torque. We can find that the results of the model with turning steering wheel rightward has 3 times higher values than the model with turning the steering wheel leftward.
- Magnitude of the steering wheel torques depends on the relation of the right and left wheel vertical movement direction. Steering wheel torque difference of the model with wheels going vertically simultaneously has 2 times higher magnitude than the model with opposing vertical wheels movement.
- Course of the steering wheel torque is not stable during the simulations with fixed position of the steering wheel. Steering mechanism kinematics and construction influences this torque change. We can find a wheel rotation around vertical vehicle axle during these simulations, too. Wheel rotation around vertical vehicle axle during the fixed steering wheel is presented in the Fig. 11.

- Maximal vehicle rotation around the vertical vehicle axle was measured during the Analysis C (left and right wheels moved simultaneously, steering wheel was turned right). Dispersion of the standard vehicle wheels rotation was measured 0,8° for left wheel and 1,8° for right wheel. Dispersion of the armoured vehicle wheels rotation was measured 0,9° for left wheel and 1,8° for right wheel.

Analysis of the load and mechanism behavior of the steering axle of the medium military truck Tatra T-810 was the purpose of this paper. Simulating multibody dynamics software was selected for the process of the analysis. The analysis were focus on the differences of the load of the basic version of the vehicle and armoured modification of the vehicle (armoured cab). The analysis of the steering axle and the other parts can be beneficial in process of vehicle modification, because common improvements bring weight increasing and driver stress increasing, too. Movement in a heavy terrain under the bad weather conditions is typical activity for the military vehicles. Protection improvements without analysis of the vehicle movement and load changes can decrease a vehicle applicability.

The next process of the work will be aimed at development of the rear axles of the vehicle Tatra T-810 (axles with leaf springs) and development of the complex vehicle model. Analysis of the complex vehicle movement in the defined terrain is the target of the further work.

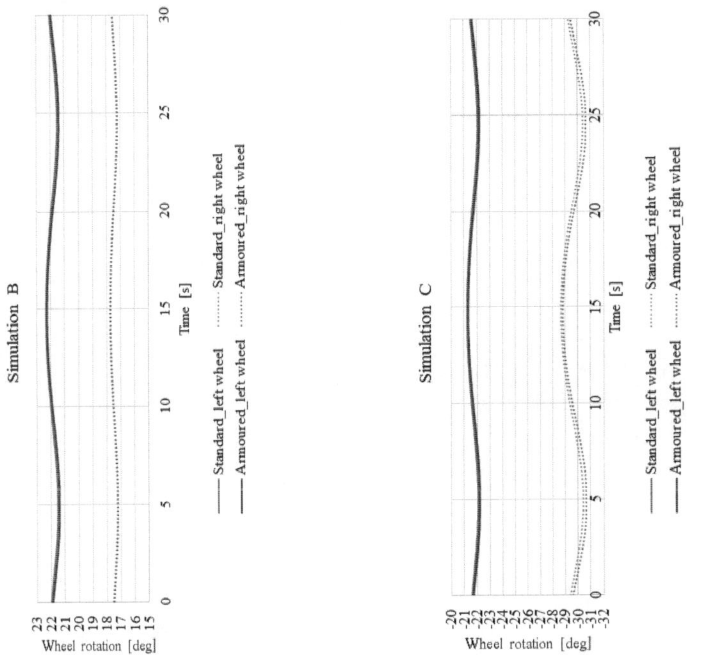

Fig. 11. Courses of the wheel rotation around vertical vehicle axle.

Acknowledgements

Presented work has been prepared with the support of the Ministry of Defence of the Czech Republic, Partial Project for Institutional Development, K-202, Department of Combat and Special Vehicles, University of Defence, Brno.

References

[1] Adams. *MSC.Software* [online] [cit. 2016-02-02]. Available at: http://www.mscsoftware.com/product/adams
[2] Adams Car. *MSC.Software* [online] [cit. 2016-02-02]. Available at: http://www.mscsoftware.com/product/adams-car
[3] T-810. *73tankovyprapor* [online] [cit. 2016-02-1]. Available at: http://www.73tankovyprapor.army.cz/fotogalerie/t-810
[4] Tatra T810 VP. *Wheelsage* [online] [cit. 2016-02-02]. Available at: http://en.wheelsage.org/tatra/t810/61229/pictures/oj61im/
[5] Medium Off-road Truck. *Army* [online]. [cit. 2016-02-1]. Available at: http://www.army.cz/scripts/detail.php?id=6631

Forum on Innovative Technologies and Management for Sustainability (ITMS'2016)

ktu 1922 panevėžys faculty of technologies and business

The 11th International Conference Intelligent Technologies in Logistics and Mechatronics Systems (ITELMS'2016), 28-29 April 2016, Panevėžys, Lithuania

Hygrothermal performance of a masonry wall retrofitted with interior insulation, the analysis of temperature and moisture distribution in wall construction

Pinigienė J.[a], Sušinskas S.[a]*

[a]*Kaunas University of Technology, Panevėžys Faculty of Technologies and Business, Nemuno str. 33, Panevėžys 37164, Lithuania*

Abstract

Renovation and retrofitting of old buildings is the best way to decrease their huge energy consumption for heating. Applying the insulation to the exterior of the façade is best way of insulating the walls. However, for historical buildings or buildings with valuable facades interior insulation remains the only feasible post-insulation technique. Applying interior insulation will however significantly modify the hygrothermal performance of the wall: the temperature in the masonry wall will decrease and also moisture level will increase. In this paper the distribution of temperature and moisture in a masonry wall with interior insulation is analyzed.

© 2016 J. Pinigienė, S. Sušinskas.
Peer-review under responsibility of the Kaunas University of Technology, Panevėžys Faculty of Technologies and Business

Keywords: interior insulation; hygrothermal performance; changes of temperature; relative humidity; excessive condensation.

1. Introduction

Most residents of Lithuania live in various old apartment buildings that were built before 1993. These buildings are cold, uneconomical and use a lot of energy for heating. Reducing the energy consumption of buildings has become increasingly imperative because of the combined demands of energy security, rising energy costs and the need to reduce the environmental damage of energy consumption. Renovation and retrofitting of old buildings is the best way to decrease their energy usage.

The best way of insulating a wall is to apply the insulation to the exterior of the façade. However, many buildings cannot be retrofitted with insulation on the exterior for reasons such as historic preservation, valuable natural stone or brick-built façade, which heritage appearance must be preserved. In such cases, the only possible alternative is an interior insulation of the wall. For people living in cold apartment buildings but disagreeing in their communities about all building renovation it is also a good way to live a bit warmer.

2. Physical processes of retrofitted wall with interior insulation

Applying interior insulation will however significantly modify the hygrothermal performance of the wall and, as a consequence, may induce a risk on interstitial condensation, frost damage, mould growth and other damage patterns. Thermal insulation lowers the temperature level in the masonry. It reduces the amount of inside heat that reaches the wall construction. On cold winter days the low temperature zone approaches inner surface of

* S. Sušinskas. Tel.: +3-706-823-9968.
E-mail address: saulius.susinskas@ktu.lt

© 2016 J. Pinigienė, S. Sušinskas.
Peer-review under responsibility of the Kaunas University of Technology, Panevėžys Faculty of Technologies and Business

wall and it can lead to frost damage. The temperature in the adjoining walls, floors and ceilings also decreases and water condensation can appear on inside surfaces. If there is no eligible interior ventilation and relative humidity is often high (>70%) the mould growth is expected which can cause health problems [1].

Interior insulation can be applied using vapor tight systems with polystyrene foam panels, mineral wool panels or ecowool in wooden frame and capillary active insulation system with polyurethane foam panels. The study [1] was performed to analyze the hygrothermal performance of the wall with vapor tight interior insulation systems and a capillary active insulation system. It was determined that vapor tight systems tend to be preferable for structures that are resistant to frost damage. For buildings sensitive to frost damage or when wooden beam ends are present, however, capillary active systems are shifted forward. Vapor tight interior insulation systems were found to perform best in respect to the potential energy savings. On the other hand, the larger thermal and vapor resistance of the vapor tight systems results in a larger risk on frost damage. The wall retrofitted with interior insulation exposed to 22 freeze-thaw cycles at -2 °C freezing temperature thresholds as compared to 9 cycles for the uninsulated wall [2].

Interior insulation causes the decrease in temperature of the masonry herewith it reduces the drying potential of precipitation moisture [3]. In order to protect walls from precipitation moisture a water-repellent impregnation or paint coat should be applied to the façade. It is recommended to improve the rain protection of the façade some time before installing the interior insulation. While the water content of the uninsulated wall drops within less than two years below the practical moisture content for brick walls, it takes more than five years for the insulated walls to dry out. Due to the higher vapor diffusion resistance, the walls insulated with polystyrene foam dries somewhat more slowly than the wall insulated with mineral wool.

3. The analysis of distribution of temperature and moisture in masonry wall construction

The 510 mm mm thick perforated ceramic brick wall of dwelling house (built before 1992) without cavity is used to analyze the distribution of temperature and moisture in masonry wall construction. Thermal transmittance (U-Value) for this type of wall is U_t = 1,05 W/(m^2·K) (R_t = 0,95 m^2·K/W) according to [4]. The thermal characteristics of various building constructions are calculated using [5]. According to [5] the maximum allowable total thermal resistance of additional interior insulation is limited in order to avoid too high relative humidity and excessive condensate at the inside surface. The maximum allowable total thermal resistance $R_{i.sum}$ (m^2·K/W) of additional thermal interior insulation is listed in Table 1. In addition to this it is emphasized that vapor barrier in insulating system should be used.

Table 1. The maximum allowable total thermal resistance $R_{i.sum}$ (m^2·K/W) of additional thermal interior insulation.

Total thermal resistance R_t (m^2·K/W) of uninsulated wall	0,7	0,9	1,2
The maximum allowable total thermal resistance $R_{i.sum}$ (m^2·K/W) of additional thermal interior insulation	1,0	1,5	2,5

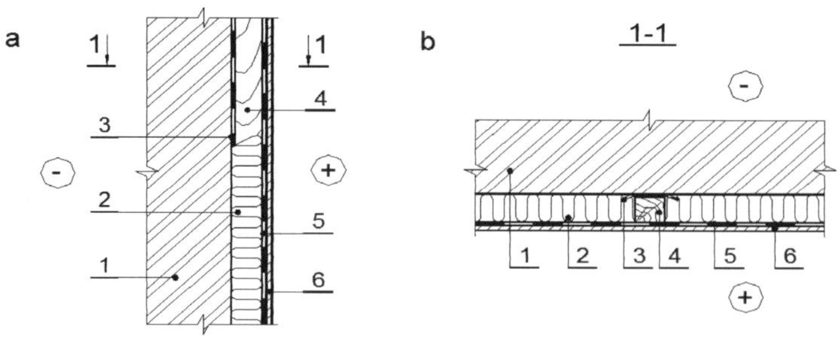

Fig. 1. (a) vertical section of the masonry wall with an interior thermal insulation system; (b) horizontal section of the masonry wall with an interior thermal insulation system. (1 – existing masonry wall; 2 - thermal insulation (mineral wool panels); 3 – flashing between wooden frame and masonry wall; 4 – wooden frame girder; 5 - vapor barrier; 6 - interior wall finish).

Vertical and horizontal sections of the masonry wall with an interior thermal insulation system are shown in Fig. 1. The analyzed masonry wall is retrofitted with interior insulation of mineral wool panels in wooden frame with water vapor barrier (Polyethylene (PE)). Gypsum boards are used for interior wall finish. Technical characteristics of materials used for additional thermal interior insulation are listed in Table 2.

Table 2. Technical characteristics of materials used for additional thermal interior insulation.

Material Number	Material	Symbol	Thermal conductivity λ_i, (W/m·K)	Thermal resistance R_i, (m^2·K/W)
1	Mineral wool in wooden frame	R_1	0,042 *	-
2	PE vapor barrier (t=0,2 mm)	R_2	-	0,040
3	Interior wall finish (gypsum boards, t=12,5 mm)	R_3	0,250	0,050

*the mean value of thermal conductivity of mineral wool with wooden frame estimated

Total thermal resistance of analyzed uninsulated masonry wall is R_t = 0,95 m2·K/W. In this case the maximum allowable total thermal resistance of additional thermal interior insulation will be $R_{i.sum}$ = 1,67 m2·K/W. This value of $R_{i.sum}$ is interpolated using data in Table 1. Formula 1 is used to calculate the total thermal resistance and Formula 2 is used to calculate the maximum allowable thickness of mineral wool panels in wooden frame:

$$R_{i,sum} = R_1 + R_2 + R_3 = 1{,}67\ \mathrm{m}^2 \cdot \mathrm{K/W} \qquad (1)$$

$$d_1 = R_1 \cdot \lambda_1 = (R_{i,sum} - R_2 - R_3) \cdot \lambda_1 = (1{,}67 - 0{,}04 - 0{,}05) \cdot 0{,}042 = 0{,}067\ \mathrm{m} \qquad (2)$$

According to the calculated data the thickness of interior insulation is selected – 60 mm thick mineral wool panels are used for retrofitting analyzed masonry wall.

Computer program THERM-7.4 and online program for calculation of U values are used to determine the changes of temperature and moisture in analyzed masonry wall with interior insulation. Data used for program analysis: outside relative humidity - 80%, outside mean daily temperature -5,1 °C. The outside relative humidity depends on the time of day and ranges from 70% to 90% in Lithuania. The mean daily maximum temperature during the heating season (October to April) is -5,1 °C [5]. The inside temperature of dwelling houses during the heating season is +20,0 °C [5] and inside relative humidity – 50%.

In contrast to a wall retrofitted with interior insulation, the uninsulated masonry wall is also analyzed (Fig. 2(a), Fig. 2(b)).

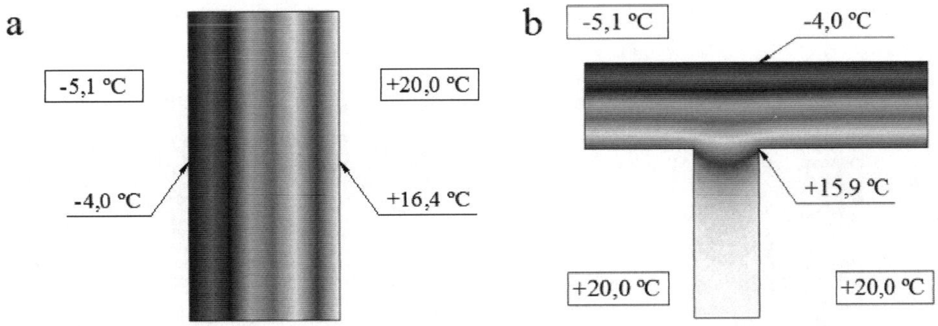

Fig. 2. (a) the changes of temperature in uninsulated masonry wall (vertical section); (b) the changes of temperature at the junction of uninsulated masonry façade wall and interior wall (horizontal section).

When the masonry wall is uninsulated the surface temperature at the inside decreases to +16,4 °C and at the junction of façade and interior walls decreases to +15,9 °C. Temperature decreases gradually through wall construction. The relative humidity at the surface is 60%. The distribution of temperature and relative humidity can be seen in Fig. 3(a). Mould formation is not expected under these conditions. However, if inside relative humidity rises to 70%, the relative humidity at the surface increases at 84%. Then mould grow is expected because most kinds of moulds start to grow at relative air humidity of 80% or more. The relative humidity in the wall construction reaches 100% (saturation point) and at this point some condensation appears (0,020 kg/m2). This amount dries in summer within 3 days (Fig. 3(b)). To avoid mould formation, the surface temperature should be increased by additional insulation.

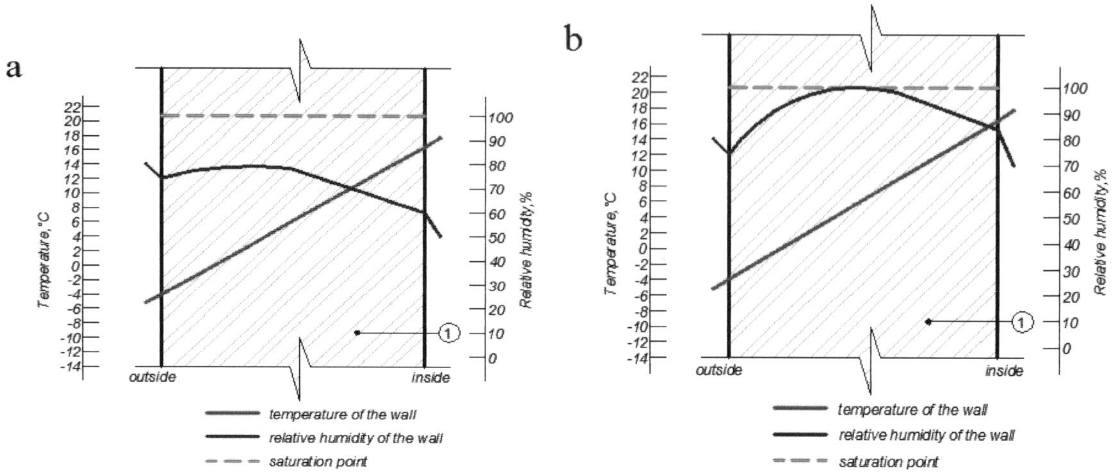

Fig. 3. (a) the distribution of temperature and moisture in uninsulated masonry wall when relative air humidity is 50%; (b) the distribution of temperature and moisture in uninsulated masonry wall when relative air humidity is 70%. (1 - ceramic brick masonry wall).

When the thermal insulation layer of 60 mm thick mineral wool panels in wooden frame and water vapor barrier is added to the interior surface of the façade walls (Fig. 1), it changes the hygrothermal performance of these constructions. Temperature decreases unevenly through wall construction (Fig. 4(a), Fig. 4(b)). The low temperature zone approaches inner surface of wall and it can lead to frost damage. Temperature decreases to +5,0 °C and vapor condensation appears (0,074 kg/m2) at the interface between the masonry wall and the insulation system. This amount of condensation dries in summer within 15 days (Fig. 5(b)). The surface temperature at the inside rises to +18,6 °C and the relative humidity is 60%. Difference between uninsulated wall and wall with thermal insulation inside surface temperature is 2,2 °C. Because of very low temperature (+10,8 °C) at the junction of façade and interior walls (Fig. 4(b)) condensation appearance and mould grow is expected in this place. This risk can be avoided by adding additional insulation to inside constructions like walls, ceiling adjoining to retrofitted façade wall (Fig. 5(a)).

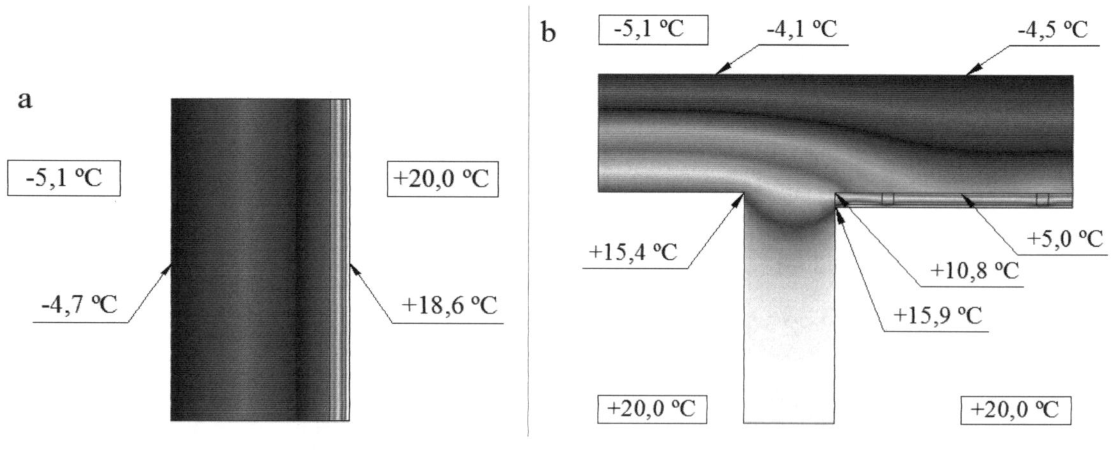

Fig. 4. (a) the changes of temperature in masonry wall with interior insulation (vertical section); (b) the changes of temperature at the junction of masonry façade wall with interior insulation and interior wall (horizontal section).

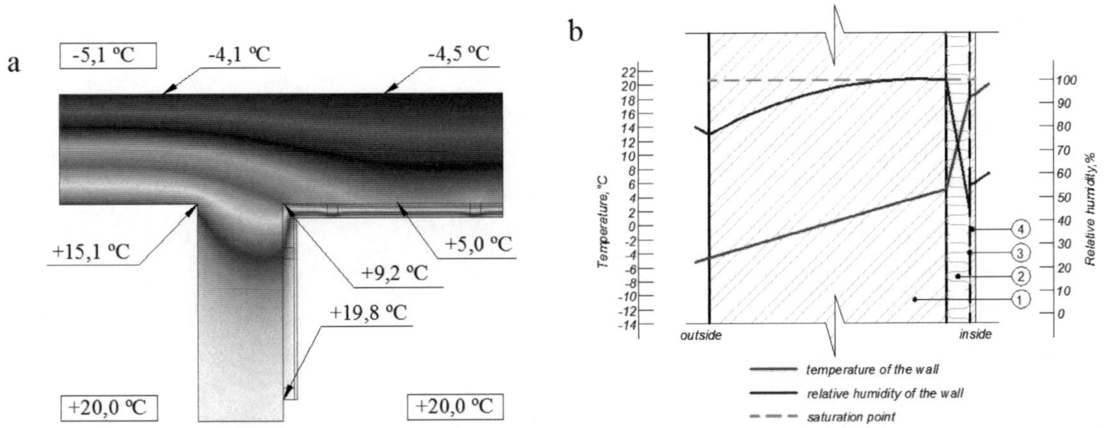

Fig. 5. (a) the changes of temperature at the junction of masonry façade wall with interior insulation and interior wall (horizontal section) when additional insulation is added to adjoining wall; (b) the distribution of temperature and moisture in masonry wall with interior insulation of 60 mm thick mineral wool, when relative air humidity is 50% (1 - ceramic brick masonry wall; 2 – mineral wool panels in wooden frame; 3 – vapor barrier; 4 - interior wall finish).

Conclusions

The distribution of temperature and moisture in masonry wall construction (510 mm thick perforated ceramic brick wall of dwelling house without cavity) is analyzed in this paper. During the heating season when inside relative humidity reaches 70% some condensation appears (0,020 kg/m2) in the uninsulated wall and mould grow is expected on the inside surface. To avoid mould formation, the surface temperature should be increased by additional insulation. If it is not possible to apply exterior insulation to the façade the interior insulation can be used. But this must be done in accordance with the conditions. The maximum thickness of mineral wool or other thermal insulation panels for interior insulation of analyzed masonry wall is 60 mm. The inside surface temperature rises when interior insulation is applied. But some vapor condensation appears (0,074 kg/m2) at the interface between the masonry wall and the insulation system that dries in summer within 15 days. To avoid water condensate and mould formation at the room corners additional insulation to inside constructions like walls, ceiling adjoining to retrofitted façade wall should be added too.

References

[1] Vereecken E, Van Gelder L et al. Interior Insulation for Wall retrofitting–A Probabilistic Analysis of Energy Savings and Hygrothermal Risks. *Energy and Buildings* 2015; 89: 231-244.
[2] Saïd MNA, Demers RG et al. *Hygrothermal Performance of a Masonry Wall Retrofitted with Interior Insulation.* CRC Press; 2003.
[3] Künzel, Hartwig M. Effect of Interior and Exterior Insulation on the Hygrothermal Behaviour of Exposed Walls. *Materials and Structures* 1998; 31 (2): 99-103.
[4] STR 2.01.09:2012 Pastatų energinis naudingumas. Energinio naudingumo sertifikavimas (Priedas 5).
[5] STR 2.05.01:2013 Pastatų energinio naudingumo projektavimas.

Forum on Innovative Technologies and Management for Sustainability (ITMS'2016)

ktu 1922 panevėžys faculty of technologies and business

The 11th International Conference Intelligent Technologies in Logistics and Mechatronics Systems (ITELMS'2016), 28-29 April 2016, Panevėžys, Lithuania

Simulation research of kinematic discrepancy in multiaxis hydrostatic drive system

Przybysz M.[a], Rubiec A.[a]*

[a]*Military University of Technology, Gen. S. Kaliskiego 2 Street, 00-908 Warsaw, Poland*

Abstract

In the article a problem of the kinematic discrepancy in multi-axis hydrostatic drive system was presented. Described vehicles model, which used to simulation research on ability of compensation discrepancies kinematic. The model was created based on the identification research of selected parameters on Unmanned Ground Vehicle, with the 6x6 suspension system and hydrostatic driving system. The simulation studies have been conducted on the incompatibility of the kinematic vehicle model created multi-axis and the results of simulation.

© 2016 M. Przybysz, A. Rubiec.
Peer-review under responsibility of the Kaunas University of Technology, Panevėžys Faculty of Technologies and Business

Keywords: hydrostatic drive system; multiaxis vehicle; kinematic discrepancy.

1. Introduction

Nomenclature

c	structural factors specific to the type of engine design hydrostatic
Δ ps	pressure drop
μ	dynamic viscosity
ns	speed hydraulic motor
c	structural factors specific to the type of construction of the hydrostatic element
qs	displacement of a hydraulic component
Qs	flow
Δ	kinematic discrepancy
ω	velocity of wheels
ρ	density of the liquid

The tasks requested from the Unmanned Ground Vehicles (UGV) during rescue activities and military missions put forward high requirements in terms of their mobility and maneuverability. They include, among others, the capacity of UGV to move on rough terrain (20-30cm), hills with the inclination angle of 40% (22°), low carrying capacity areas=150kPa, overcoming typical field obstacles, i.e.: debris, logs, ditches [1,2]. The above requirements are possible by applying a hydrostatic drive system and suspension ensure well distributed of pressure wheels on the ground [3].

* Corresponding author. E-mail address: arkadiusz.rubiec@wat.edu.pl

© 2016 M. Przybysz, A. Rubiec.
Peer-review under responsibility of the Kaunas University of Technology, Panevėžys Faculty of Technologies and Business

Compliance with these requirements is all the more important, given the problems of vehicle control system teleoperation. These problems are presented in the papers [4,5]. Due to such a requirement propulsion system in Unmanned platforms land is one of the most important systems responsible for providing high mobility. For this reason, when designing a much attention is paid to the understanding of the phenomena occurring in the powertrain driving. Often transferred experience with conventional vehicles [6,7,8,9].

Driving with UGVs on very rough terrain or with different values of dynamic radiuses of road wheels, which may be caused by uneven load or different value of pressure in tires, results in wheels rotating with different rotary speeds (Fig. 1). The phenomenon is known as kinematic discrepancy and it is defined by a degree of kinematic discrepancy. The degree of kinematic discrepancy Δ of a system is the ratio of the result of substracting the numbers of wheels rotation ω', ω'' performed on a given road section (while rotating without slipping), to the number of rotations of the wheel rotating with greater speed [10,11]

$$\Delta = \frac{\omega' - \omega''}{\max(\omega', \omega'')} \tag{1}$$

Kinematic discrepancy of the drive system may be examined between the wheels located on the same axis or between different axes. This study presents the issue of kinematic discrepancy between different axes.

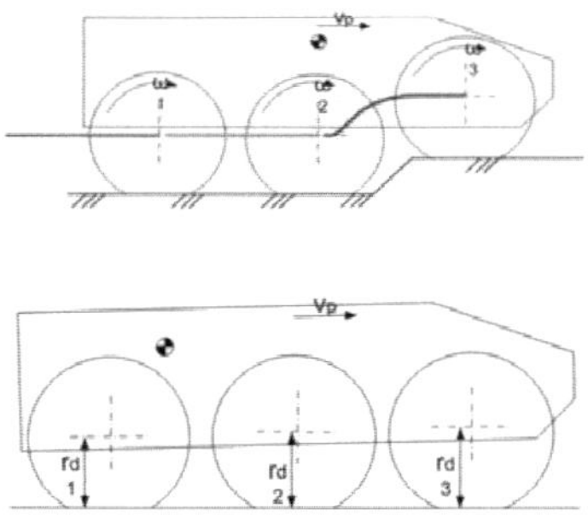

Fig. 1. Kinematic discrepancy in six-wheels vehicle: a-drive over obstructions terrain b- different radius of wheels.

Hydrostatic drive systems are characterized by great kinematic rigidity, which is associated with small compressibility of the working agent, i.e. a medium in HDS that ensures power transfer from the pressure generator to the receivers, which are the hydraulic motors responsible for the road wheels drive. This feature results in the fact that HDS provides a great kinematic accuracy of performed movements, but it is also sensitive to the occurring kinematic discrepancies. The accuracy depends mainly on internal leaks resulting from the construction features of the used elements and pressure drops on particular elements [12,13,14]. The leaks enable, to some extent, the differentiation of rotational speed of wheels necessary for covering the kinematic discrepancy - which is defined as kinematic flexibility. On the other hand, the capacity of covering the kinematic discrepancy by means of ensuring kinematic flexibility of the system is called kinematic discrepancy compensation.

Thus, the impact of leaks in the system on the capacity of differentiating the speed of wheels, enabling the provision of appropriate cooperation of the wheel with the road bed when driving with large values of kinematic discrepancy, avoiding the development of the circulating power phenomenon, should be examined. for this purpose, a model of the vehicle together with the propulsion system allows you to investigate the effect of discrepancies kinematic.

2. Identification studies

Identification tests were performed using a multi-axis unmanned ground vehicle (weight 4200 kg) with a hydrostatic drive system. The distance between the axes of the wheels is 1 meter. Hydrostatic drive system consists of two main displacement pump with variable capacity (3). Each pump supplies the hydrostatic motors (gerotor) (1) belonging to a its own side of the vehicle. The system also used Flow dividers (2) to ensure consistent rotational speed of the wheels (Fig. 2b).

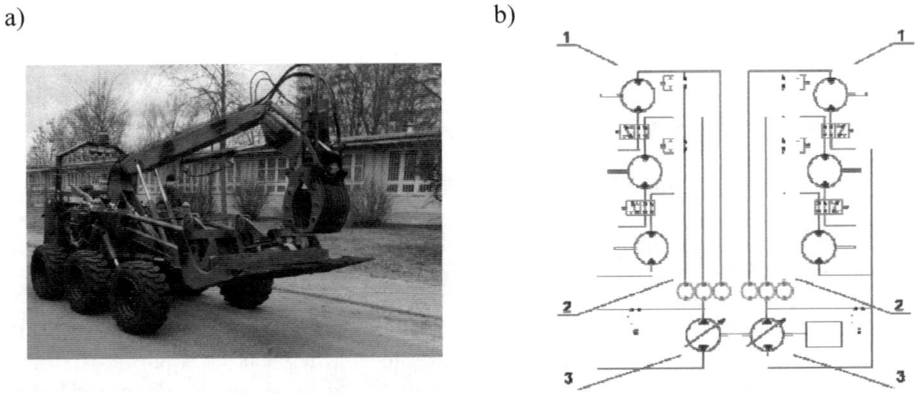

Fig. 2. Unmanned Ground Vehicle "Marek": a - view of the vehicle; b - hydrostatic drive system: 1 - hydrostatic motors; 2 - flow dividers; 3 - variable displacement pump [15].

Research discrepancies kinematic drive system occurring during the overcoming of embankments specified beating embankment (fig. 3) with a height h_2=80 cm, l_1=2m - has been erected without tamping.

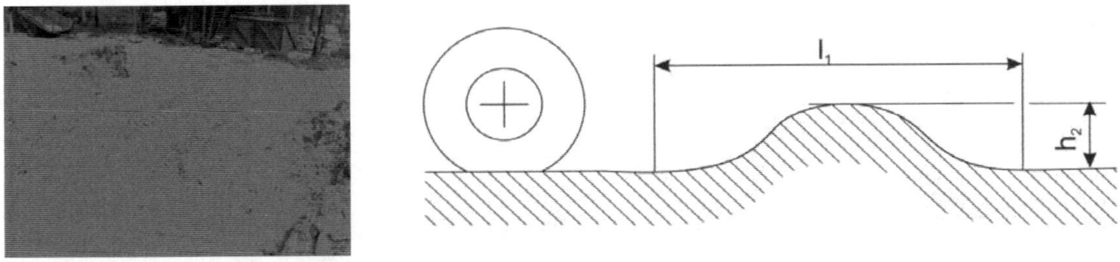

Fig. 3. Embankment: used in identification tests of kinematic discrepancy.

Based on the study, we identified the parameters necessary to develop a simulation model of the vehicle. This model will be used to investigate the ability of compensation of discrepancies kinematic. Summary of the results of the test vehicle and the simulation is shown in Figure 5. Comparative parameters shown in Table 2.

3. Model simulation model of the vehicle for testing kinematic discrepancy

The study simulation was prepared simulation model of the vehicle. It consists of a dynamic model of the vehicle body and the suspension structure. In addition, using a model susceptible wheel and its impact on the ground and model of the hydrostatic drive system (Fig. 1).

3.1. Dynamic model of the vehicle and the suspension structure

The geometry of the vehicle model has been developed in the Catia v5 and then imported into the program serving MSC Adams multibody simulation. To model the following assumptions:

- The model is a multi-mass system consisting of rigid bodies by 17 degrees of freedom;
- The body of which the vehicle consists connected by means constraint ideal (did not include the friction and clearances);
- The stiffness of the suspension springs replaced with constant elastic parameters k1, k2, k3 and damping c1, c2, c3;
- The substrate is non-deformable shape corresponding inequalities field used during testing polygon.

Parameters of model vehicle shown in Table 1, whereas the suspension structure diagram of Figure 4.

Table 1. Masses and parameters of vehicle model

Mass of vehicle [kg]		3300
suspension mass [kg]	Front	40
	Middle	60
	rear	40
Mass of wheel [kg]		20
Stiffness k1,k2,k3 [N/mm]		900
Damping c1,c2,c3 [N*s/mm]		90

Model (without wheels) comprises four rigid bodies connected to each other by means of rotational kinematic constraint of the longitudinal axis perpendicular to the plane of the drawing, located at points A-L corresponding to the pattern of the suspension of the physical object. Model (without wheels) comprises four rigid bodies connected to each other by means of rotational kinematic constraint of the longitudinal axis perpendicular to the plane of the drawing, located at points A-L corresponding to the pattern of the suspension of the physical object. Points A-F show joints mounting the actuators rocker arms suspension, with coefficients of stiffness $k_{1\text{-}3}$ and damping $c=_{1\text{-}3}$.

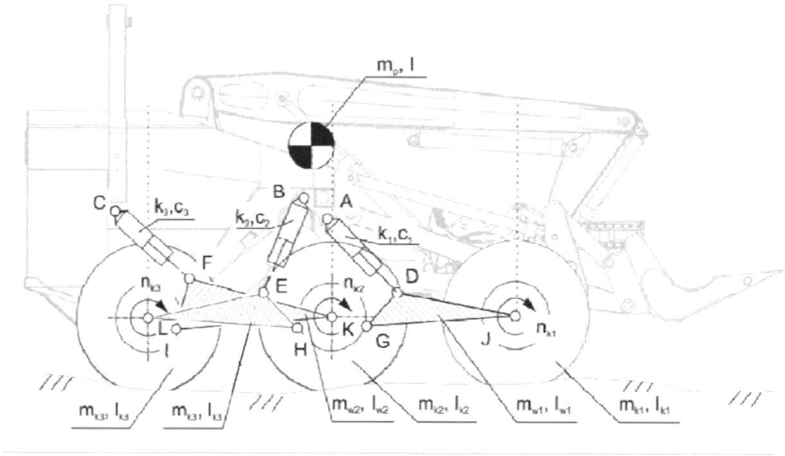

Fig. 4. Model of vehicle.

In the model of vehicle was applied a flexible wheel model [2] developed in the Institute of Mechanical Engineering WAT. Model consists of finite elements forming the circuit of the wheel. Each of elements has a conferred stuffiness and damping force related to a wheel rim in directions longitudinal, lateral and radial. Each of the elements is given with respect to stiffness and damping of the wheel rim in the longitudinal, lateral and radial direction. Developed model allows to calculate the value of the driving force in function of wheel slip and engaging positively the wheel relative to the curvature of the ground.

3.2. Model of hydrostatic drive system

Simulation studies of circulating power is assumed to use components of the hydrostatic drive system running on different characteristics of volumetric efficiency. For this purpose, they assumed the use of motors hydrostatic of structures (piston and gerotor) and dividers flow with different accuracies distribution of the working medium (a divide gear with an accuracy of dividing a flow of 2% and 5% in nominal conditions of operation, and slide divider enabling the division of flow in the range of 20% -50%, or free flow).

The model of the hydrostatic drive system includes both the characteristics of efficiency (volumetric and mechanical) for hydraulic motor and flow divider. It takes into account elements such as design features modeled hydraulic components, the properties of the working fluid and the rigidity of the hydraulic lines. Volumetric efficiency of the model of the hydraulic motor is determined on the basis of equation[16]:

$$\eta_v = \frac{1}{1 + c \cdot \mu \cdot \dfrac{\Delta p_s}{2 \cdot \pi \cdot \mu \cdot n_s} + c_r \cdot \dfrac{1}{n_s} \cdot \sqrt{\dfrac{2 \cdot \Delta p_s}{\rho}} \cdot \sqrt[3]{q_s}^{-1}} \tag{2}$$

The efficiency of the mechanical-hydraulic model hydraulic motor is determined by the equation:

$$\eta_{m-h} = sign(\Delta p_s \cdot \frac{1}{1+\cdot\frac{1}{3\cdot n_s}}, n_s) \qquad (3)$$

Model flow divider was developed on the basis of data on the accuracy of the distribution of the stream of the working medium. The starting point is to balance the pressure drop generated on the flow divider, which is the kinematic connection between the lines of the hydraulic system.

In the balance of this it follows that the individual sections of the divider can work in a hydraulic motor (pressure drop), or as a pump (increase in pressure) in a line after the divider. In both states there are leaks dependent on the pressure drops, resulting in compensation for discrepancies kinematic.

Therefore, the divider sections are coupled together in a kinematic way:

$$\begin{cases} Q_{fd1} = n_{fd} \cdot q_{fd} \cdot \eta_{v1} \\ Q_{fd2} = n_{fd} \cdot q_{fd} \cdot \eta_{v2} \\ Q_{fd3} = n_{fd} \cdot q_{fd} \cdot \eta_{v3} \end{cases} \qquad (4)$$

values of the volumetric efficiency determined from the equation:

$$\eta_v = \frac{1}{1 + c \cdot \frac{\Delta p_s}{Q}} \qquad (5)$$

Developed in this way, the hydraulic system allows to carry out research on the impact of leaks different components volumetric efficiencies on the ability to compensate for discrepancies kinematic

To verify the assumptions made when creating the model of the vehicle and hydrostatic system were carried out the verification tests and then comparing the results with results from studies of identification.
The study consisted in overcoming obstacles in the form of off-road ditch and recording speeds and drive torques to the individual running wheels belonging to one side. Summary of the two waveforms in Figure 5.

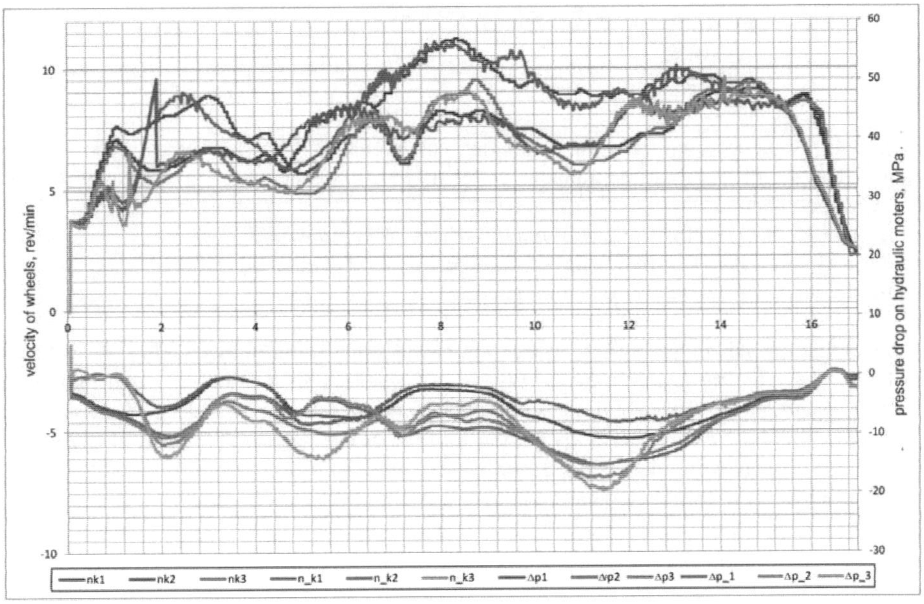

Fig. 5. Wheel speeds and drive torques to the wheels obtained during testing identification and simulation.

To assess compliance waveforms obtained during simulation tests with the results of identification used indicators: standard deviation, correlation coefficient, mean absolute percentage error.

Table. 2. Compare the results of identification and simulation

	Parameter to compare					
	n_kl	n_k2	n_k3	Mom_1	Mom_2	Mom_3
standard deviation	2,5%	1,9%	2,3%	36,1%	12,6%	5,0%
correlation coefficient	92%	84%	87%	85%	91%	87%
mean absolute percentage error	0,21	0,14	0,15	1,44	0,88	0,39

4. Simulation tests of the power circulating in the hydrostatic drive system

Research of kinematic discrepancy and circulating power in hydrostatic drive system, included the simulated terrain obstacle crossing vehicle and to record the selected parameters (such as the wheel speed, driving torque and flow of hydraulic oil). Drive was carried out at a constant speed of the wheels, corresponding to the vehicle speed 0.4 m / s. For the simulation, the following components of the transmission:

- Hydrostatic motors: (A- without leaks, B- hydrostatic piston motors, C- hydrostatic gerotor motor);
- Flow divider (I- without leaks and II- with a precision distribution of flow 2% and III- 5%, IV-20% V 50%).

Simulation tests were performed for all possible combinations of components of the hydraulic system.

The shape of the obstacles of adopted in the study was developed based on the profiles of the obstacles that are on the training ground used for experimental studies of Unmanned Ground Vehicle.

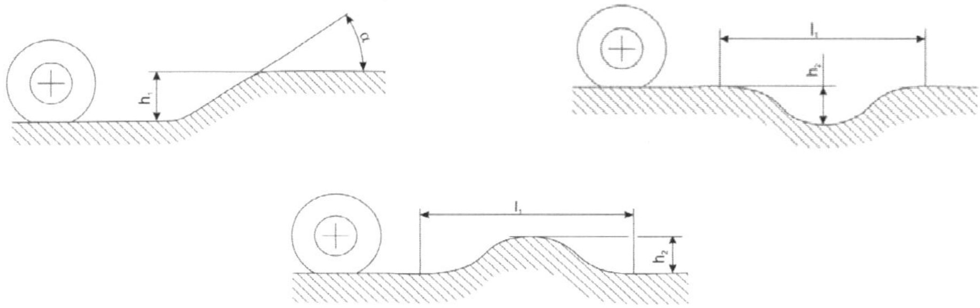

Fig. 6. Obstacles in simulation tests: elevation, trench, embankment.

Examples of the results obtained from the simulations overcome obstacles off-road (type embankment), is given in Figure 7 and 8. In this case, the arrangement consisted of a piston motors and gear flow divider (B-II). These results represent them:

- speeds wheels: nk1- front wheels, nk2- middle wheel, nk3- rear wheel;
- torque of wheels: Mk1- front wheel, Mk2- middle wheel, Mk3- rear wheel;
- slip of wheels::sk1 front wheel, sk2- middle wheel, sk3- rear wheel.

Due to the fact that the vehicle is moving on uneven terrain, both on the slopes facing up and down, so that the wheels of the vehicle at any given moment must drive (if the entrance to the hill) and inhibit (Downhill). Therefore, it was necessary to apply condition allowing determine the torque and its direction giving rise to circulating power in the system. For this purpose, the average of the determined value of the vehicle driving torque. In the case of the driving torque occurring with a different sign compared to the average value of torque, the assumed time giving rise to the phenomenon of power circulating in the system.

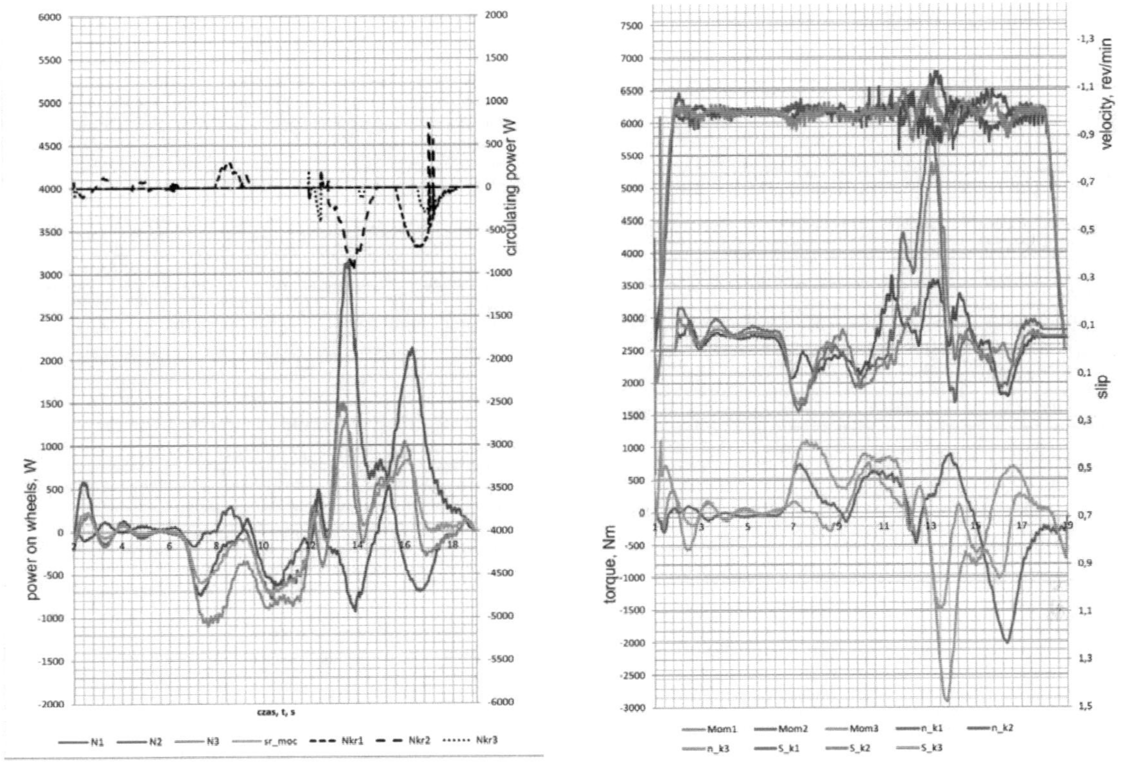

Fig. 7. Results of simulation research simulation for system in configuration B-II.

Analyzing the simulation charts, we see that the motion begins (t = 2s) and ends (t = 18s). At the time lasting between t = 2-3 S followed by acceleration of the vehicle. This results in an increase of the torque to the wheels and the increase in wheel slippage relative to the ground. At time t = 3-6s following stabilize the vehicle by damping motion. Followed driveway obstacle. In this time velocity of wheels have the same value as evidence of low value discrepancies kinematic. At this time there are no significant differences in the speed and drive torque result from the dynamics of acceleration and the rolling resistance of the wheels.

At t = 6,4s raid front wheel follows an obstacle, resulting in increased resistance to drive vehicle, the deviation of the vehicle in a reverse direction and moving the center of gravity toward the rear of the vehicle. This causes a momentary dociążenie central and rear wheels, and then unloading the wheel center. At the time course of driving torque, this effect appears as an increase in driving torque and wheel slippage and power circulating in the drive system of the 500W. Because of their high volumetric efficiency, the system is unable to sufficiently compensate for discrepancies kinematic resulting in the occurrence of too low a speed of the front wheels.

At t = 9,6s followed raid front wheel on top of the obstacles to the resulting alignment wheel speeds, and thus decrease the drive torques. Then, at t = 10s followed by raid the central wheels on the obstacle. At the moment when the wheel is on top of the hill at the time t = 12s is blocked due to the suspension of limitation of the actuator stroke suspension system. At time t = 13s followed by downhill wheel center of the obstacles and the rapid inclination of the vehicle in the forward direction. This increases laden front wheels. The phenomenon of power circulating about the value of 1kW at the center circle. Then, the downhill of the rear wheels and stop the vehicle.

5. Results

The tables below contain the results of the simulations. Table 1 shows greatest value of the sum of the power at the wheels of the vehicle registered at overcoming obstacles. Table 2 shows the power of circulating registered overcoming the obstacles

Table 3. The maximum value of the sum under the wheels during simulation tests

		obstacles								
		elevation			trench			embankment		
the accuracy of the distribution of stream flow divider%		A	B	C	A	B	C	A	B	C
	I	1164	1164	776	1320	1100	880	1500	1250	1000
	II	1067	921,5	727,5	1210	1045	825	1375	1187,5	937,5
	III	970	776	679	1100	880	770	1250	1000	875
	IV	485	388	242,5	550	440	275	625	500	312,5
	V	194	97	97	220	110	110	250	125	125

Table 4. Power values of circulating occurring during the overcoming obstacles

		obstacles								
		elevation			trench			embankment		
the accuracy of the distribution of stream flow divider%		A	B	C	A	B	C	A	B	C
	I	3104	3395	3104	3520	3850	3520	4000	4375	4000
	II	3007	3104	2910	3410	3520	3300	3875	4000	3750
	III	2910	3007	2813	3300	3410	3190	3750	3875	3625
	IV	2619	2619	2619	2970	2970	2970	3375	3375	3375
	V	2231	2231	2231	2530	2530	2530	2875	2875	2875

6. Conclusion

Assessment ability of compensation of kinematic discrepancy is particularly important at the stage of design of vehicle propulsion extreme mobility. The reason for this assumption is the possibility of driving significant uneven terrain gives rise to a significant kinematic discrepancy and the possibility of significant power in the drive system.

Particularly important is circulating power which adversely affects the efficiency of the entire power train. Research has shown that the power circulating in the system can be up to 40% of the power needed to drive around obstacles.

In considered system the greatest impact on reducing the phenomenon of power circulating was accurate distribution of the flow divider. Comparing the change in the distribution of the flow divider at constant hydrostatic motor efficiency, a reduction in the power of the circulating 1.5kW to 250W. Not while there was such a significant effect to reduce the power of circulating considering the change in hydrostatic motors.

The analysis results also showed a significant effect of wheel slippage with decrease of volumetric efficiency of the components used with the hydraulic system of 0.9 to 1.5. This effect is noticeable in the same manner by changing the volumetric efficiency of both hydraulic motors and flow dividers. The adverse effect was recorded during the reduction in volumetric efficiency due to hydraulic motors that are caused to decrease speed of the vehicle in comparison with the change of volumetric efficiency of flow dividers. This value is a maximum value was 12%.

Acknowledgements

We would like to thank of National Centre for Research and Development for founding grant number LIDER/007/424/L-4/12/NCBR/2013 within have been conducted investigation described in this paper.

References

[1] Jaskolowski M, Konopka S, Krogul P, Spadlo K. Capabilities to negotiate terrain obstacles by articulated tracked platforms. *Intelligent Technologies in Logistics and Mechatronics Systems ITELMS 2015*. Kaunas: Kaunas University Technology Press; 2015. p. 158 – 165

[2] Dabrowska A, Jaskolowski M, Krogul P, Spadlo K. Mobility evaluation of a lightweight four-wheel unmanned ground vehicle with articulated steering system. *Intelligent Technologies in Logistics and Mechatronics Systems ITELMS 2015*. Kaunas: Kaunas University Technology Press; 2015. p. 95 – 101

[3] Bartnicki A, Łopatka MJ, Muszyński T. Stiffness evaluation of Fire Rescue Robot suspension with hydropneumatic components. *Mechatronic systems, mechanics and materials II. Solid State Phenomena*, vol. 210, Zurich: Trans Tech Publications Ltd.; 2014. p. 301-308

[4] Typiak A. Ladar - Aided Navigation System for Unmanned Ground Vehicles. *Intelligent technologies in logistics and mechatronics systems - ITELMS'2013*. Kaunas: Kaunas University Technology Press; 2013. p. 254-261.

[5] Cieslik K, Przybysz M, Rubiec A, Typiak R. Concept for a vision system for an extreme mobility unmanned ground vehicle. *Intelligent technologies in logistics and mechatronics systems - ITELMS'2015*. Kaunas: Kaunas University Technology Press; 2015. p. 75-79

[6] Bartnicki A, Dąbrowska A, Łopatka MJ. Experimental research on directional stability of articulated tractors. *Mechatronic systems. mechanics and materials II Book Series: Solid State Phenomena* Volume: 210, 2013; p. 77-86

[7] Konopka S, Sprawka P, Muszynski T. Investigating Resistance of Turn Six-Wheel Skid-Steer Vehicle. *Intelligent technologies in logistics and mechatronics systems - ITELMS 2013*. Kaunas; 2013. p. 107-115

[8] Bartnicki A, Łopatka MJ, Muszyński T, Jaskółowski M. The Researches of Stability Evaluation Articulated Loader while Working on Slopes. *ITELMS '2013*. Kaunas; 2013.

[9] Łopatka MJ, Muszyński T, Rubiec A. Loads analysis of skid steer robot drive system. In: *Method and Models in Automation and Robotics Międzyzdroje 2013, IEEE*. New York; 2013. p. 362-367

[10] Wong JY. *Theory of ground vehicle, August, 2008*. New York; 2008.

[11] Studziński K. *Teoria I sterowanie. Samochód.* WKŁ, Warszawa, 1980.

[12] Budny E. *Napęd i sterowanie układów hydraulicznych w maszynach roboczych*. ITE, Radom; 2001.

[13] Bartnicki A, Sprawka P. *Zastosowanie hydrostatycznych układów napędowych we współczesnych maszynach i pojazdach lądowych.* LOGITRANS, Szczyrk; 2008.

[14] Stryczek S. *Napęd hydrostatyczny. Tom I. Wydawnictwa Naukowo* . Techniczne. Warszawa; 2005.

[15] Łopatka MJ. Praca zbiorowa, Inżynieryjny Robot Wsparcia IOD/EOD – usuwania ładunków niebezpiecznych. *Sprawozdanie z projektu rozwojowego Nr OR00001205/PBR*, Warszawa; 2011.

[16] Kollek W. *Podstawy projektowania napędów i sterowań hydraulicznych.* Politechnika. Wrocławska; 2013.

Forum on Innovative Technologies and Management for Sustainability (ITMS'2016)

panevėžys faculty of technologies and business

The 11th International Conference Intelligent Technologies in Logistics and Mechatronics Systems (ITELMS'2016), 28-29 April 2016, Panevėžys, Lithuania

Application of Capacitive micromachined ultrasound transducers for detection of specific interaction of materials

Sapeliauskas E.[a]*, Barauskas D.[a], Pelenis D.[a], Vanagas G.[a], Mikolajūnas M.[a], Viržonis D.[a]

[a]Panevėžys Faculty of Technology and Business, Kaunas University of Technology, Nemuno str.33, LT-37164 Panevėžys, Lithuania

Abstract

We present here our results of designing, testing and researching of capacitive micromachined ultrasound transducers (CMUT) as detectors in biosensing. In our work we used two types of CMUT structure, which correspond to two types of detection: resonance and Scholte waves. We found that biosensing is more efficient when the interdigital (IDT) CMUT structure is used for Scholte waves excitation and receive, since IDT structure enables the real-time biosensing in fluid environment CMUT IDTs for 10 MHz operation in water, with 146 μm spaced double fingers were designed and fabricated using the surface micromachining technique. Fabricated CMUTs were tested for their resonance in air and for Scholte-type wave transmission in alcohol/deionized water solutions. The amplitude and phase velocity of the excited and received Scholte waves were measured in a 200 μm height microchannel, capped with a thick layer of soft polymer, which suppressed the production of non-informative guided waves. Experimental data were also used to verify the adequacy of the finite element model. It was determined that for the analyzed conditions the sensitivity of the measurement channel is expected to be no worse than 2 kHz GPa^{-1} in terms of the Scholte wave and CMUT IDT resonance frequency. This leads to a positive conclusion on the feasibility of the new sensor type. Additionally we explored the advantage of the CMUT for liquid mixing and pumping.

© 2016 E. Sapeliauskas, D. Barauskas, D. Pelenis, G. Vanagas, M. Mikolajūnas, D. Viržonis.
Peer-review under responsibility of the Kaunas University of Technology, Panevėžys Faculty of Technologies and Business

Keywords: CMUT; biosensing in fluid; Scholte wawe; microchannel.

1. Introduction

With global increase of chronic, neurodegenerative and lifestyle related diseases the need for early diagnosis with portable, fast and relatively inexpensive diagnostic systems is becoming the top priority. The most effective way to diagnose the mentioned diseases is biomolecular diagnostics. A biosensor is biomolecular diagnostics is the main tool of detection. Generally, the biosensor structure consists of two parts – a bioactive surface on which biochemical interactions takes place and another part - detector which converts this interaction into electrical signals. A number of different molecular diagnostics tools are being developed, but due to the complexity of the methods used, analytical chip fabrication costs and the specific requirements of the analyte preparation makes most of these tools not suitable for use in out of the laboratory conditions.

We believe that using CMUT (capacitive micromachined ultrasound transducers) structure can make biosensors and molecular diagnostics to be faster, cheaper and generally more accessible.

2. Resonance frequency measurements in immersion

The main and straightforward application of CMUT is use of its sensitivity to mass loading of its structures.

* Corresponding author. Tel.: +370-698-80721.
E-mail address: e.sapeliauskas@gmail.com

© 2016 E. Sapeliauskas, D. Barauskas, D. Pelenis, G. Vanagas, M. Mikolajūnas, D. Viržonis.
Peer-review under responsibility of the Kaunas University of Technology, Panevėžys Faculty of Technologies and Business

In our earlier experiments CMUT membranes were modified with bovine leukemia virus (BLV) gp51 protein antigen (Ag) and we have studied this protein interaction with gp51-specific antibody (Ab), without real time monitoring of sensor resonant frequency in air changes [1]. As for as the most important biochemical reactions take place in liquids, real-time sensing requires detector operation in immersion. We have made experiments with different density fluids to test the sensor working conditions with liquid analyte. For testing we used: isopropanol with the density of 786 kg/m^3 and the transformer oil (Neste TRAFO 10X), density 895 kg/m^3. Results of the real-time immersion resonance measurements are shown in Fig. 1. We can see that the change of CMUT immersion resonance frequency due to oil is grater that in case of isopropanol [2].

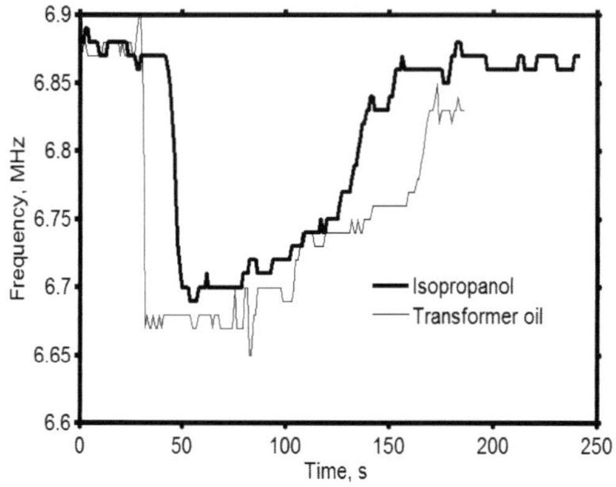

Fig1. Real-time plot of the oscillator output frequency during the drop loading experiment [2]

When CMUT is immersed into fluid it dissipates a lot of energy into the environment. This loss of energy increases measurement noise and significantly reduces sensor sensitivity and accuracy.

3. Investigation of surface waves generated with CMUT structure

The best way for CMUT structure to get in contact with liquid is by fabricating a microchannel. Optimization of microchannel was done and we found that decreasing its height from 400 um to 50 um CMUT structure operation was intact.

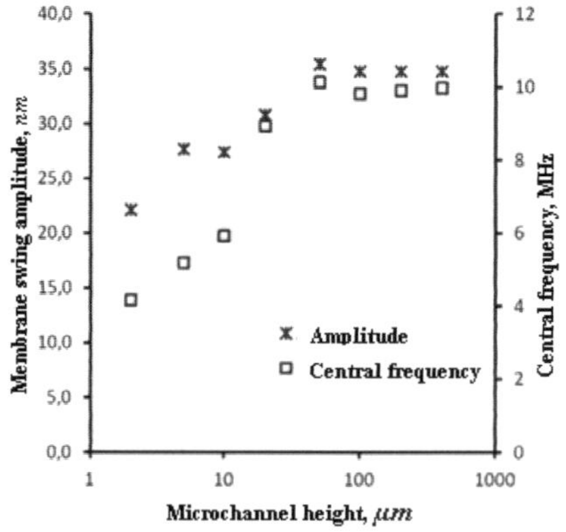

Fig. 2 The dependency of informative CMUT parameters over the microchannel height [3]

Further decreasing of the microchannel hight cause decrease of amplitude and frequency. This is due to dynamic interactions between the ultrasound wave reflections inside the microchannel. The data are shown in Fig. 2 [3] .

If CMUT device is used in the resonance detection regime, the quality of vibrations determines the sensitivity of the sensor for detecting changes in mechanical biochemical sample properties on the CMUT surface [4]. At the same time, while analyzing transient processes in microchannel system we can also detect other types of acoustic oscillations, one of which are Scholte waves. They have good application potential for sensing in immersion as they propagate at the sensor surface boundary with the liquid [5, 6]. To excite these waves a special interdigital (IDT) electromechanical transducer structure is needed, which can resonate with the interfacial wave and and the resonance quality will depend not only from the amount of CMUT structure vibration energy, but also from how many finger couples are in the path of the propagating wave [7]. CMUT with IDT structures is shown in Fig. 4.

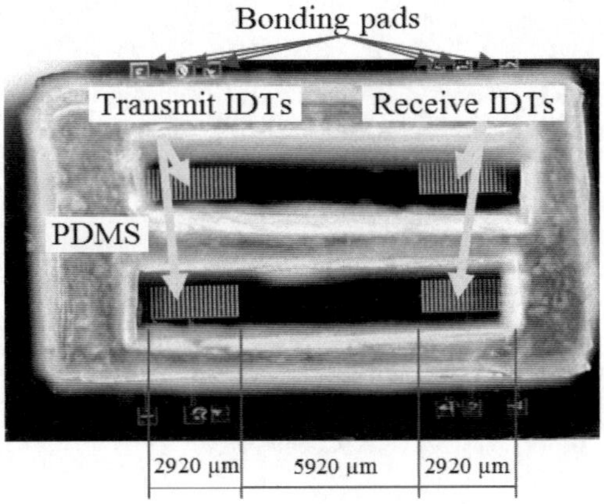

Fig. 4 Photo of two channel CMUT IDT delay lines assembled with microchannel. The capping wall of the microchannel is removed for better general representation and (in some cases) easier access to the microchannel contents during experimental work [7]

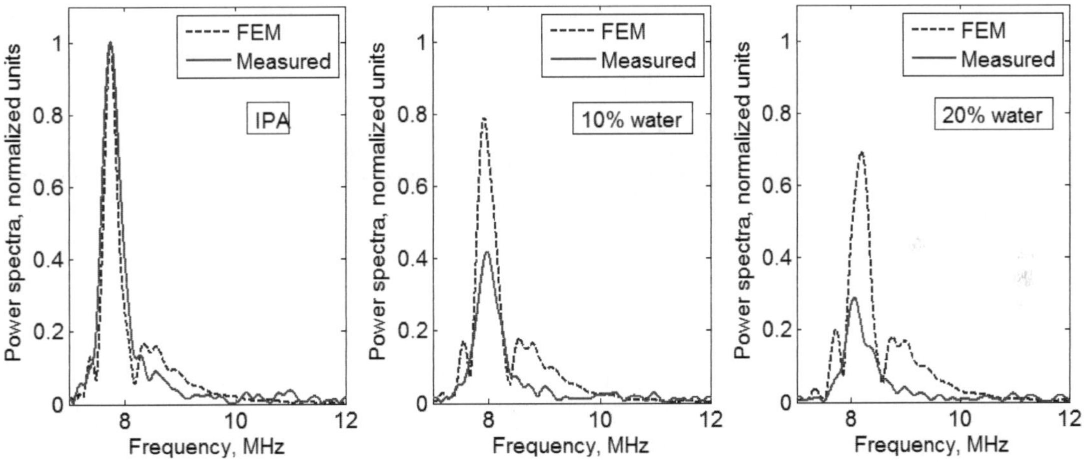

Fig. 5 Finite element model output and verification with experimental data: (a) time plots of the model output obtained at receiving points; (b) zoomed-in area of the time plots from 1.2 to 1.8 µs with Scholte wave arrival time indicated; (c) power spectra of the model output plotted with corresponding experimental data for comparison [7]

During finite element analysis it was assumed that the 200μm high PDMS-capped microchannel was filled with the liquid with the properties corresponding to IPA (Isopropanol) and two different water solutions in IPA with homogeneous volume proportions of 10% and 20%. The output data of transitional analysis of the FEM are presented in figures 5(a) and (b). Frequency spectra of the simulated and experimental data are presented for comparison in figure 5(c). Very good fit in terms of the resonance frequency can be observed and the sensitivity of the measurement channel is expected to be no worse than 2 kHz GPa$-^{1}$ in terms of the Scholte wave and CMUT IDT resonance frequency At this point we are able to conclude that our model for finite element analysis is adequate enough to capture the transitional processes in the described measurement channel, related with the structural resonance of CMUT IDTs and Scholte waves.

4. Fluid mixing in microchannel

When liquid is flowing inside the microchannel with the height of 200 um its flow is laminar, which is a serious problems from the use of microchannel for mixing. Since CMUT electromechanical structure in principle can be used for fluid mixing, we have researched that possibility by changing the shift in CMUT IDT excitation phase. We have conducted the experiment with colored IPA passing it through the microchannel and after 50 seconds changing it back to colorless. By analyzing the image fragments with MATLAB we found that phase shift significantly changed the dynamics of the fluid mixing.

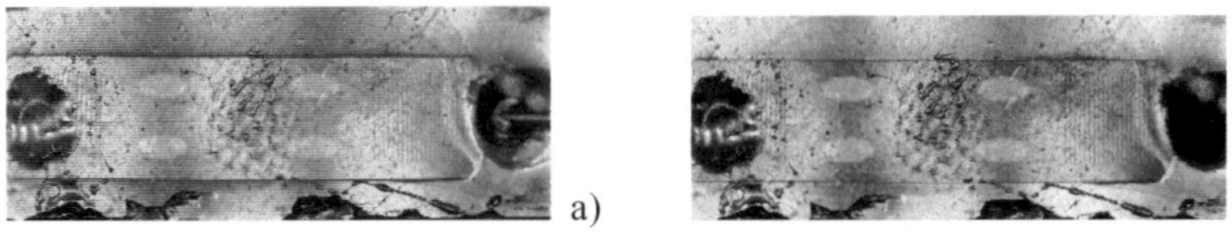

Fig. 6. a) image of microchannel, b) image of microchannel with colored IPA at its maximum color intensity

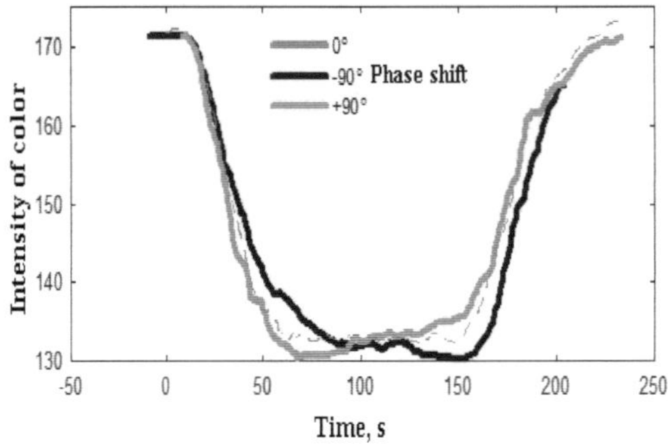

Fig. 7. Dashed line shows the fluid mixind dynamics when the phase shift is zero. Light blue line corresponds to the positive phase shift, and the dark blue line – to the negative phase shift

In the fig. 7. we can see that when the phase shift is positive, in the same direction of the auxiliary pumped flow, maximum color intensity is reached in 60 seconds. While the phase shift opposes direction of the flow, we can see that the maximum intensity of color is reached only after 150 seconds. This shows us the possibillity to use the microchannel with CMUT stuctures as a micropump. Also, by changing phase during sensing process, one can easily disturb the laminar flow and induce fluid mixing, which will provide better conditions for specific interaction of materials.

Conclusions

Our experiments show the applicability of CMUT structure for sensing of very low changes of the liquid composition. Phase velocity of Scholte wave can be measured with CMUT IDT structure and provide sensitivity of 2,9 m/s per 1 % of the water/IPA solution concentration change within the range of 0 – 20 % concentrations.

We also determined that real-time reading of the resonance between CMUT IDT structure and the Scholte wave is much more informative and have greater sensitivity potential than real-time reading of the CMUT membrane resonance in immersion.

Also we determined that sensing with CMUT structure in microchannel immersion can be further enhanced by estabilishing the fluid mixing, which breakes the laminar flow of a liquid and enables better conditions for specific interaction between the analyte and sensor structure.

References

[1] Ramanaviciene A, et al. Capacitive micromachined ultrasound transducer (cMUT) for immunosensor design. *Analyst* 2010;135: 1531-1534.

[2] Vanagas G, et al. Analysis and testing of liquid loaded CMUT oscillator. In *Intelligent technologies in logistics and mechatronics systems (ITELMS'2013)*, Panevezys, Lithuania; 2013. p. 267-273

[3] Vanagas G, et al. Study of the CMUT operation in microfluidic application. In *2012 IEEE International Ultrasonics Symposium*; 2012. p. 1814-1817

[4] Sapeliauskas E, et al. CMUT as a sensor in microfluidic environment. *Sensor Letters*, submitted.

[5] McLean J, Degertekin FL. Directional scholte wave generation and detection using interdigital capacitive micromachined ultrasonic transducers. *Ultrasonics, Ferroelectrics and Frequency Control, IEEE Transactions on* 2014; 51: 756-764.

[6] Sapeliauskas E, et al. Surface micromachined CMUTs for liquid phase sensing. In *2014 IEEE International Ultrasonics Symposium*; 2014. p. 2580-2583

[7] Sapeliauskas E, Barauskas D, Mikolajunas M, Pakenas E, Pelenis D, Jukna T, Virzonis D. Design, simulation and testing of capacitive micromachined ultrasound transducer- based phospholipidic biosensor elements. *Journal of Micromechanics and Microengineering* 8 June 2015; 25: 8.

Forum on Innovative Technologies and Management for Sustainability (ITMS'2016)

ktu 1922 panevėžys faculty of technologies and business

The 11th International Conference Intelligent Technologies in Logistics and Mechatronics Systems (ITELMS'2016), 28-29 April 2016, Panevėžys, Lithuania

Study of the mechanical properties components made by SLM additive technology

Śnieżek L.[a], Grzelak K.[a], Torzewski J.[a]*, Kluczyński J.[a]

[a]*Wojskowa Akademia Techniczna, Wydział Mechaniczny, Instytut Budowy Maszyn,ul. gen. Sylwestra Kaliskiego 2, 00-908 Warszawa 49, Poland*

Abstract

The paper presents the results of research strength properties of metal components produced one of the modern methods of additive manufacturing, which is the technique of selective melting using a laser beam SLM (Selective Laser Melting). An essential part of the strength tests were preceded by designing the additive manufacturing process of samples on the machine SLM 125 HL. Dimensions of Samples were adjusted to machine working area and the guidelines contained in the test standards. Roughness measurements were made on the metal surfaces after processing. Tensile and fatigue properties were determined by means of the Instron 8802 servohydraulic test system using samples prepared by melting 316L austenitic stainless steel powder. Fatigue tests were carried out under the conditions of tension (R = 0.1). The test results are shown in the Wöhler diagrams. In addition, the study of the influence of heat treatment on the fatigue life of the manufactured parts was conducted.

© 2016 L. Śnieżek, K. Grzelak, J. Torzewski, J. Kluczyński.
Peer-review under responsibility of the Kaunas University of Technology, Panevėžys Faculty of Technologies and Business

Keywords: additive manufacturing; selective laser melting; mechanialp properties.

1. Introduction

Since 2010, there is the significant increase in using additive manufacturing techniques (commonly referred to as three-dimensional printing) in many areas of the commercial industry. One of the most promising methods for additive shaping is a SLM technique involving selective melting metal powder with a high power laser beam of high power (100-1000W). SLM technique enables the production of parts with very complex geometries without using of additional tools. Most materials used in the preparation of this method are alloys based on iron, titanium, aluminum, and nickel [1-3].

This technique is most common in companies involved in the production: dentures and dental prostheses, molds for injection molding machines, individual components or low-volume production of parts with complex geometry.

The process of SLM manufacturing starts by loading specially prepared CAD model into machine software and the setup of parameters of the process. After starting the process, special recouter applies a layer of powdered metallic material having a thickness of 30 microns. Distributed layer is melted by laser in places where is the section of manufacturing element. After that platform with part is lowered by the thickness of the layer and it is possible to put next powder layer. The described cycle is repeated until the full manufactured component geometry. This process is shown in Figure 1.

* Corresponding author. E-mail address: janusz.torzewski@wat.edu.pl.

© 2016 L. Śnieżek, K. Grzelak, J. Torzewski, J. Kluczyński.
Peer-review under responsibility of the Kaunas University of Technology, Panevėžys Faculty of Technologies and Business

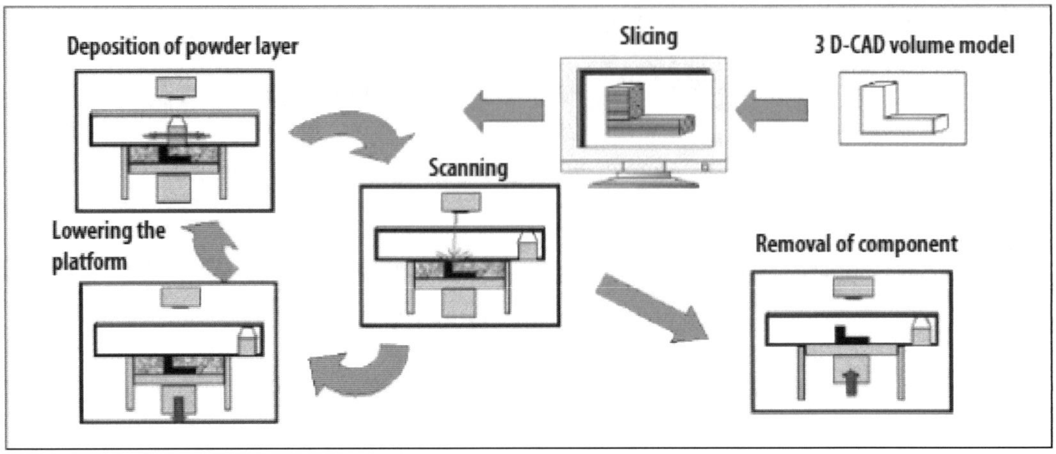

Fig. 1. Visualization of the SLM manufacturing process [6].

A characteristic feature of SLM is an obligation for using support material for hanging structures. This is due to the specific nature of the production process, during which the temperature gradient is large (in the working chamber, the temperature is about 40 ° C, while the melting point of the for example stainless steel powder is above 1400 ° C). It is giving stresses rise which deforms the model during manufacturing process. Furthermore, the support structure allow for easier removal of the model of the work platform. Removing the structure is easy and can be done using simple and publicly available tools (grinder, cutter, etc.).

Since November 2014 the SLM manufacturing machine SLM125HL has been launched in the Additive Manufacturing Laboratory in the Institute of Machine Building Military University of Technology. The main directions pursued by a research team is to analyze the possibilities of improving the performance of components made using additive manufacturing. In addition, there is investigations which are consisting on manufacturing gossamer and lattice - structured elements to allow reducing the weight of the final product and maintaining high strength properties of the parts.

The article includes the results of mechanical properties research of model elements produced using one of the modern additive manufacturing method, which is the technique of selective laser melting (SLM). An essential part of the strength tests were preceded by preparing the additive manufacturing process of specimens on the machine SLM 125HL. Test parts were adapted to the manufacturing process using that machine. Research is based on the literature analysis of material microstructure and surface roughness own specimens measurement. Tensile and fatigue properties was determined with using Instron 8802 servohydraulic test system. Specimens were made of 316L austenitic stainless steel and they were angled of 60° to the base unit. The results of fatigue tests conducted by a factor of asymmetry cycle of R=0.1 is shown in a Wöhler graph. Furthermore, the research include the effect of heat treatment on fatigue life prepared specimens.

2. Material and specimen preparation

SLM125HL working space and Instron material testing machine parameters required designing appropriate dimensions and geometry of the specimens. SLM manufacturing machine offers unlimited possibilities of elements configuration: parts arrangement on the working platform, the exposure time, exposure method, temperature of the work platform, exposure energy etc. The proper selection of these parameters requires a whole series test batch of products studies of a particular size including the additional procedures, such as, for example heat treatment or surface treatment.

2.1. The material and geometry of the specimens

The geometry of the specimens is based on ASTM E466 96. The dimensions were adjusted to the recommended load that can be accomplished using an Instron 8802 servohydraulic test system. The recommended minimum force during test cannot be lower than 17kN, so the minimum cross section of the specimens was determined from the equation:

$$A_0 = \frac{F_m}{R_m} \quad (1)$$

Symbol explanation: A_0 - speciment minimum cross-section area,
F_m - minimum tensile strength,
R_m - 316L steel tensile strength.

Standard dimensions of specimens based on the equations (1) and ASTM E466 96 are given in Table 1.

Table.1. Specimens dimensions.

Measurement length [mm]	Width [mm]	Thickness [mm]	Cross- section area [mm^2]	Max. lenght [mm]
25	9	3	27	105

Basing on the dimensions listed in Tab.1 the specimen was design in CAD software - CATIA V5R21 (Fig. 2), the specimen geometry was saved as STL file.

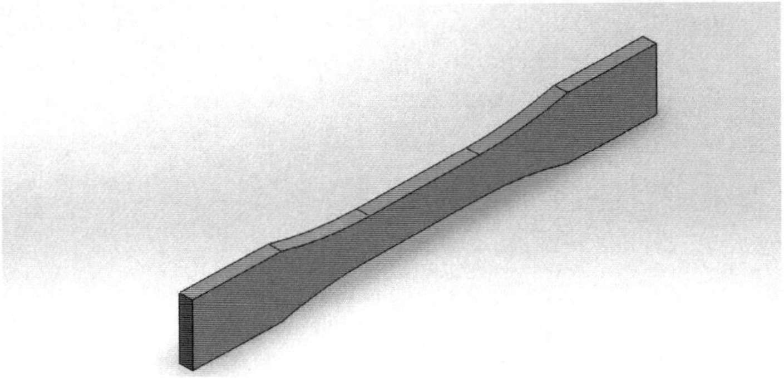

Fig.2. Isometric view of the specimen.

Samples were manufactured from the austenitic stainless steel 316L powder, whose chemical composition is given in Table 2.

Table.2. The chemical composition of the 316L steel powder.

C	Mn	Si	P	S	N	Cr	Mo	Ni
% weight								
max. 0,03	max. 2,0	max. 0,75	max. 0,045	max. 0,03	max. 0,10	16,0-18,0	2,00-3,00	10,0-14,0

2.2. Specimens preparation process

Each element manufactured using SLM technique requires the creation of a support structure which allows easy detachment part from the working platform. Supports cannot be too weak. It have to fix element so strongly that it does not break away from the surface of the platform. Performing too strong supports makes difficult to remove produced part of the working platform.

2.2.1. Support structure generation

Support structure was created using Magics 19 software. Apart from the supports there was used pins with a diameter of 1 mm, allowing for stronger attachment elements to the working platform.

Besides generating supports the specimens were angled of 60° to the working platform surface, and it has been rotated through an angle of 31o to the vertical axis of the coordinate system (OZ) in order to avoid damaging the wiper which distribute the powder. Perpendicular direction of arrangement of the specimens would result in the wiper cutting – after cut the wiper the process have to be stopped and the wiper need to be replaced. Specimens placement on the working platform is shown in Fig. 3.

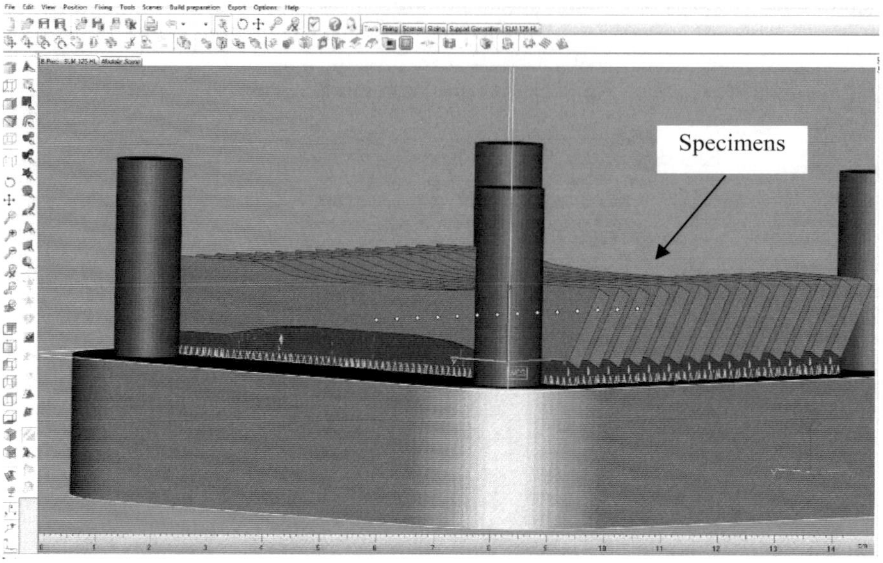

Fig. 3. Specimens with support structure on the working platform - the image of the program interface Magics 19.

2.2.2. SLM manufacturing process parameters

Input files for machine software SLM 125 HL are prepared in the program Magics 19 equipped with a module SLM Metal Build Processor it ensures slicing STL models into layers and setting up the manufacturing parameters such as exposure power, laser speed, vectors of laser paths configuration etc.

Exposure energy density, vectors spacing and vectors hatch space, laser speed and layer thickness for a given material is determined by the machine manufacturer. These parameters can be changed, of course, but it involves a change of the mechanical properties of manufactured parts, which significantly differ from those given by the SLM 125 HL machine producer. Laser energy density for 316L stainless steel should be 58 J / mm3, for the supports should be about 50 J / mm3. The value of this parameter is determined from the equation:[4]

$$E_{density}\left[\frac{J}{mm^3}\right] = \frac{Laser_{power}[W]}{Laser_{speed}\left[\frac{mm}{s}\right] \cdot Hatch_{space}[mm] \cdot Layer_{thickness}[mm]} \quad \# \; \# \; \# \; \# \; \# \quad (2)\#$$

There is possibility of using two types of exposure stripes and chess. These two configurations are shown in Fig.4.

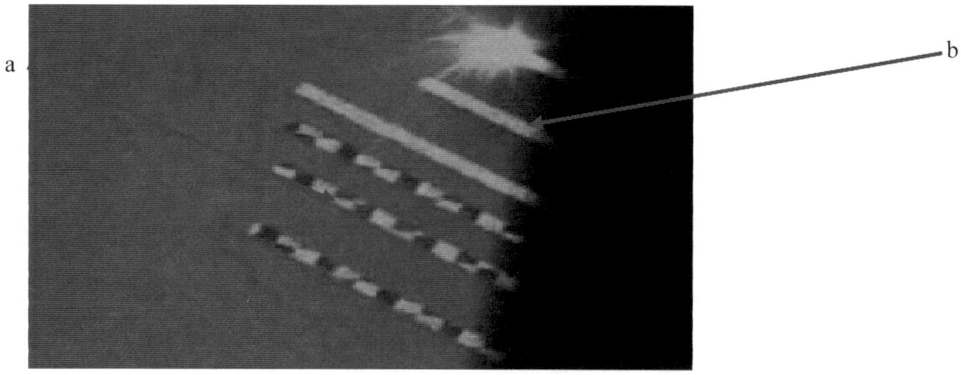

Fig.4. The layer of powder exposed in a chess (a), and stripes (b) configuration.

For the purposes of this research specimens were exposured in a linear configuration, and formed in combination of parameters:

- Power irradiation: 178W
- Laser Speed: 850 mm / s
- Distance between vectors: 0.12mm

- Layer thickness: 0.03mm
- Minimum layer time: 10 seconds
- Layer from which began the process: 10
- Number of exposures of the first layer: 1
- Temperature of the working platform: 130 ° C
- Argon pressure in the working chamber: 12mbar

Approval of the settings will start the file preparation process of slicing models into layers and save the file with the extension ".slm". File is exported to the machine software, where after loading it is possibility to set up the minimum exposure time, a layer of which the process begins, the number of exposures of the first layer and the other process parameters. The minimum layer time is assumed to 10 seconds, in order to prevent too fast start of exposuring another layer, which would result in the formation of thermal stresses. Start the process from 10 layers allowed for better attachment supports and pins to the platform. The use of more than one layer of the number of exposures is used when for some reason the process was interrupted and the surface temperature of the created model significantly decreased. The temperature of the working platform has an impact on the support and pins fixing strength to the platform. The value of 130 ° C is tested and recommended by the machine producer. Very patient and careful process preparation and justified introduction of the parameters allowed to start the manufacturing process, which resulted in the successful production of specimens shown in Fig. 5

Fig.5. Manufactured specimens on working platform.

3. Results

3.1. The microstructure of the material and surface roughness of specimens

An image of the microstructure was captured using a confocal microscope OLYMPUS OLS4100. The resulting image of the microstructure allowed an in-depth analysis of it.

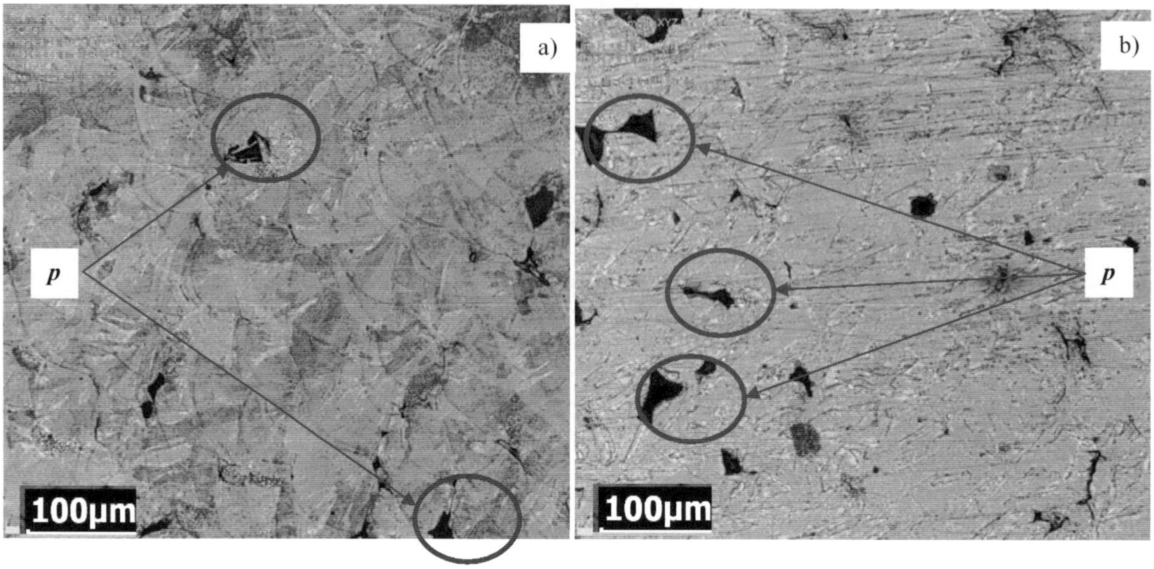

Fig.6. The metallographic section of 316L steel of specimen produced using SLM technique - marked pores (p): a) without heat treatment, b)with heat treatmen.

The presence of pores (marked outline in Fig. 6) is unavoidable during manufacture SLM technique. However, it can be reduced the amount of up to 1% of the total volume of using isostatic pressing process. The porosity of the structure does not differ from the elements produced by casting.

A characteristic feature of parts produced using additive manufacturing techniques is high elements surface roughness. In SLM technique there is a way to reducing surface roughness by interfering with the manufacturing process (eg. to change the wavelength of the laser beam). In order to check the roughness of the samples were subjected to measurements of the geometric structure of the surface using a confocal microscope OLYMPUS OLS4100. The measurement results are shown in Fig. 7.

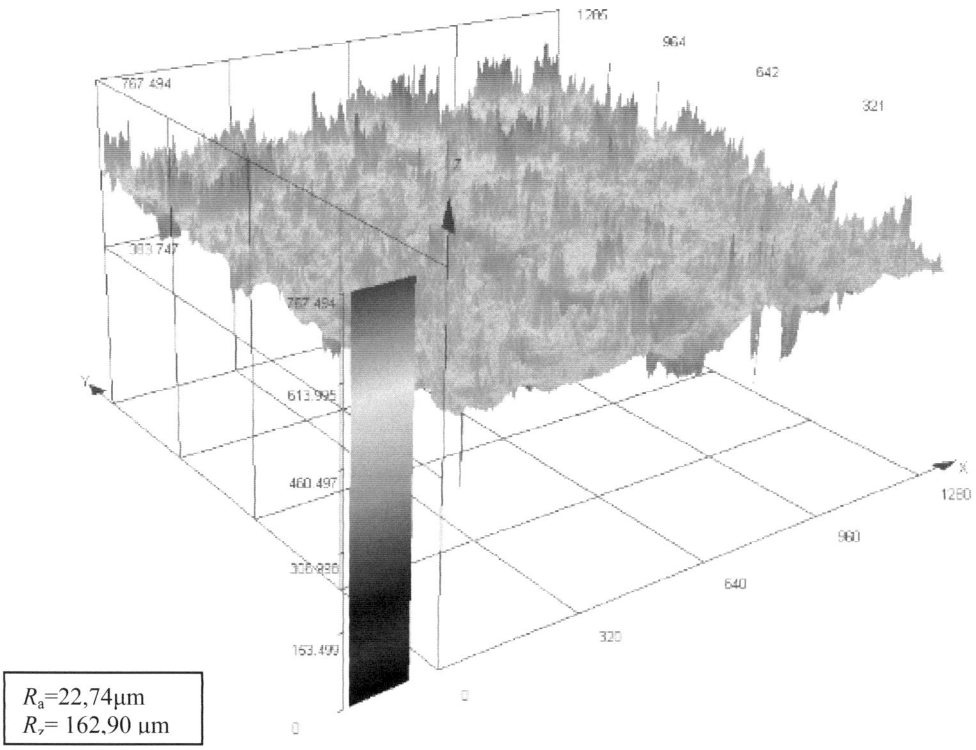

Fig. 7. Roughness of the surface metal parts made using SLM technology.

The surface roughness research allowed the determination of specimens roughness parameters R_a and R_z: R_a = 22,74μm and R_z = 162,90 μm. This is a relatively low surface roughness, taking into account the elements prepared by conventional techniques, e.g. casting process. Compared to other additive manufacturing methods - produced by SLM elements are characterized by relatively low roughness. Obviously surface of the SLM technique produced may be subjected to various types of machining.

3.2. Tensile properties

Examined steel was fabricated in selective laser melting (SLM) process. A specimen angled to working platform of 60° and a linear type of exposuring was tested for static tensile testing using an Instron 8802 servohydraulic test system. On the base of results of the static tensile test the stress–strain curve was plotted (Fig. 8).

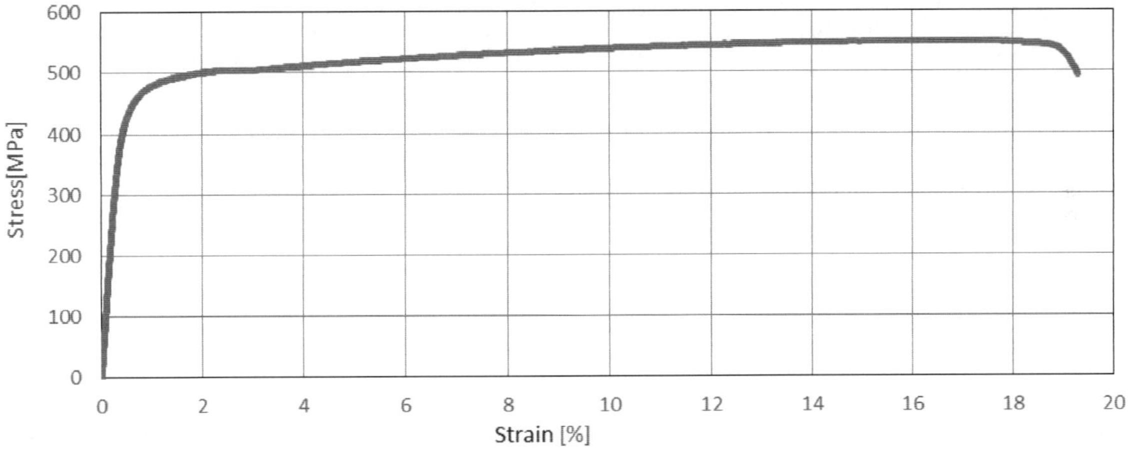

Fig. 8 Stress–strain curve for specimen made of 316L stainless steel.

The research results correspond to the results of the elements 316L produced by conventional and additive manufacturing techniques (Table 3).

It should be noted that in Table 3 relate to the strength properties of samples not subjected to heat treatment or machined surface. The statement includes data sheets for steel 316L, the data companies SLM Solution and EOS, and the results of research conducted by the Institute for Rapid Product Development St Gallen (IRPD)

Table 3. List of the mechanical properties of steel 316L produced using various manufacturing techniques.

Material	**Source**	**Yield strength [MPa]**	**Tensile strength [MPa]**	**Elongation [%]**
316L	Casted steel [6]	290	558	50
	SLM- Solutions [7]	550	654	35
	SLM 125 HL (MUT)	**470**	**550**	**19**
	EOS [8]	310	620	30
	IRPD [8]	640	760	30

The results clearly show that elements prepared using the SLM technology are characterized by a similar tensile strength and a higher yield strength in comparison to the specimens made from 316L stainless steel which were produced using conventional manufacturing techniques. The decrease of elongation is due to lack of any thermal or mechanical treatment of the material after the formation of the part.

3.3. Fatigue properties

The fatigue properties were characterized on the basis of the carried out research for the specimens produced using SLM technique. The specimens were exposed to variable levels of maximum stress and stable cycle asymmetry ratio $R = 0{,}01$ ($R = \sigma_{min} / \sigma_{max}$). The load was sinusoidal with frequency $f = 10$ Hz. The results are show on the Wöhler fatigue chart (Fig. 9). This graph was prepared for specimens directly after melting process, without subsequent heat treatment. Stress values at which the test was carried out fatigue life was determined on the basis of the graph prepared in advance static tensile test.

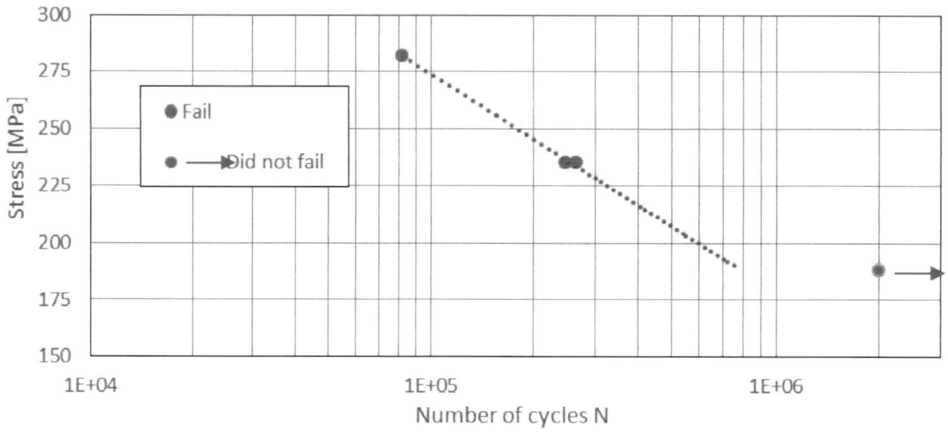

Fig. 9. Wöhlers' chart of specimens manufactured using SLM technique.

The specimen without heat treatment reached unlimited fatigue life at maximum stress σ = 188 MPa . Two samples tested at σ = 235 MPa failed after a very similar number of cycles (N = 247 166 cycles and N = 264 480 cycles), which may indicate the repeatability high-cycled fatigue properties of produced components. SLM-manufactured parts can be subjected to any heat treatment, thermo-mechanical and chemical treatment, which is used for machining parts made using conventional manufacturing techniques.

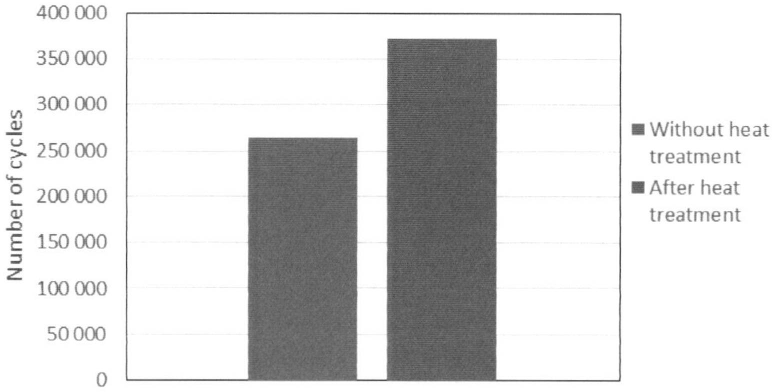

Fig. 9. Effect of heat treatment on the fatigue life of components manufactured using SLM technique.

Metal powders melting using a laser beam is similar micro welding (powder particle size of about 15 microns). The paper attempts to pre-determine the effect of heat treatment in the form of normalization annealing at 950°C on the fatigue life of manufactured components. Applied annealing temperature exceeds approx.50°C 316L steel austenitizing temperature. The results of the number of cycles of each specimen are shown in the bar graph (fig. 9).

Conclusions

The use of SLM- additive manufacturing technology allows to obtain elements with very similar mechanical properties to parts produced by conventional manufacturing techniques. The fundamental factor giving SLM technology advantage is the possibility of obtaining almost any shape without the use of complicated and expensive tools. This method of manufacture avoids the high costs of single or low-volume parts with very complex geometries.

In addition to the SLM machine having the necessary software can be relatively quickly (compared to the time of delivery of the component from the other end of the country or the world) to perform an essential element for the correct operation of the device which is damaged. This fact makes the additive manufacturing technologies of an increasingly more important in the wider industry. Found during conducting the research high surface roughness and a significant share of pores in the structure of the material had a negative impact on designated fatigue life. However, these problems can be eliminated by machining the surface, which indicate the authors of the few works devoted to this subject [8]. The pores in the structure could be reduced using the isostatic pressing as in the case of castings. The use of various additional mechanical and thermal treatments can affect the properties of the parts produced using SLM technique close to the properties of their counterparts produced by conventional manufacturing techniques.

Preliminary own research showed the positive effect of the applied heat treatment on the fatigue life, manifesting itself in increasing the sustainability element of the test for a limited fatigue life by 30%.

References

[1] Song B, Dong S, Liu Q, Liao H, Coddet Ch. Vacuum heat treatment of iron parts produced by selective lasermelting: Microstructure, residual stress and tensile behaviour. *Materials and Design*; 2014. p. 727-733.

[2] Gu DD, Meiners W, Wissenbach K, Poprawe R. Laser additive manufacturing of metallic components: materials, processes and mechanisms. *International Materials Reviews* 2012;. 57, 3: 133-64.

[3] Brandl E, Heckenberger U, Holzinger V, Buchbinder D. Additive manufactured AlSi10Mg samples using Selective Laser Melting (SLM): Microstructure, high cycle fatigue, and fracture behavior. *Materials & Design* 2012; 34: 159–169.

[4] Bremen S, Meiners W, Diatlov A. Selective Laser Melting, A manufacturing technology for the future?, *2012 WILEY-VCH Verlag GmbH & Co. KGaA, Weinheim, Laser-journal.de;*2012. p. 1-4

[5] SLM Solutions GmbH. *Introductory Training: Transition from AutoFab to SLM Build Processor.*

[6] http://www.aksteel.com/pdf/markets_products/stainless/austenitic/316_316l_data_sheet.pdf.

[7] http://www.stage.slm-solutions.com/download.php?f=0c4e94b4e06f805e65f3698bf1ff391e.

[8] Spierings A. B., Starr T. L., Wegener K. Fatigue performance of additive manufactured metallic parts. *Rapid Prototyping Journal* 2013; 19 Iss 2: 88 – 94.

Forum on Innovative Technologies and Management for Sustainability (ITMS'2016)

ktu 1922 panevėžys faculty of technologies and business

The 11th International Conference Intelligent Technologies in Logistics and Mechatronics Systems (ITELMS'2016), 28-29 April 2016, Panevėžys, Lithuania

Experimental study on ballistic AA 2519 / Ti6Al4V laminate according to STANAG 4569 Level 1

Szachogluchowicz I.[a*], Sniezek L. [a], Mierzynski J. [a], Koperski W. [a]

[a]Military University of Technology, gen. S. Kaliskiego 2 str., 00-908 Warsaw, Poland,

Abstract

The aim of the paper is to investigate ballistic layered materials selected light alloys under various configuration and thickness of the protective plates. The samples layers panel join by explosive bonding made of developed material AA2519 and Ti6Al4V with the intermediate layer alloy AA1050.

The paper describes the impact of the bullet impact energy structure of the panel, Al-Ti, depending on the parties shelling and thickness of layers. The analysis of samples carried out in accordance with the approved program and methods of investigation, according to current standards, including standards STANAG 4569 Level 1. The research program included testing on basis materials and layers panels made by an explosive bonding AA2519 and Ti6Al4V plates. The results of the research were compared to the standard material used for ballistic shields and base materials.

© 2016 I. Szachogluchowicz, L. Sniezek, J. Mierzynski, W. Koperski.
Peer-review under responsibility of the Kaunas University of Technology, Panevėžys Faculty of Technologies and Business

Keywords: explosive welding; layered composites; mechanical properties of the composites.

1. Introduction

Adopted a more practical work included examination of selected mechanical properties of layered composite material composed of alloys: Ti6Al4V / AA1050 / AA2519 with the resulting diffusion layer Al_3Ti type [1,4,10]. It was assumed that such material will be characterized by unique properties that combine the beneficial properties of titanium and aluminum alloys (high strength, high plasticity and low specific gravity) and alloys, intermetallic matrix Ti-Al (high hardness and stiffness) [2,3,5,13] . The composite layer made of base materials in the form of aluminum alloy AA2519 (AlCuMgMn + ZrSc) and the titanium alloy Ti6Al4V. Chemical composition and mechanical properties of base materials are provided in Table 1 and Table 2.

Table 1. Mechanical properties and chemical composition of the Ti6Al4V alloy.

Mechanical properties			Chemical composition (wt %)							
Rp0,2 (MPa)	Rm (MPa)	A (%)	O	V	Al	Fe	H	C	N	Ti
950	1020	14	<0,20	3,5	5,5	<0,30	<0,0015	<0,08	<0,05	rest

* Corresponding author. Tel.: +48-261-837-285; fax: +48-261-837-211.
E-mail address: ireneusz.szachogluchowicz@wat.edu.pl

© 2016 I. Szachogluchowicz, L. Sniezek, J. Mierzynski, W. Koperski.
Peer-review under responsibility of the Kaunas University of Technology, Panevėžys Faculty of Technologies and Business

Table 2. Mechanical properties and chemical composition of the AA2519 alloy

Mechanical properties			Chemical composition (wt %)								
Rp0,2 (MPa))	Rm (MPa)	A (%)	Si	Fe	Cu	Mg	Zn	Ti	V	Zr	Sc
312	335	6,5	0,06	0,08	5,77	0,18	0,01	0,04	0,12	0,2	0,36

Combining explosive was realized at the Department of Technology Energetic "EXPLOMET". Cooperation with the plant allowed to undertake research aimed at obtaining and determining the performance of a construction material in the form of composite Al-Ti layer with specific properties, which does not have a solid material [6-12].

The density change and different textures base material should have an impact on how to change the energy dissipation. The use of materials with significantly different mechanical properties, offers the possibility of selecting the type of the form of energy dissipation. A significant impact on them is the ratio of the thickness of the materials used, and set the order of the layers panel in relation to fire. The method and energy absorption affects not only the rapid deceleration of the projectile, but also on the number of fragments generated. In layered materials, there is a risk of delamination of the material sought by external energy.

Nomenclature

A	elongation sample
Rm	maximum rupture resistance
$Rp_{0,2}$	ductility limit

2. Procedures and subject research

The concept of research of dynamic (firing) base materials and panels made of composite material developed assumed testing sandwich panels made by putting together panels AA2519-Ti6Al4V combined method of explosive. Shelling samples followed from the titanium-type missiles M193 rifle cartridge 5,56x45 mm in conditions complying with the class I STANAG 4569 (Table 3). The cartridge is in the moment of impact energy has 1300-2300 J.

Table 3. The ammunition used in the test.

Ammunition used	Type of projectile	Velocity
Indirect missile – 5,56 x 45 mm	M 193	937 m/s

Research ballistic base materials and developed sandwich panels was conducted in the following order:

- The trials ballistic laminate base material in the form of: aluminum alloys AA2519 and titanium Ti6Al4V.Test specimens made of aluminum alloy AA2519 have been prepared in the form of the casting. Samples of titanium Ti6Al4V were made of titanium rod cold drawn with a diameter of 100 mm.
- Execution of the distinctive firing a bullet cartridge M193 5,56x45 mm rifle three-layer panel made of laminate AA2519-Ti6Al4V of the interlayer alloy AA1050. The distance between the plates was 15 mm. Panels are configured with layered laminate with dimensions of 200x120x11 mm (Fig. 1).

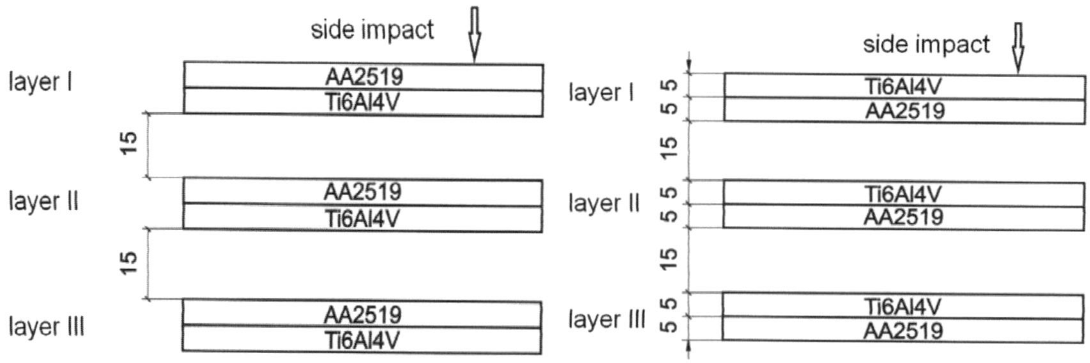

Fig. 1. Variant shelling plates from the alloy Ti6Al4V and the distance between the plates of 15 mm.

- Firing missiles M193 5,56x45 mm rifle cartridge dual-layer laminate panels of AA2519-Ti6Al4V about the distance between the plates of 15 mm. The panels used in the second stage of the research of a laminate layer with dimensions of 200x120x10 mm.
- Missiles firing a rifle cartridge M193 5,56x45 mm six-layered laminate panels of AA2519-Ti6Al4V about the distance between the plates of 15 mm. The panels used in the second stage configured studies were layered laminate with the dimensions 200x120x6 mm.
- Shelling multilayer panels to compare the results of research. Shelling multilayer panels made of alloys AA2519 and Ti6Al4V armor steel Armox 500. Comparative material were panels: 6 layer alloy AA2519 with dimensions of 200x120x10mm, Ti6Al4V alloy with dimensions of 200x120x10mm and a three-layer panel 200x120x6mm.

2.1. Section headings

Stand for testing ballistic resistance basis materials and developed layers panels shown in Fig. 2. and the photographs (Fig. 3. and Fig. 4.). This position used to fire the samples in the laboratory missiles B32 API (7,62x54R mm rifle cartridge).

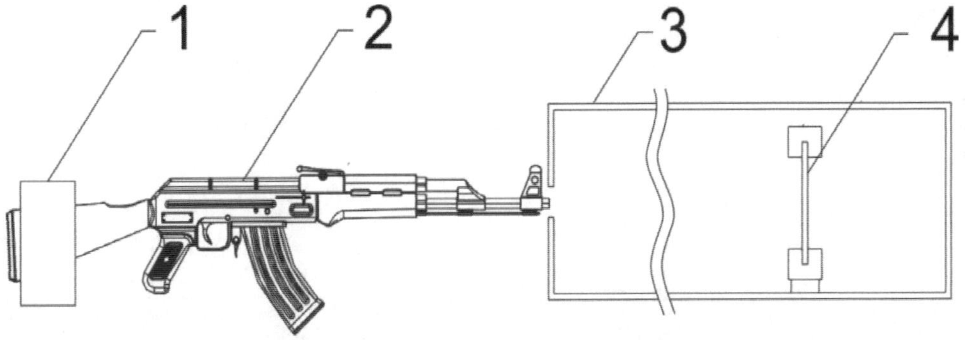

Fig. 2. Schematic stand for testing ballistic base materials and developed layers panels: 1 - bracket for mounting weapons, 2 - -throwing device (suitable caliber firearm), 3 - butt, 4 - the test sample mounted in a holder.

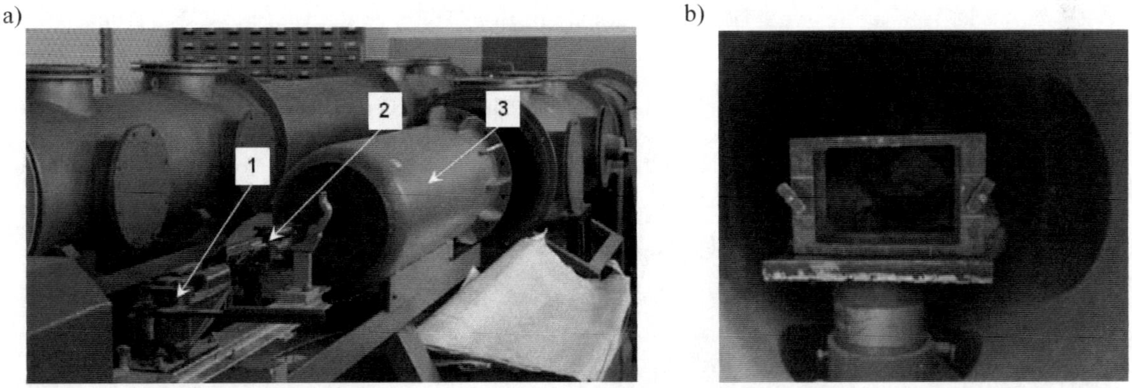

Fig. 3. a) Stand for testing ballistic resistance basis materials and developed layers panels: 1 - bracket for mounting weapons, 2 - throwing device (suitable caliber firearm), 3 - butt b) handle layers panels placed in the butt.

Fig. 4. The weapon mounted in the handle portion of a shooting during ballistic tests.

Rifle 5.56 mm model 1944 was placed on the bench in order to ensure that you get hit in the designated area of the sample. Before each test checks for proper mounting arms and putting the barrel axis relative to the sample. During the shot distance of the muzzle of the sample was 2.5 m.

3. Procedures and subject of research

The results of the samples of aluminum alloy AA2519 are shown in Fig. 5. The cross section area of the sample in aluminum alloy AA2519 in the mold. The missile has penetrated the material to a depth of 14 mm on the alloy surface around the inlet, formed crater formed by ejection of the material by a projectile. The core Remained in the bullet sample. The diameter of the opening at the entry location of the projectile is 5 mm, and a depth of 10 mm. The Result test sample made from titanium alloy Ti6Al4V shown in Fig. 6. no piercing projectile formed a recess with a diameter of approx 13 mm. The depth of the fingerprint was 4 mm.

Fig. 5. Sample No. 02 - face sample of aluminum alloy AA2519 in the mold heat-treated after firing a missile M 193.

Fig. 6. Sample No. 03 - face sample of titanium alloy Ti6Al4V after firing a missile M 193.

Results of the laminate AA2519-Ti6Al4V consisting of three layers of the laminate with the dimensions 200x120x10 mm protection (distance between layers of 15 mm) are shown in the photographs (Fig. 7-10). The missile was launched from the alloy AA2519. He stopped at the second layer by penetrating it to a depth of about 3mm. The hole in the first layer is characterized by irregular shape simillar to the oval. Titanium sections within the aperture indicate a change in alignment with respect to the flight path of the projectile upon impact Ti6Al4V alloy that is significantly harder than the AA2519 alloy. Firing from the Al alloy increases occurrence precession and nutation, which preferably a improves the ballistic resistance of the panel.

Fig. 7. Sample No. 6 - surface area the front panel with a shot projectile M 193 5,56x45mm from the alloy AA2519.

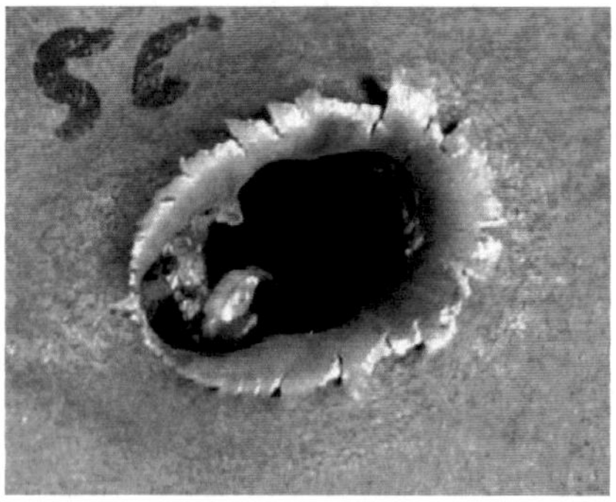

Fig. 8. Sample No. 6 - place inlet bullet M 193 5,56x45mm.

The missile launched from the titanium alloy stopped on the second layer. The projectile damaged to a depth of about 0.5 mm. The inlet opening is characterized by a sharp edge of the recess material. This shows the occurrence of plugging process, which will squeeze the material in the shape of the plug. An alloy layer of aluminum has also been pushed, and she had also significantly dissipate energy.

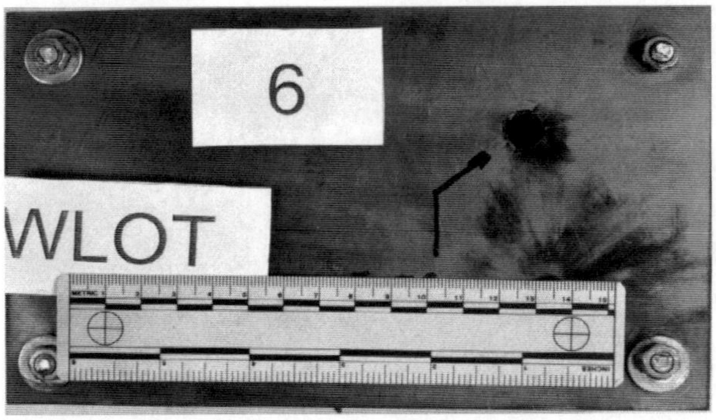

Fig. 9. Sample No. 6 - surface area the front panel with a shot projectile M 193 5,56x45mm from the alloy Ti6Al4V.

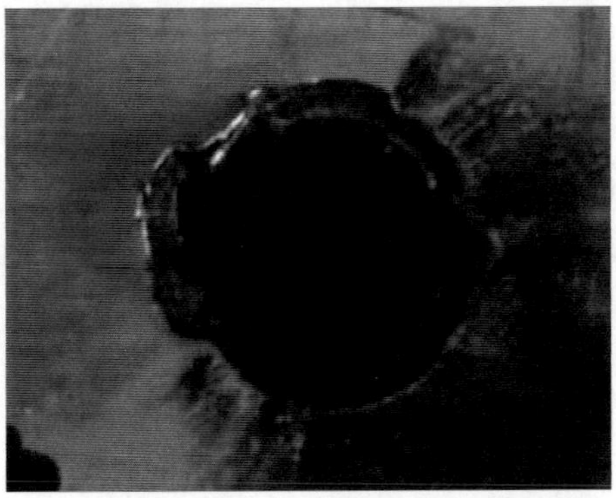

Fig. 10. Sample No. 6 - place inlet bullet M 193 5,56x45mm.

M 193-type missiles firing a rifle cartridge 5,56x45mm six-layered laminate panels of AA2519-Ti6Al4V about the distance between the plates of 15 mm. The panels used in the second stage configured studies were layered laminate with the dimensions 200x120x6 mm. (Fig. 11 - 3.14). Missile penetrated the panel 2 layers. He stopped on the third deforming her spot at a depth of about 2 mm. The resulting crater has an irregular edge. The left side has the shape of a circle, while the right shape oval. This demonstrates the buckling projectile position relative to its flight path. The effect of changing its orientation is to increase the contact surface of the projectile to the panel. This increases the possibilities of dispersion energy panel and reducing the generation of debris.

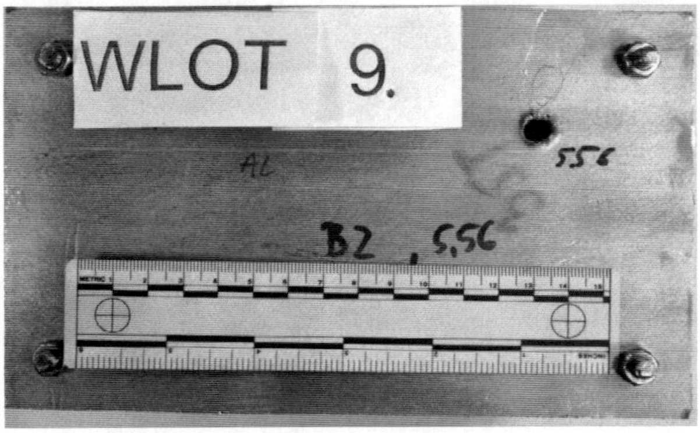

Fig. 11. Sample No. 9m - surface area the front panel with a shot projectile M 193 5,56x45mm from the alloy AA2519.

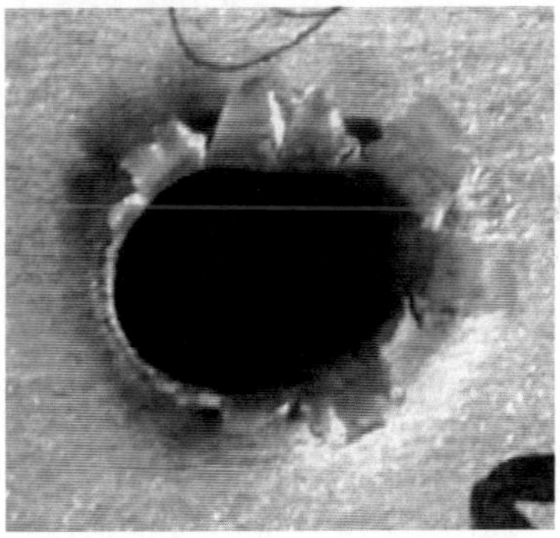

Fig. 12. Sample No. 9 - place inlet bullet M 193 5,56x45mm.

For fire from the titanium alloy bullet stopped in August for the third layer. The hole has a sharp edge which proves that has been cut, thereby generating the fragments. The energy generated by the projectile focused spot. In this process, the following dulling or breaking up the projectile. Energy density is so high that the material is not capable of absorbing energy and phenomenon occurs to break the stopper. Produced plug takes over from the missile a significant amount of energy thus becoming part of destruction. A positive development is to increase the contact area allowing for greater heat dissipation, and greater opportunity to absorb it by the protective panel.

Fig. 13. Sample No. 9 - surface area the front panel with a shot projectile M 193 5,56x45mm from the alloy Ti6Al4V.

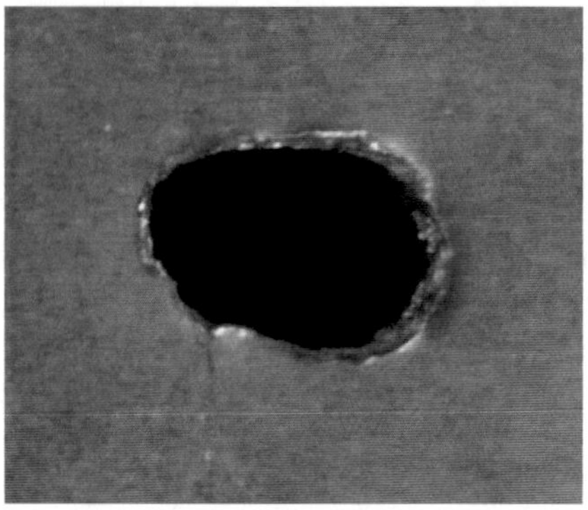

Fig. 14. Sample No. 9 - place inlet bullet M 193 5,56x45mm.

Test results AA2519-Ti6Al4V laminate consisting of two layers of the laminate with dimensions of 200x120x20 mm are shown in the photographs (Fig. 15-18). For the embodiment in which the panel was fired from the aluminum alloy projectile stopped in the first layer. Has pierced only the aluminum alloy. The entrance opening has a regular circular shape having a diameter of about 6 mm. The bullet stopped on a combination of the two alloys orientation perpendicular to the trajectory. Changing his position was at the moment of impact of the bullet titanium alloy.

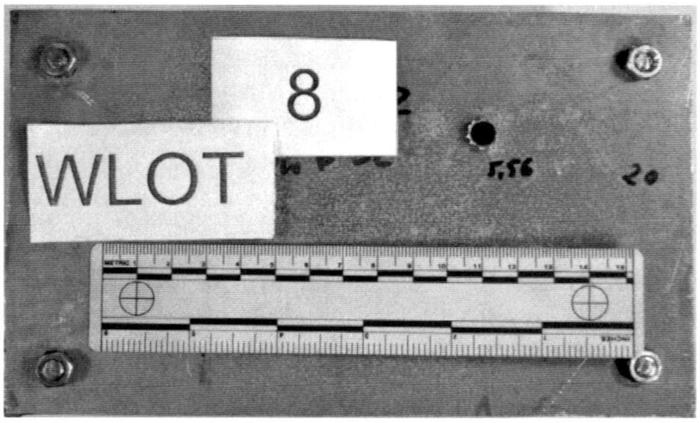

Fig. 15. Sample No. 8 - surface area the front panel with a shot projectile M 193 5,56x45mm from the alloy AA2519.

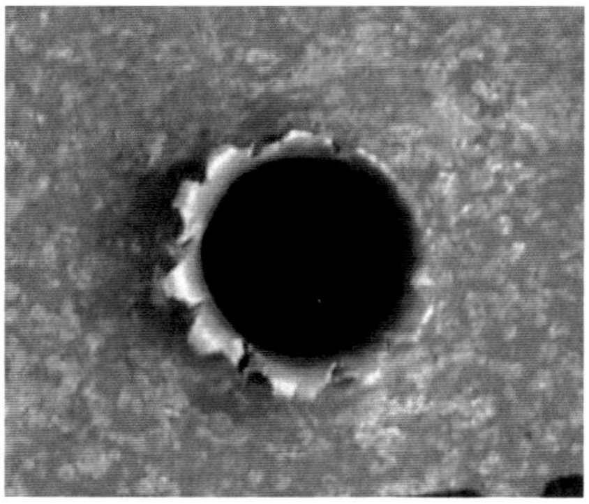

Fig. 16. Sample No. 8 - place inlet bullet M 193 5,56x45mm.

Variant in which the firing was the site of titanium as the first bullet caused damage to a depth of about 2-3 mm. The recess in the material has an irregular shape. The missile hit a hard titanium alloy and has been defragmented. The resulting fragments of deformed bullet hole created. Some of the edges have been broken off and discarded or deformed by creating crisp sheets, after which the missile is only slipped.

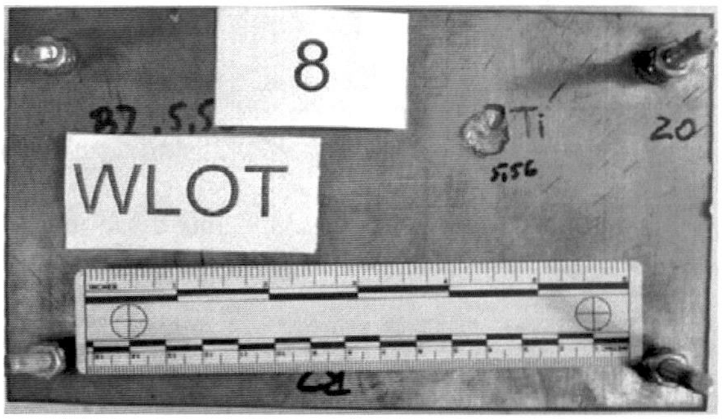

Fig.17. Sample No. 8 - surface area the front panel with a shot projectile M 193 5,56x45mm from the alloy Ti6Al4V.

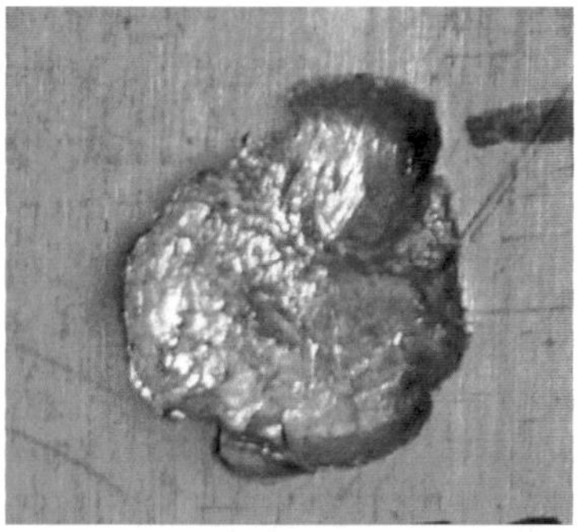

Fig. 18. Sample No. 8 - place inlet bullet M 193 5,56x45mm.

The test results of fire panel made of steel armor Armox 500 consisting of six layers of protective shown in the Fig. 19-26. The bullet stopped at the first layer which stuck. Seeder caused deformation of the material to a depth of about 6 mm. On the outer pane at the impact point of the projectile generated to a plurality of pieces of steel originating Armox 500. In the areas of concentration of the impact energy from a plurality of visible break of the material

Fig. 19. Sample No. 10 - the front face a panel made of stainless antitank Armox 500 after shelling a missile M 193 5,56x45mm.

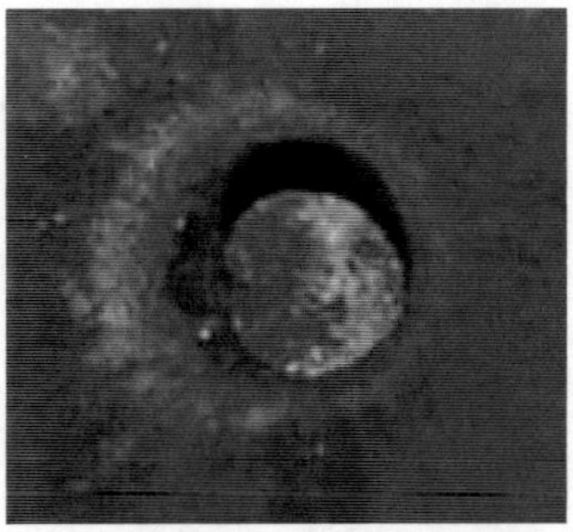

Fig. 20. Sample No. 10 - place inlet bullet M 193 5,56x45mm.

Fig. 21. Sample No. 10 - the lateral surface of the tunnel made of stainless antitank Armox 500 after shelling a missile M 193 5,56x45mm.

Fig. 22. Sample No. 10 - place inlet bullet M 193 5,56x45mm.

The bullet penetrate the two layers. Material stopped at the third being limited to about 0.5 mm to form a cavity having a diameter of about 10 mm.

Layers panel have not been bent. At the edge of the hole formed a caldera, providing a significant strength of the material. AA2519 alloy underwent pushing allowing better distribution of energy through the material. The density of the material does not cause significant blunting or crushing missile, but alters the arrangement of the bullet relative to its flight path.

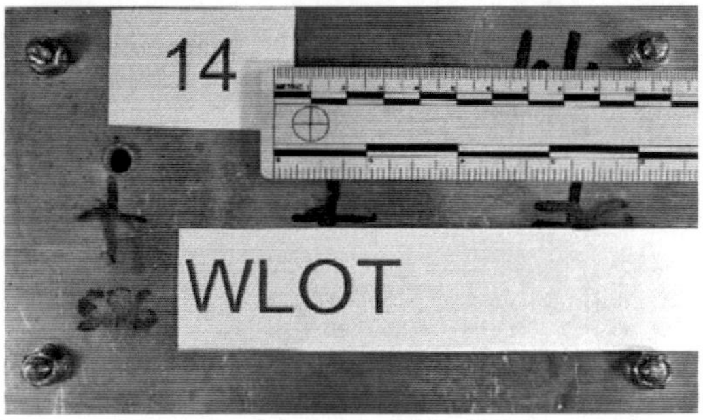

Fig. 23. Sample No. 15 - the front face panel made of alloy AA2519 after shelling a missile M 193 5,56x45mm.

Fig. 24. Sample No. 15 - place inlet bullet M 193.

The test results firing a panel made of titanium alloy Ti6Al4V consisting of three protective layers (distance between layers 15 mm) shown in the Fig. 25-26.

The bullet penetrated the first layer of material. The second layer was not deformed. The hole has a beveled edge, and a circular hole with a smooth side. The second layer has a recess of about 1mm 5mm. This is the result of hitting a bullet with a cut stopper. The thickness of 10 mm influenced the panel good stiffness and absorption of the energy.

Fig. 25. Sample No. 15 - the front face panel made of alloy Ti6Al4V after shelling a missile M 193 5,56x45mm.

Fig. 26. Sample No. 15 - place inlet bullet M 193.

The density of aluminum is about 2.0 g/cm3 density of titanium of 4.5 g/cm3 and the density of armor steel 7.8 g/cm3. The armor steel armox 500 despite the lack of stopping the penetration creates large deformation and generated numerous shards that may pose a threat as dangerous as a bullet. The dispersion of fine debris can cause, in fact, much more damage than a bullet. Analyzing the impact of the projectile in base materials and layered materials the differences are considerable. The projectile striking the AA2519 alloy in the initial state it has penetrated to a depth of 8mm. During the shelling of the base material in the form of alloy Ti6Al4V missile plunged into 3mm while firing panels made of titanium interface layer has a thickness of 10 mm each bullet pierced a total of 11mm. The panel of AA2510 alloy layers having a thickness of 10 mm 23 mm Total raised. The panel layer during firing from the alloy AA bullet pierced a total of 13 mm whereas when firing from the titanium alloy 11. Shelling Al -t laminate thickness of 20 mm projectile basins to 3 mm and from the aluminum to 10 mm. When firing from the Ti and AA2519 material behaved the same way as in the base material firing layer 6x6 mm from the Al 18 and from the t 12 firing layered titanium and a laminate from the side of Titanium effect is the same - namely, the missile has penetrated after 11 mm laminate firing the titanium side panel 20 mm thick aluminum layer increase the rigidity of the material and improve the depreciation of impact of the projectile, which is energy dissipation. For 6 mm aluminum layers will not carry out its functions too well, because it is pushes to the side and does not slow the bullet. It is worth noting that even for a 10mm projectile stop layer on the second layer, that is, broke alloy Titanium 5mm thickness aluminum alloy. The layer 20 millimeter projectile penetrated only a thickness of about 2-3 mm.

Acknowledgements

The project is carried out under Project PBS2/A5/35/2013 funded by the National Research and Development Centre.

References

[1] Hutsaylyuk V, Sulym H, Pasternak Ia, Turchyn I. Transient Plane Waves Propagation in Non-homogeneous Elastic Plate. *Composite Materials: The Great Advance. Proc. of the 19th Int. Conf. Composite Materials, ICCM19;*Montreal, Canada, 28 July – 2 August 2013. p. 8890–8897

[2] Sniezek L, Szachogłuchowicz I, Gocman K. The mechanical properties of composites AA2519-Ti6Al4V obtained by detonation method. *Intelligent Technologies in Logistcs and Mechatronics Systems ITELMS'2014, Proc. of The 9TM International Conference edited by Z. Bazaras and V. Kleiza;* 2013. p. 214–219

[3] Sniezek L, Szachogłuchowicz I, Hutsaylyuk V. Research of Property Fatigue Advanced Al/Ti Laminate. *Intelligent Technologies in Logistcs and Mechatronics Systems ITELMS'2014, Proc. of The 9TM International Conference edited by Z. Bazaras and V. Kleiza*; 2014. p. 232–238

[4] Lysak VI, Kuzmin SV. *Explosive welding of metal layered composite materials* / Ed. B. E. Paton. Kiev: E.O. Paton Electric Welding Institute of NASU; 2003.

[5] Abrahamson GR. Residual periodical deformations of surface under action of moving jet. *Proc. of ASME, Ser. E, Appl. Mechanics* 1961;. 28 (4): 45–55.
[6] Cowan GR, Holtzman AH. Flow configuration in colliding plates: explosive banding. *J. Appl. Phys* 1963; 34 (4): 928–939.
[7] Cowan GR, Bergmann OR, Holtzman AH. Mechanism of Bond Zone Wave Formation in Explosion Clad Metals. *Metallurgical Transactions* 1971; 2 (11): 3145–3155.
[8] Godunov SK, Kiselev SP, Kulikov IM, Mal, VI. Numerical and Experimental Simulation of Wave Formation during Explosion Welding. *Proceedings of the Steklov Institute of Mathematics* 2013; 281: 12–26.
[9] Achenbach JD. *Wave propagation in elastic solids*. New York: Amer. Elsievier Publ. Co.; 1973.
[10] Sulym H, Hutsaylyuk V, Pasternak Ia, Turchyn I. Stress-strain state of an elastic rectangular plate under dynamic load. *Mechanika* 2013;19 (6): 620–626.
[11] Gorskov AG, Medvedsky AL, Rabinsky LN, Tarlakovsky DV. *Waves in continuum media*. Moscow: Fizmatlit; 2004.
[12] Sneddon I, Fourier transforms. – New York: McGraw-Hill, 1951. 542 p.
[13] Totten GE, Mackenzie DS. *Handbook of Aluminum: Alloy Production and Materials Manufacturing*. CRC Press; 2003.

Forum on Innovative Technologies and Management for Sustainability (ITMS'2016)

panevėžys faculty of technologies and business

The 11th International Conference Intelligent Technologies in Logistics and Mechatronics Systems (ITELMS'2016), 28-29 April 2016, Panevėžys, Lithuania

Structure and magnetic properties of Fe_6C_2 particle

Tamuliene J.[a]*, Barkaline V.[b]

[a]*Vilnius University, Institute of Theoretical Physics and Astronomy, Sauletekio al. 3, LT-10222, Lithuania*
[b]*Belarusian National Technical University,65 Nezalezhnasci Praspekt, 220013 Minsk, Belarus*

Abstract

Fe_6C_2 compounds were investigated by using state-of-the-art computational ab-initio methods. The total energy calculations for the various structures different from point of view of the symmetry were performed. The most stable geometric structure of the particle is obtained. The electronic structure of the most stable particle and its isotropic total susceptibility were investigated, too. The results could shed some light the Fe_3C particles formed during carbon nanotube production.

© 2016 J. Tamuliene, V. Barkaline.
Peer-review under responsibility of the Kaunas University of Technology, Panevėžys Faculty of Technologies and Business

Keywords:

1. Introduction

Carbon nanotubes were first discovered in arc discharge from a carbon rod containing transition metal catalysts [1,2] Currently, substantial progress has been made to synthesize nanotubes by using other methods such as laser-ablation, chemical vapour deposition, pyrolysis and etc. [3-5]. However, these new methods are based on metal, or a metal compound catalyst. Among various metals, Fe, Co, and Ni are frequently used. Metal or a metal compound can mix with the nanotubes to affect their electronic and magnetic properties. Moreover, nanoparticles could be formed during nanotube growth [6]. On the other hand, the fabrication of carbon-encapsulated magnetic nanoparticles has attracted much attention due to their potential applications in magnetic data storage, magnetic toners, drug delivery and magnetic resonance imaging, since the carbon shell provides the necessary chemical stability and magnetic isolation for the core magnetic material.

The effect of interactions between the Fe and C atoms on the magnetic properties of Fe-filled nanotubes by comparing with their free-standing wires was studied by Yong-Ju Kang [7]. This study exhibited that the magnetic properties of encapsulating Fe nanowires strongly depend on nanowire thickness: the magnetic moments was greatly reduced for thicker nanowires due to interactions between the Fe atoms and the C nanotube. Moreover, x-ray diffraction studies of the C-coated Fe wires showed the presence of both body-centered cubic (bcc) α-Fe and face-centered cubic γ-Fe structures together with graphite [8]. T. Mühl et al. measured the coercivity of Fe-filled multiwalled carbon nanotubes with temperature for both the parallel and the perpendicular field orientations [9]. The results obtained exhibited that the coercivity increased almost linearly with decreasing temperature for both field orientation. Grobert et al. obtain similar results [10]. However, the coercivity values obtained in these two studies mentioned were significant different [11]. Moreover, Prados et al. found a considerable change in the magnetization at low temperatures, but the results was not proved by experimental measurements performed in other laboratories.

It was noted that the degree of graphitization exhibited by the nanotube walls could be extremely high. Referring to results presented, we predicted that the above discrepancies could occur due to presence of the different number and size of Fe_3C particles that were formed due to interaction of the Fe atoms and carbon nanotubes.

* Corresponding author. Tel.: 370-6-89-12-133.
E-mail address: Jelena.Tamuliene@tfai.vu.lt

© 2016 J. Tamuliene, V. Barkaline.

Peer-review under responsibility of the Kaunas University of Technology, Panevėžys Faculty of Technologies and Business

The Fe_3C compounds are belonged to the most common and important strengthening phases in carbon steels and white cast, which plays a critical role during the process of heat treatment (such as aging, tempering, annealing etc.) and hot working (such as forging, rolling etc.) in steels. It is known that slight changes in the amount and distribution of the carbon content can completely mask any changes due to grain size, stress, and the effects of other impurities. The studies of the susceptibilities, of the amorphous carbon films containing Fe_3C nanograins show an increasing behaviour with temperature until a maximum was reached in both conditions of zero field cooling and a field cooling [12]. Moreover, the increasing of susceptibility gradually from low temperature to a maximum indicated a no constant size of grains existing in the films. Focusing at the films with pure Fe_3C grains, the grain increases slightly with decreasing carbon concentration from 85% to 72% and both coercivity and saturation magnetization, increase with increasing grain diameters.

In this paper, we presented the first results obtained aiming to shed some light on the structure and their magnetic properties of the Fe_3C nanoparticles that could form during production of the carbon nanotubes.

2. Method of investigation

The structure of the molecule and its fragments has been studied by the Becke's three-parameter hybrid functional applying the non-local correlation provided by Lee, Yang, and Parr (B3LYP) [13], – a representative standard DFT method. The most advantage of the DFT method is a significant increase in computational accuracy without the additional increase in computing time [14]. The present DFT method is derived for obtaining total energies as the function of the nuclei position and is often the method of choice for reaction calculations because the electron correlation energy is accounted, while one of the major significant deficiencies of the Hartree-Fock method is ignoring of the electron correlation. However, the correlation energy, as well as the total energy of the system investigated, is highly dependent on the basis set used. Thus, our investigation was performed with the 6-311G(d) basis set [15,16]. The particles under investigations were created used the most stable structures of Fe_3C particles obtained by M. V. Ryzhkov [17]. The total number of the symmetrically different particles investigated was 24. The structure parameters of the molecules under study were optimized with no symmetry constraint. We computed force constants and the resulting vibrational frequencies. The zero-point energy was included into calculations of the total energy of the particles. Referring to the results obtained, the most stable particles were found. The structural changes due to temperature of the most stable particle were investigated, too. The suspensibility of the structures obtained was calculated with presence of temperature to describe dependence of the magnetic properties of the compounds investigated on the temperature. The Gaussian program packages were applied [18].

3. Investigation Results

The view of the most stable Fe_6C_2 particle is presented in Fig.1, while the detailed description of it takes place in Table 1.

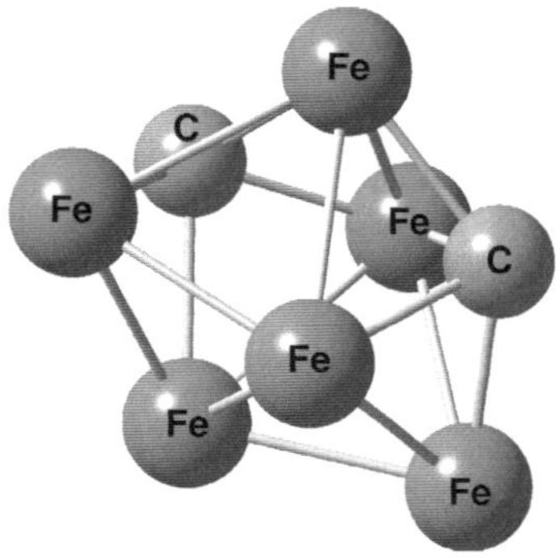

Fig.1 The view of the most stable Fe_6C_2 particle obtained.

Table 1. The coordinates (Å) of the most stable particle obtained.

Atoms	X	Y	Z
Fe	-0.439933	0.826385	0.131352
Fe	1.359317	0.303097	2.797758
Fe	1.747895	0.256331	0.640606
Fe	-0.439933	1.583615	2.278648
Fe	1.359317	2.106903	-0.387758
Fe	1.747895	2.153669	1.769394
C	0.115549	2.440696	0.795672
C	0.115549	- 0.030696	1.614328

The total energy of the particle with zero point energy is equal to -7657.689 a.u. The total energy of other particles, that could be name as one of the most stables, is equal to -7657.674 a.u or -7657.676 a.u., and it is more than 0.38 eV higher than those of the most stable particle under investigation. The total energy of other particles investigated, but unmentioned is much more higher than 0.38 eV. It is evident that the structure of the Fe_6C_2 particle presented is favourable. HOMO-LUMO gap is equal to 1.45 eV and indicate that the particle is semiconductor. Moreover, the relative large HOMO-LUMO gap indicates high chemical stability of the particle under investigation.

The shape of the most stable particle is irregular, but the particle is symmetric, the point group of which is C_2. It is interesting, that the particle consist of C atoms that not located as much as possible closer to each other, but they are not as far as possible from each other, too. The particles, in which carbon atoms form C-C bond, are much more unstable.

Bond lengths, atom charge and overlap population are listened in Table 2. The overlap population represents the number of electrons shared between the pairs of atoms and characterized bonded and non-bonded pairs of atoms. The bonded pairs of atoms are presented in Fig.2. The Fe-C bond lengths obtained coincide well with results of the investigation of Fe_3C compound presented by M.V. Ryzkov et al. [17]. However, distances between Fe-Fe atoms obtained by us are significantly smaller than that presented in [17], but coincide well with value of 2.198 A measured by L. Fohlmeister et al. [19]. The short Fe-Fe bond exists in a related mixed valence amidinate bridged complex and indicates the presence of Fe-Fe multiple bonding. It is of note that the Fe–Fe interaction in this high-spin complex is ferromagnetic in nature, as opposed to the more common antiferromagnetic interactions seen in related dimers, where the Fe–Fe bonds length is equal to 2.515 A. We would like to pay attention to the fact that the short Fe-Fe bonds take place in the Fe_6C_2 particle under investigation, which the isotropic total susceptibility of 212.148 calculated reveals the particle to be paramagnetic. It allows us to predict that the Fe-Fe multiple bonding is present in the Fe_6C_2 particle under investigation.

Table 2. The bond length, condense atom to all electrons and atoms charge of the most stable particle. The atoms numbers are presented in Fig. 2.

Bond	Bond length, A	Overlap population
Fe1-C7	1.831	0.196
Fe1-C8	1.801	0.276
Fe2-Fe3	2.192	0.080
Fe2-Fe6	2.152	0.361
Fe2-C8	1.749	0.170
Fe3-C5	2.152	0.361
Fe4-C7	1.801	0.276
Fe4-C8	1.831	0.196
Fe5-Fe3	2.152	0.361
Fe5-Fe6	2.192	0.080
Fe5-C7	1.749	0.170

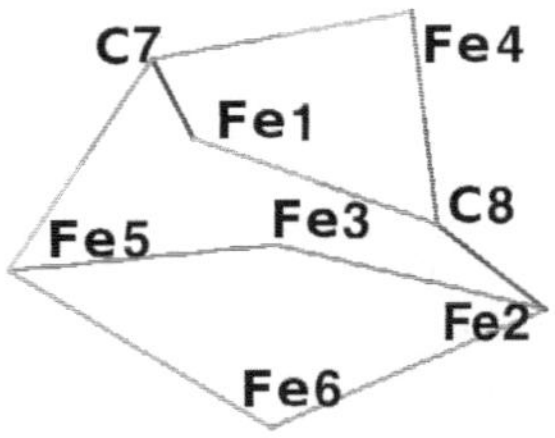

Fig.2. The bonded pairs of atoms obtained in the most stable Fe_6C_2 particle under investigation.

The presence of the Fe-Fe multiple bonding indicates overlap population, too. It is clearly seen from Table 2., that values of the overlap population of the shortest Fe-Fe bonds is the largest (0.361), although that of other Fe-Fe bonds is the smallest (0.080). Referring to the data and the Mulliken charges, we speculate that the oxidation states of the Fe atoms of the Fe_6C_2 particles forming different length and type bonds are different (Table 3).

Table 3. Mulliken charges obtained. The atom numbers are presented in Fig. 2.

Atom	Fe1	Fe2	Fe3	Fe4	Fe5	Fe6	C7	C8
Charge	0.50	0.35	0.03	0.50	0.35	0.03	-0.89	-0.89

Moreover, it allows us to consider that, of the metal-based valence electrons in the active space of the particle occupied bonding molecular (p-type), anti-bonding (p*-type), and nonbonding (of d and d* character). This gives rise to a different overlap population, and yields some unpaired electrons, and, as results, unpaired spin.

The magnetic susceptibility calculations of the particle under investigations showed that it is largely independent of temperature between 323 and 573 K, remaining at 212.148. Hence, the electronic structure of the particles in the temperature range is not changed, i.e. the multiple Fe-Fe bonds are not destructed, althoug we predict that this destruction could occurred at higher temperature or due to Fe interaction with C atoms. Hence, the results of the investigation indirectly confirm that 'the magnetic properties of encapsulating Fe nanowires strongly depend on nanowire thickness: the magnetic moment was greatly reduced for thicker nanowires due to interactions between the Fe atoms and the C nanotube'.

Conclusions

The following results of our investigation were obtained

- The geometric and electronic structure of the most stable F_6C_2 particle is obtained;
- The multiple Fe-Fe bonds are observed;
- The magnetic susceptibility is evaluated.

The results of the investigation indirectly confirm that the magnetic properties of encapsulating Fe nanowires depend on the interactions between the Fe atoms and the C nanotube that could lead to both the destruction of the multiple Fe-Fe bond and formation of new chemical composition particles.

Acknowledgements

The authors greatly appreciate the contribution of the Supercomputing centre of MIF of the University of Vilnius for resources and support provided. The computations were carried out at the Computer Center of University of Bialystok, too. The work is supported by TAP-LB-14-011 project.

References

[1] Iijima S, Ichihashi T. Single-shell carbon nanotubes of 1-nm diameter. *Nature* 1993; 363: 603-5.

[2] Bethune DS, Kiang CH, de Vries MS, Gorman G, Savoy R, Vazquez J, Beyers R, Cobalt-catalysed growth of carbon nanotubes with single-atomic-layer walls *Nature* 1993: 363: 605-7.

[3] Thess A, Lee R, Nikolaev P, Dai HJ, Petit P, Robert J, Xu CH, Lee YH, Kim SG, Rinzler AG, Colbert DT, Scuseria GE, Tomanek D, Fischer JE, Smalley RE, Crystalline Ropes of Metallic Carbon Nanotubes. *Science* 1996; 273: 483-7.

[4] Kong J, Cassell AM, Dai H. Chemical vapor deposition of methane for single-walled carbon *nanotubes. Chem. Phys. Lett.* 1998; 292: 567-į74.

[5] Li WZ, Xie SS, Qian LX, Chang BH, Zou BS, Zhou WY, Zhao RA, Wang G, Large-Scale Synthesis of Aligned Carbon Nanotubes. *Science* 1996; 274: 1701-3.

[6] Li DCh, Dai L, Huang Sh, Mau AWH, Wang ZL. Structure and growth of aligned carbon nanotube films by pyrolysis. *Chem. Phys. Lett.* 2000;316:349-355.

[7]. Kang YJ, Jin Choi, Moon ChY, Chang KJ. Electronic and magnetic properties of single-wall carbon nanotubes filled with iron atoms. *Physical Review B* 2005; 71: 11544 -7.

[8] Karmakar S,. Sharma SM, Teredesai PV, Sood AK. Pressure-induced phase transitions in iron-filled carbon nanotubes: X-ray diffraction studies. *Phys. Rev. B* 2004; 69:165414-5.

[9] Mühl T, Elefant D, Graff A. Kozhuharova R, Leonhardt A, Mönch I, Ritschel M, Simon P, Groudeva-Zotova S, Schneider C M. Magnetic properties of aligned Fe-filled carbon nanotubes. *Journal of Applied Physics*. 2000; 393: 7894-7896.

[10] Grobert N, Hsu WK, Zhu YQ, Hare JP,. Kroto HW,. Walton DRM. Terrones M, Terrones H, Redlich Ph, Ruhle M, Escudero R, Morales F. Enhanced magnetic coercivities in Fe nanowires. *Applied Physics Letters* 1999; 75: 3363-3365.

[11] Prados C, Crespo P, Gonzalez JM, Hernando A, Marco JF, Gancedo R, Grobert N, Terrones M, Walton RM, Kroto HW. Hysteresis shift in Fe-filled carbon nanotubes due to y-Fe. *Phys. Rev. B* 2002; 65: 113405-4.

[12] Lee YH, Han TC, Huang JCA. Magnetic properties of Fe_3C nanograins embedded in carbon matrix. *Journal of Applied Physics* 2003; 93:8462-8464.

[13] Becke AD. Density-functional thermochemistry. iii. the role of exact exchange. *J. Chem. Phys.* 1993; 98:5648-52.

[14] Zhao Y, Pu J, Lynch BJ, Truhlarand DG. Tests of Second-Generation and Third-Generation Density Functionals for Thermochemical Kinetics. *Phys. Chem. Chem. Phys.* 2004; 6:673–676.

[15] Raghavachari K, Binkley JS, Seeger R, Pople JA. Self-Consistent Molecular Orbital Methods. 20. Basis set for correlated wave-functions. *J. Chem. Phys.*, 1980;72: 650-54.

[16]. Petersson GA, Al-Laham MAA complete basis set model chemistry. II. Open-shell systems and the total energies of the first-row atoms. *J. Chem. Phys.* 1991; 94:6081-90.

[17]. Ryzhkov MV, Ivanovskii AL, Delley B. Electronic structure and geometry optimization of nanoparticles Fe_2C, FeC_2, Fe_3C, FeC_3 and Fe_2C_2. *Chemical Physics Letters* 2005; 404(4-6):400-408.

[18] Gaussian 03, Revision C.02, Gaussian, Inc., Wallingford CT; 2004.

[19] Fohlmeister L, Liu S, Schulten Ch, Moubaraki B, Stasch A, Cashion JD, Murray KS, Gagliardi L, Jones C. Low-Coordinate Iron(I) and Manganese(I) Dimers: KineticStabilization of an Exceptionally Short Fe-Fe Multiple Bond. *Angew. Chem. Int. Ed.* 2012; 51: 8294 – 8298.

Forum on Innovative Technologies and Management for Sustainability (ITMS'2016)

ktu 1922 panevėžys faculty of technologies and business

The 11th International Conference Intelligent Technologies in Logistics and Mechatronics Systems (ITELMS'2016), 28-29 April 2016, Panevėžys, Lithuania

Spectra of N-(2,4,6-trinitrophenyl)-*1H*-1,2,4-triazol-3-amine

Tamuliene J.[a]*, Sarlauskas J.[b], Bekesiene S.[c]

[a]*Vilnius University, Institute of Theoretical Physics and Astronomy, Sauletekio al. 3, LT-10222, Lithuania,*
[b]*Vilnius University, Institute of Biochemistry, Sauletekio al. 7,LT-10222 Vilnius, Lithuania*
[c]*The General Jonas Žemaitis Miltitary Academy of Lithuania, Silo g. 5A, LT-10322 Vilnius*

Abstract

A paper presents the investigation of the *N*-(2,4,6-trinitrophenyl)-1*H*-1,2,4-triazol-3-amine(HM-I) spectrums – the new explosive material. The IR and UV-Vis spectra measurements are presented. These spectra are calculated in various solvent, too. The significant difference of the shape of the spectra is obtained, although the common features are also specified. The obtained results indicate that HM-I molecule optic properties are solvent dependant. It allows us to conclude that the detection of the new hazardous material by spectroscopic method is rather difficult.

© 2016 J. Tamuliene, J. Sarlauskas, S. Bekesiene.
Peer-review under responsibility of the Kaunas University of Technology, Panevėžys Faculty of Technologies and Business

Keywords: explosive materials; explosive-based terrorism; explosive detection; spectroscopy; UV spectra; IR spectra; calculations; DFT method.

1. Introduction

The detection of traces of explosive materials is of utmost importance in the present time because explosive-based terrorism has grown enormously in recent years. This growing is presented because explosive-based weapons are simple, easy deployed, and can cause enormous damage [1]. Several methods, including ion mobility spectroscopy, mass spectroscopy, terahertz spectroscopy, infrared spectroscopy, laser-induced breakdown spectroscopy, cavity ring down spectroscopy, electrochemical sensors and immunosensors, have been implemented [2]. However, all these methods have their limits such as poor, incapable detection, portability of instruments, and false alarms due to the inherent lack of selectivity [3,4]. The other major obstacle in the identification of explosives is their elevated spectral similarity with those compounds that share an analogous elemental composition. The camouflage of explosive devices and the complexity of the composition of suspicious objects complicate this identification. Detecting of explosives is a challenging task even due to frequent introduction of novel explosive compositions. Therefore, the explosive detecting technologies must be specially adapted for the problems of explosives detection. The adaption requires a lot of knowledge on various properties of explosive materials. In order to assess the feasibility of explosive-detecting techniques and design viable detection systems, molecular properties and spectroscopic signatures of the target molecules are needed.
The study of the N-(2,4,6-trinitrophenyl)-1H-1,2,4-triazol-3-amine (HM-I) aiming to examine both the stability and explosive performances confirms that the material is explosive, non sensitive, with good explosive characteristics [5]. The new explosive material is also synthesized. However, there is no evidence how to detect the new material. Hence, the spectroscopic properties of HM-I are investigated taking into account the both facts:

* Corresponding author. Tel.: 370-6-89-12-133.
E-mail address: Jelena.Tamuliene@tfai.vu.lt

© 2016 V. Adomavičius, J. Valickas, G. Petrauskas, L. Pušinaitis. **T428F0023** . MEDIMOND
Peer-review under responsibility of the Kaunas University of Technology, Panevėžys Faculty of Technologies and Business

- optical spectroscopy, widely used for measurement of many different species at trace levels or in complex backgrounds, is applied as the basis for possible detection systems;
- for any detection scheme, the essential information required is data on wavelengths and intensities corresponding absorption and/or emission features of the target compounds.

In this paper, a theoretical method was used to predict both the IR and UV-Vis spectra of HM-I. The theoretical results obtained were compared with the experimentally measured data aiming to check the theoretical predictions made. Hence, the goal of our studies was to elucidate the possibility to use spectroscopy to detected HM-I material.

2. Method of investigation

The structure of the molecule and its fragments has been studied by the Becke's three-parameter hybrid functional applying the non-local correlation provided by Lee, Yang, and Parr (B3LYP) [6], – a representative standard DFT method. The most advantage of the DFT method is a significant increase in computational accuracy without the additional increase in computing time [7]. The present DFT method is derived for obtaining total energies as the function of the nuclei position and is often the method of choice for reaction calculations because the electron correlation energy is accounted, while one of the major significant deficiencies of the Hartree-Fock method is ignoring of the electron correlation. However, the correlation energy, as well as the total energy of the system investigated, is highly dependent on the basis set used. Thus, our investigation was performed with the cc-pVTZ basis set [8]. The structure parameters of the molecules under study were optimized with no symmetry constraint. We computed force constants and the resulting vibrational frequencies. The force constants were determined analytically, too. To describe energetic properties of the compounds investigated, the detonation velocity and oxygen balance were calculated. An excited state calculation using the time-dependent DFT method was performed, too [9]. The Polarizable Continuum Model (PCM) using the integral equation formalism variant was applied to performed simulations in the presence of a solvent by placing the solute in a cavity within the solvent reaction field [10]. The Gaussian program packages were applied [11].

UV/Vis spectra measurements of N-(2,4,6-trinitrophenyl)-1H-1,2,4-triazol-3-amine were spectrophotometrically carried out applying Lambda 25 (Perkin Elmer, Waltham, MA, USA) UV-VIS spectrophotometer using 1.0 cm optical path cells and recording spectra with 0.4 nm accuracies in ethanol medium. IR spectra were recorded in KBr disc on a Perkin-Elmer spectrophotometer (FT-IR Spectrum BX II) at accuracies 1 cm^{-1}.

3. Investigation Results

3.1. IR spectra

Let us remind, that the spectra of the HM-I molecule is both measured and calculated. The spectrum measured is used to compare the experimental and theoretical spectra. The comparison allows us to determine precision of the results obtained and to foresee some properties of the molecules that is important performance of hazardous materials. The spectra of HM-I take place in Fig. 1 and 2, while more precise description is presented in Table 1.

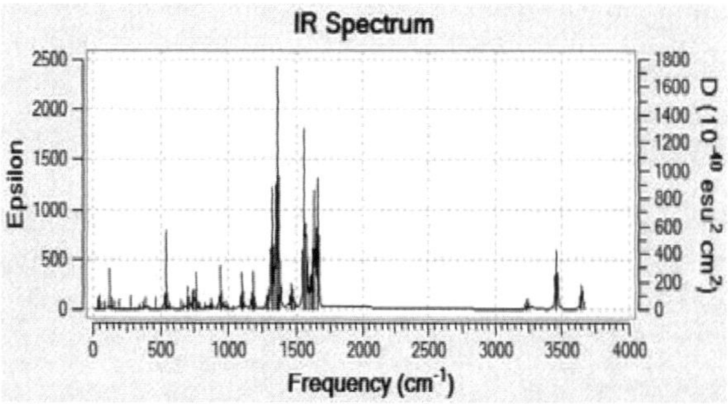

Fig. 1 The calculated spectrum of HM-I

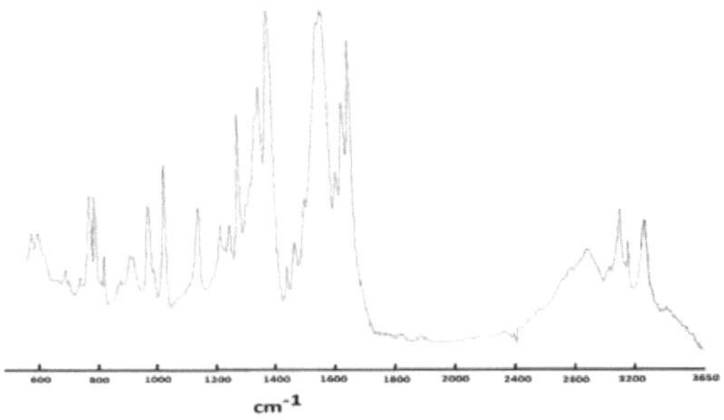

Fig.2 The measured spectrum of HM-I

The comparison of the spectra indicates that the spectrum calculated is slightly shifted to larger wave number region. However, the shift is not greater than the potential uncertainties, i.e. the accuracy of the theoretical investigation is high enough. It is emphasized, that in both spectra the lines correspond to the N-H and C-H bond stretching are present. The good agreement between theoretical and experimental results as well as the studies of the other author [12] is achieved, too. Hence, the IR spectrum measured or calculated allows one to detect any compounds that share the analogous elemental composition of HM-I.

We would like to pay attention that some of the vibrations, that are not the most intensive (Table 2), are related with the intensive stretching of -NO_2 and -NH, joined the benzene ring and the -$C_2H_2N_3$ group.

Table 1. The most intensive line of the IR spectra of HM-I molecule.

Wave length measured, cm^{-1}	Wave length calculated, cm^{-1}	Intensity calculated	Vibration spectra assignment
1311.82	1324.44	292.7016	ν , N-C
1344.69	1361.78	596.7952	ν , C-H, N-H
1527.62	1560.75	508.0189	ν , N-H
1595.83	1635.63	349.9145	ν , N-H, C-H
1619.07	1658.91	380.7052	ν , N-H, C-H

The IR spectrum in the various solvents such as water, acetonitrile and ethanol was calculated to foresee the solvent influence. The spectra calculated are presented in Tables 2. Referring to results presented, it is possible to see that IR spectrum of the molecules under investigations calculated in vacuum changes significantly when simulations are performed in the presence of a solvent by placing the solute in a cavity within the solvent reaction field. The spectra calculated in the presence of the solvents and vacuum are different due to both the intensity of the lines and the number of it. As examples: the intensity of the lines at 1361 cm^{-1} and 1560 cm^{-1}increases in solvent, but this intensity is not the largest in comparison with those of the spectra calculated in vacuum; the intensity of the wavelength of 1324cm^{-1} is one of the largest in the presence of solvent, while that in the vacuum is the smallest; intensity of 1635 cm^{-1} and 1658 cm^{-1} is change slightly. The simplest explanation of the phenomena is different attraction of the molecules under study with solvent molecules, formation of hydrogen bonds and etc.

However, referring to results presented in Table 2 it is possible to see that IR spectra of the molecules calculated in different solvent are not different significantly.

Table 2. IR spectra of the HM-I calculated in the presence of a solvent by placing the solute in a cavity within the solvent reaction field. The wave length bolded are obtained in the IR spectra calculated in vacuum.

Acetonitrile		Water		Ethanol		Vibration spectra assignment
Wave length. cm^{-1}	Intensity	Wave length. cm^{-1}	Intensity	Wave length. cm^{-1}	Intensity	
1183.06	267.328	1183.69	270.7771	1182.72	256.8517	ν
1291.98	243.7403			1292	228.2786	ν
1311.76	**723.0984**	**1311.01**	**826.4378**	**1311.87**	**731.4279**	N
1339.39	877.5563	1338.81	957.5727	1339.32	853.0158	ν . Σ
1347.99	**753.2162**	**1347.23**	**747.3596**	**1348.37**	**735.2247**	N
1374.34	232.2083	1372.96	268.3246	1373.17	250.0698	ν . Σ
1451.37	260.9656			1451.83	250.3838	N
1534.64	**597.334**	**1531.3**	**736.3055**	**1535.38**	**626.9658**	N
1542.58	757.1702	1539.71	645.8711	1543.29	696.55	N
1569.64	382.7652	1565.33	409.2672	1570.6	356.9159	ν . Σ
1592.28	**262.2246**	**1582.35**	**277.8787**	**1593.19**	**265.4388**	N
1616.21	**511.2168**	**1612.14**	**514.8691**	**1616.51**	**510.8566**	N
1646.03	604.2924	1644.86	604.9508	1646.31	607.4036	Σ
3284.64	640.7571	3276.84	654.3227	3297.51	619.1577	N
3454.48	401.334	3441.01	447.6203	3457.27	392.651	N

In this case, the main difference is absence of the lines of 1292 cm^{-1} and 1451 cm^{-1} in the spectrum calculated in the presence of water. However, the intensity of these lines in other spectrum calculated taking into account solvent is relatively small. It implies, that shape of the spectra of HM-I molecule measured in various solvents should be the same. Hence, from first sign, the IR spectrum measured in solvent could be used to detect HM-I. However, taking into account the fact that IR spectra of amines are greatly influenced by solvents [13], we speculate that IR spectrum features described should be obtained in the spectra of compounds that share the analogous elemental composition of HM-I. It allows us to conclude that identification of the new explosive materials by using spectroscopic methods could be rather difficult due to spectra-shape-change in solution.

3.2. UV-Vis spectra

The UV-Vis spectra of the materials under study were both measured and calculated, too. The spectrum of HM-I was measured only in ethanol solvent due to the low solubility of the molecule, although the spectrum is calculated in the presence of different solvents (Figs. 3-5).

The both measured and calculated spectra indicate presence of the 250.30 nm and 354.50 nm lines, although the intensities of the lines are not coincided, i.e. intensity of 250.30 nm calculated is too low in comparison with intensity of this line measured. The uncoincidence is not a surprise, because ethanol and water, other solvent of this study, are protic solvents having O-H or N-H bonds. The protic solvents can participate in hydrogen bonding, which is a powerful intermolecular force. Additionally, these O-H or N-H bonds can serve as a source of protons (H+). The phenomena are insufficiently described by PCM model used, although the results obtained is enough accurate to foresee the main features of the solution influence to the optic properties of HM-I.

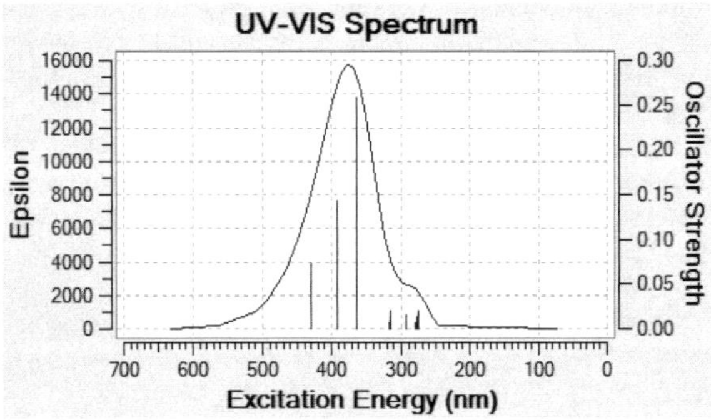

Fig.3 The UV-Vis spectrum of HM-I calculated in the presence of ethanol.

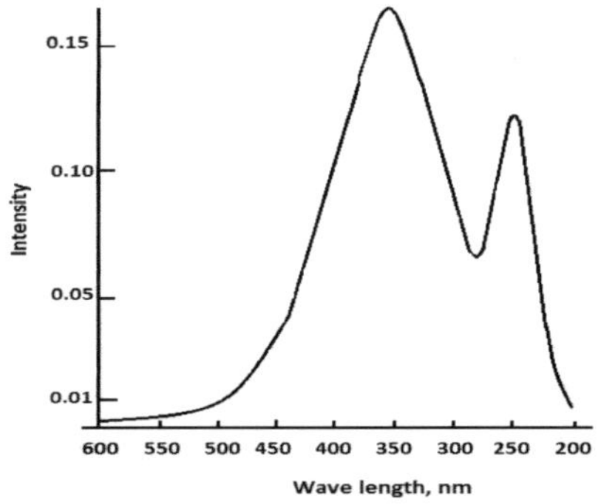

Fig.4 The spectrum of HM-I measured in ethanol.

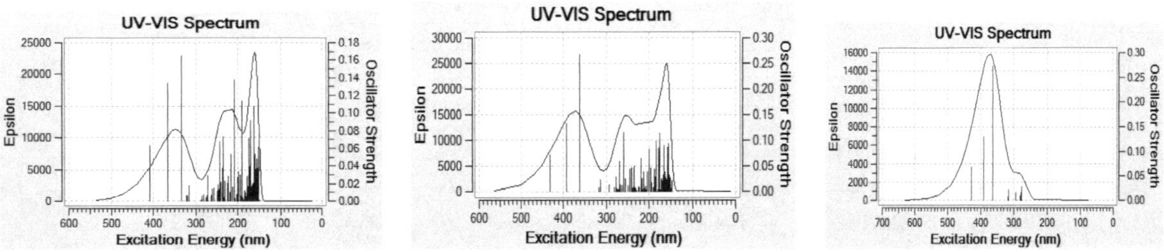

Vacuum **Acetonitrile** **Water**

Fig.5. The UV-Vis spectra of HM-I calculated in vacuum (on the left), acetonitrile (in the middle), and water (on the right).

As it is evident from Fig. 5, the lines of 250.30 nm and 354.50 nm are present in all spectra calculated in different solvents and vacuum, while their intensity is different. It implies, that the shapes of the UV-Vis spectra of HM-I calculated in vacuum, acetonitrile and water are different, although one of these spectra calculated in water and ethanol are similar. We pay attention, that investigation is performed in the presence of the solvent that belong to following groups:

- polar aprotic (acetonitrile);
- polar protic (ethanol, water).

Hence, the difference of the shape is present due to both the solvent and solvent properties. It allows us to conclude that the detection of the new hazardous material by UV-Vis spectroscopic method could be rather difficult, too.

Conclusions

The following results of our investigation were obtained:

- The IR and UV-Vis spectra of the new hazardous material are both measurement and calculated in the presence of various solvent and vacuum;
- The general features, i.e. wavelengths and wave numbers that are obtained in all spectra investigated, are achieved;
- The differences of the IR and UV-Vis spectra calculated in various solvents and vacuum are exhibited.

The above observations allow us to foresee that the detection of the new hazardous material by spectroscopic method could be rather difficult.

Acknowledgements

This work was conducted within the framework of the LMA scientific project "A theoretical and experimental investigations of new potentially explosive materials using quantum mechanical methods (NSPROG-I4)". The authors are thankful for the high performance computing resources provided by the Information Technology Open Access Center of Vilnius University

References

[1] Senesac L, Thunda TG. Nanosensors for trace explosive detection. *Materials Today* 2008; 11: 28–36.
[2] Hwang J, Namhyun Choi N, Aaron Park A, Park JQ, Chung JH, Baek S, Cho SG, Baek SJ, Choo J. Fast and sensitive recognition of various explosive compounds using Raman spectroscopy and principal component analysis. *Journal of Molecular Structure* 2013;1039: 130–6.
[3] Moore D. Instrumentation for trace detection of high explosives. *Rev. Sci. Instrum.* 2004; 75: 2499–2512.
[4] Izake E L.Forensic and homeland security applications of modern portable Raman spectroscopy. Forensic Sci. Int. 2010; 202: 1–8.
[5] Tamuliene J, Sarlauskas J, Bekesiene S, Kleiza V. ITELMS'2014 : Proceedings of the 9th international conference, May 23-24, 2014, Panevėžys, Lithuania. Kaunas: Technologija; 2014.
[5] Steinfeld J I, Wormhoudt J, Explosives detection: A Challenge for Physical Chemistry. Annu. *Rev. Phys. Chem.* 1998; 49:203–32
[6] Becke A.D.Density-functional thermochemistry. iii. The role of exact exchange. *J. Chem. Phys.* 1993; 98:5648-52.
[7] Zhao Y, Pu J, Lynch B J, Truhlarand D G. Tests of Second-Generation and Third-Generation Density Functionals for Thermochemical Kinetics. *Phys. Chem. Chem. Phys.* 2004; 6:673–676.
[8] Kendall R A, Dunning Jr. T H, Harrison R J.Electron affinities of the first-row atoms revisited. Systematic basis sets and wave functions. *J. Chem. Phys.* 1992; 96:6796–6806.
[9] Scalmani G, Frisch M J, Mennucci B, Tomasi J, Cammi R, Barone V. Geometries and properties of excited states in the gas phase and in solution: Theory and application of a time-dependent density functional theory polarizable continuum model. *J. Chem. Phys.* 2006;124:094107: 1-15.
[10] Cossi M, Rega N, Scalmani G, Barone V. Energies, structures, and electronic properties of molecules in solution with the C-PCM solvation model. *J. Comp. Chem.* 2003; 24:669-81.
[11] Gaussian 03, Revision C.02, Gaussian, Inc., Wallingford CT; 2004.
[12] Silverstein R M, Bassler G C, Morrill T C, *Spectrometric Identification of Organic Compounds*. 4th ed. New York: John Wiley and Sons; 1981
[13] Xie M, Liu Y, Deng Z. Influence of substituents on IR spectrum of aromatic amines in different solvents. *PubMed.* 2000; 20(6):819-21

Forum on Innovative Technologies and Management for Sustainability (ITMS'2016)

panevėžys faculty of technologies and business

The 11th International Conference Intelligent Technologies in Logistics and Mechatronics Systems (ITELMS'2016), 28-29 April 2016, Panevėžys, Lithuania

Load spectrum development tractor during tillage

Tchórz T.[a]*, Śnieżek L.[a], Grzelak K.[a]

[a]*Military University of Technology, Sylwestra Kaliskiego str. 2, 00-908, Warsaw, Poland*

Abstract

This article includes load spectrum methodology development for agricultural tractor with engine power 38 kW. The article presents different ways of determining loads acting on the tractor's transmission from analysis through to experimental studies. Based on the analysis of available literature in this area, defining the initial load participation with occurs during various agricultural works and transportation used for agricultural tractors. In order to verify the assumptions made
in basis of literature, there have been done fuel consumption measurements during various works execution. The measurements analysis show that the tractor consumed the biggest quantity of fuel during ploughing, which means that this work generates the greatest load for the power train.
The rest of this article describes loads methods registration that occurs during ploughing. The resistance ploughs movement course was finally determined. It represents the largest share of total tractor load during ploughing. For this purpose test stand has been built. It is the plough which leg has been pivotally mounted. The possibility of rotation was blocked after force sensor installation, thanks to that forces were recorded between legs and frame of a plough. In the next step numerical model in MSC Adams software has been built based on the real geometrical dimensions of the plough. The model was used to determine forces value which act on the tractor during ploughing. In the last step the tractor load spectrum has been developed which based on forces acting on the tractor during ploughing which obtained from simulation.

© 2016 T. Tchórz, L. Śnieżek, K. Grzelak.
Peer-review under responsibility of the Kaunas University of Technology, Panevėžys Faculty of Technologies and Business

Keywords: farm tractor; load spectrum; tillage; ploughing.

1. Introduction

Measurement and loads analysis of tractor that occur during agricultural works are essential in the design stage of the transmission and motor selection for a particular type of tractor. It is therefore necessary to record information on the most commonly used gear, speed achieved by the tractor and the driving forces behind the work of cultivation and transportation. Research in this area suggests that the greatest load in the power train of the tractor exist during such operations as plowing and soil cultivation [1].

All of these agricultural works belong to the hard work group, but it is necessary to select a procedure that generates the greatest load for tractor's drive train. One of the parameters determining the tractor load is fuel consumption. Fuel is tractor energy source and its consumption determines machine performance the work resistance value of cooperating implements that the tractor must overcome. Moreover, fuel consumption during a single treatment is affected by many factors such as soil texture, moisture, terrain, tractor type (two or four-wheel drive), tractor size of a and a method of selecting implements to the tractor (if they are not too large or too small). For the fuel consumption when using the same implement can vary significantly [2].

* Corresponding author. Tel.: +22-261-839-419; fax: +22-261-837-211.
E-mail address: tomasz.tchorz@wat.edu.pl, lucjan.sniezek@wat.edu.pl

In order to investigate the initial loads contribution that occur during individual agricultural works performance fuel consumption was mesured. The object of the research is tractor Ursus C-360 with engine nominal power of 38 kW and a carb weight of 2830 kg. The fuel consumption results for each works shown in Fig. 1.

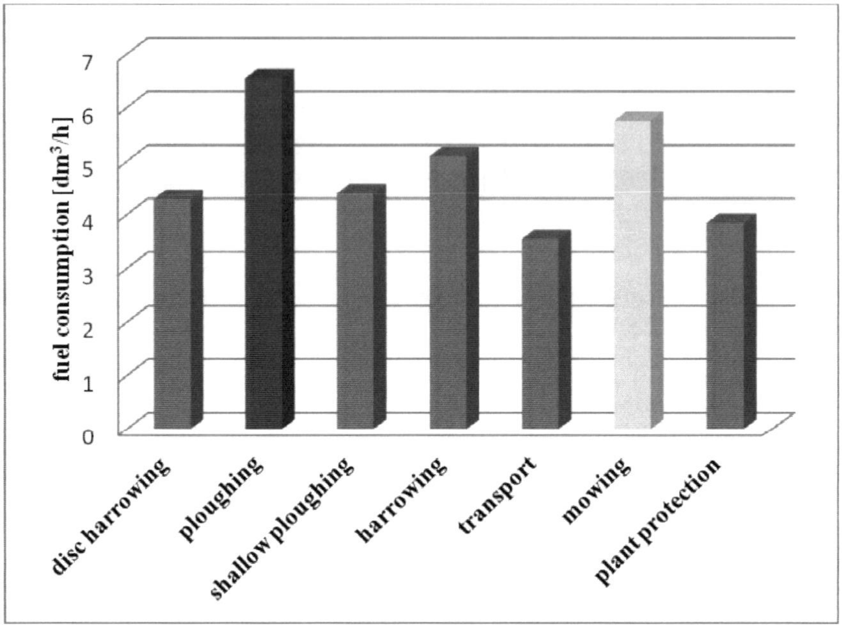

Fig. 1. Fuel consumption of the tractor Ursus C-360 which performs various field works

Based on the measurements it can be explicitly stated that the highest fuel consumption occurs during such operations as ploughing and mowing. It is remarkable, however, that when working with the mower tractor's transmission does not transfer the total load because the mower is driven by the tractor PTO. Therefore, bar corresponding to the fuel consumed during mowing has been marked in yellow. The highest fuel consumption, which was observed during the tests occurred during ploughing, when the total torque generated by engine is transmitted by the drive system. Bar in the graph corresponds to the fuel used in this case is marked in red. Later in the article authors treated loads methods and analysis during the ploughing.

Nomenclature

F_P	plough resistance [N]
R_S	sensor reaction [N]
G_P	plough weight [N]
G_{TR}	tractor rear axle weight [N]
R	soil reactions [N]
R_{TL}	top link reaction [N]
R_{LA}	lift arms reactions [N]
P_X	traction effort [N]
O	leg plough axle rotation
f	rolling resistance coefficient
m	mass of the plough [kg]
g	acceleration of gravity [m/s^2]
k	specific resistance of the soil against deformation [N/m^2]
d	working depth [m]
w	working width [m]
ε	coefficient, depending of the kind of mouldboard and soil [$N\,s^2/m^4$]
v	ploughing speed [m/s]

2. Methods for determining resistance forces during the ploughing

The main tractor load in the agricultural performance is the cultivation implements resistance work power. However, the tractor also subject to such forces as tractor wheels rolling resistance, co-operating machines rolling resistance, inclination terrain resulting resistance. The sum of these forces is the strength of resistance for the tractor is transmitted through the drive train on the cultivation performance.

In order to achieve tractor's maximum efficiency its work should be carried out using the maximum drawbar power, which is the product of actual speed and traction force [9]. With the increase of drawbar power, wheel slip increases due to loss of grip. The result is a reduced possibility of using the maximum engine power what causes fuel and time waist.

2.1. Plough work resistance analytical calculation

The tractor's tractive effort value must be greater than the sum of soil cutting resistance occurring while ploughing. The sum of soil cutting resistances are a plough work resistance, which value is the result of plowshares geometry and mouldboards, soil structure, soil moisture and tractor speed. Knowing these basic parameters approximate value of the horizontal component of traction effort which must have a tractor can be determined in an analytical way, which will be aggregated with the plough. For this purpose, the following formula may be used [3]:

$$F_{PX} = (f \cdot m \cdot g) + (k \cdot d \cdot w) + (s \cdot d \cdot w \cdot v^2) \qquad (1)$$

In order to determine the exact parameters values which appear in equation 1 it is to be performed research working plough elements in the soil in the same conditions. The plough resistance value during ploughing calculated analytically does not reflect the exact dynamics system.

While tillage occur temporarily increases the work implements resistance, which cause an overload for the engine and transmission. In addition, slip value increased, it may even reach 100%. Therefore, it is essential to ensure the power supply and keep the tractor load while working at 60-70% of its maximum load [4]. To ensure this condition it is necessary to know the acting loads nature on the tractor transmission during the toughest works which include the ploughing.

2.2. Plough resistance work experimental determination

Most of the tillage implements could be attached to the tractors rear three-point linkage, where there are two lift arms and one top link. While working, the lift arms are stretched while the top link is usually compressed. It makes it difficult to measure the total work force the implement on the tractor. One work resistance measuring during ploughing is towing the tractor, when attached to a plough, using another tractor [5]. This method diagram is shown in Fig. 2, this solution advantage is the easy way to register force occurring in the line connecting the tractor towed with the drive.

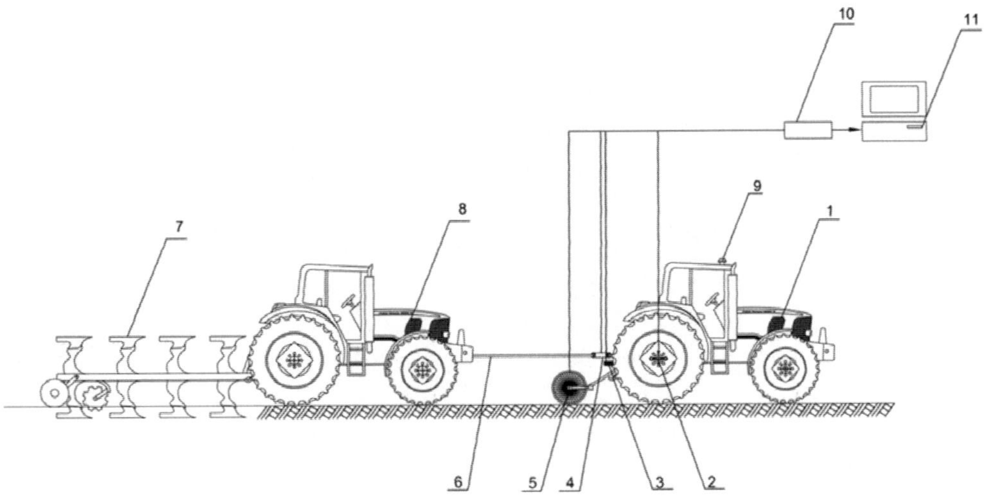

Fig. 2. Measuring the resistance of work the plough using tractor towed and driving. Labels: 1-test tractor, 2-sensor for theoretical speed, 3-sensor for PTO speed, 4-dynamometer, 5-travel speed sensor, 6-link, 7-implement load, 9-GPS, 10-data acquisition, 11-computer.

The method main disadvantage is work plough total resistance sum measuring and the tractor towed rolling resistance one. In order to isolate the resistance which generates plough there have also to be measured the tractor towed rolling resistance with raised implement and then calculate work plough resistance using the following equation:

$$F_P = F_{P+T} - F_T \tag{2}$$

The calculated resistance value F_P may not be accurate because the dependence from which it was calculated does not include the tractor down force axle, which can occur when working with the mounted implement.

3. Work plough resistance measurement

Studies was done on 10.11.2016 in Lublin province in the municipality of Stary Zamość in the field to growing an area of 0.36 hectares, which dominates the soil with clay surface. The exact parameters to the conditions the test was carried out and the devices information used are shown in Table 1.

The loads particular nature during ploughing done tractor Ursus C-360 with a total mass of 2830 kg. with rear-wheel drive was determined on the basis of the following research. During the study a plough having three furrow was used, where the leg pivotally mounted to the frame. The rotation possibility was blocked by the mounting force sensors ZEPWN CL-140, with which the forces acting on the leg were recorded. Force sensor mounting and the plow frame leg is shown in Fig. 3.

Table 1. The values of occurring parameters during research.

Parameter name	Value	Unit
Soil moisture	12	%
Depth of the ploughing	0,25	m
Driving speed during the ploughing	1,3	m/s
The average tractor wheel slip	27	%
Engine power	38	kW
Load front axle	9	kN
Load rear axle	19,3	kN
Own weight of the plough	0,3	kN
Working width of the plough	0,9	m
Cutting angle of the plowshare	30	°

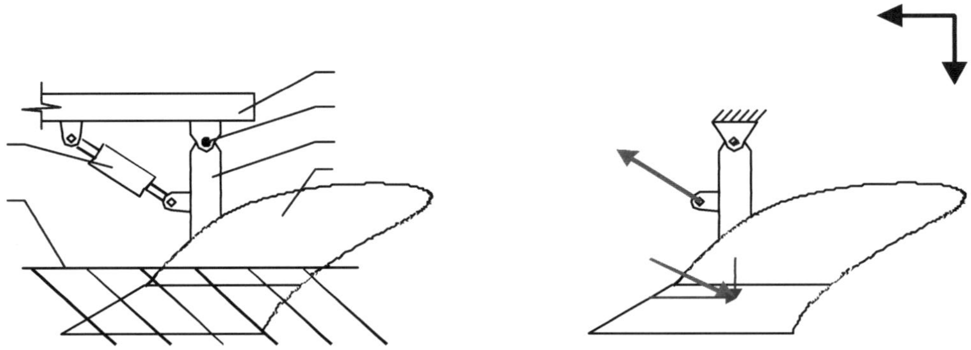

Fig. 3. Construction of the measuring plough (a) method of assembly force sensor and the leg to the plough frame. Labels: 1-frame, 2-pivot, 3-leg of plough, 4-mouldboard, 5-force sensor, 6-surface of soil, (b) simplified system of forces that occur during ploughing.

Plough working elements are characterized by helical geometry due to the work nature they perform, their job is to separate the upper layer of soil and flip it to cover the plants. The plowshares and mouldboards shape complexity level make it impossible to clear and precise pressure soil distribution on the working elements during tillage. Therefore, first and most important criterion is to consider simplifying the system: tractor - machine as a flat forces system. The resultant forces acting on each of the working plough's elements calculation was performed based on the assumption that these forces points are present on the plough bodies in the ploughing middle depth [6]. It is also assumed that the force points are under axes, whereby the vertical component does not affect the torque value of relative to the rotation axis. The forces acting on the object of research during tillage shown in Fig. 4.

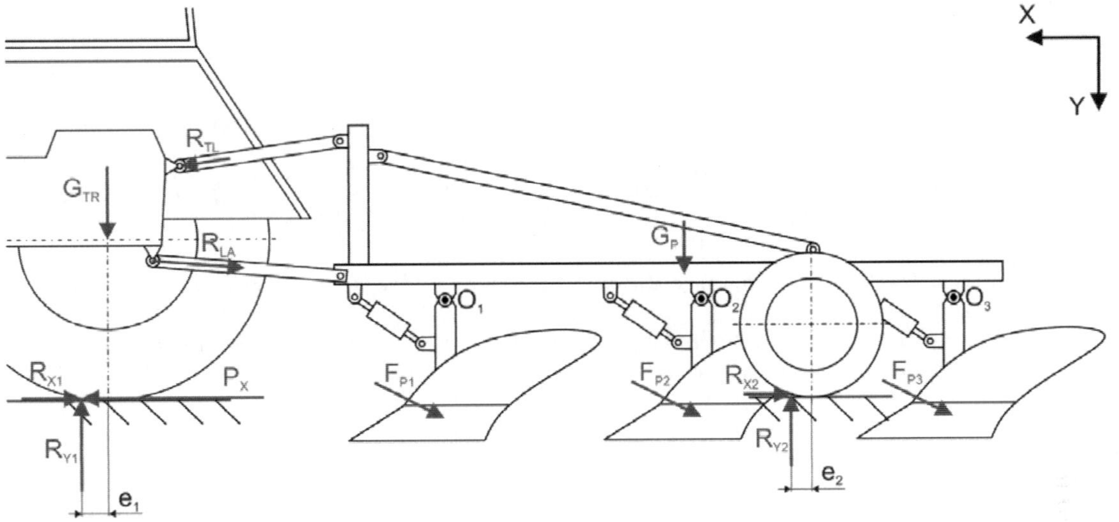

Fig. 4. Distribution of forces acting on the plough and tractor during tillage.

On the basis of plough and tractor geometry precise measurements, it was built numerical model in the MSC Adams, which was introduced force in the form of waveforms of the forces F_{P1}, F_{P2} and F_{P3},which was recorded during research [10]. Waveforms of these forces shown in Fig.5.

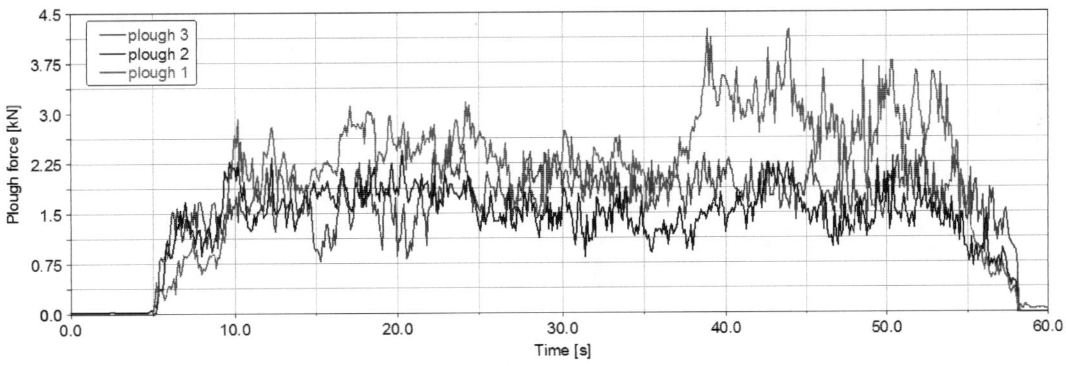

Fig. 5. Waveforms horizontal forces acting on the plough bodies. Labels: plough 1-F_{P1}, plough 2-F_{P2}, plough 3- F_{P3}.

During ploughing the plowshares penetrate the soil causing its cutting and inverting and as a result the plough working elements are additionally loaded by vertical component of forces F_{P1}, F_{P2} and F_{P3}. In order to reflect the actual working conditions and determine the force which presses the plough to the ground, the vertical work resistances components have also been included in the analytical model. The forces values for tillage implements with an angle of plowshare cutting 30 ° is about 40% of the horizontal component values [7].

Fig. 6 shows the waveforms forces acting on a tractor during ploughing, which have been obtained from simulation. The color blue represents the traction effort that tractor generates during tillage, however red color signified the sum of the vertical reaction components which were occurring in the three point linkage.

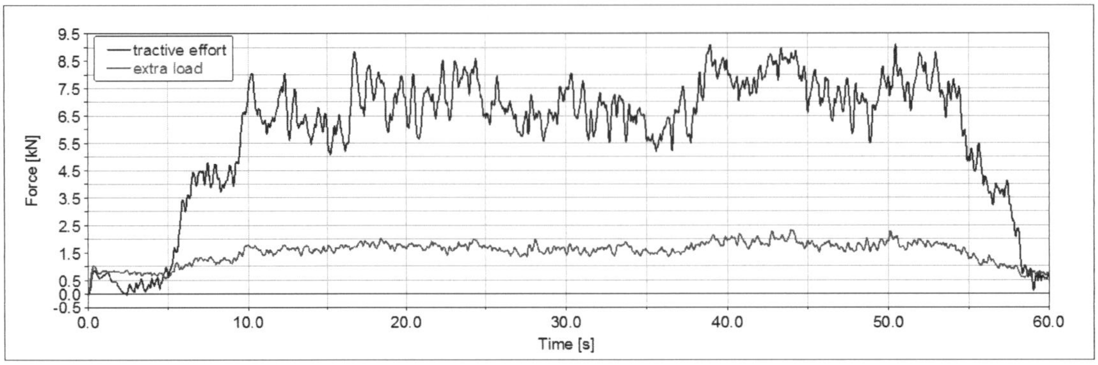

Fig. 5. Waveforms forces acting on a tractor during ploughing. Labels: tractive effort - P_X, extra load - $\Sigma R_{TCY} R_{LCY}$.

While analyzing the forces shown in Fig. 6 it can be observed that the greatest force values occur in the time interval from 10 to 55 sec. In other time intervals analyzed forces reach much lower value due to plough deepening beginning, which lasted for 10 sec. and decrease the driving speed tractor to a halt after 55 sec.

The average traction effort value during ploughing in the range from 10 to 55 sec. is 7,13 kN, while the maximum force value, occurred in the analyzed interval is approx. 9 kN which represents 47% tractor load rear axle. This means that the maximum adhesion coefficient value reaches the 0,47, on the assumption that tractor has not been additionally loaded during tillage. In order to verify tractor work with a plough impact on the extra load rear axle, there was presented vertical components waveform sum reaction occurring in the three point linkage tractor (Fig. 6). In the time interval from 10 to 55 sec. the sum average forces resulting extra rear tractor load is 1,73 kN while the maximum value exceeds 2 kN which represents about 10% of the total load on the rear axle. Increasing the load rear axle may be important for the tractor driveline components durability. This is due to the fact that, for high-gear ratio in driveline the tractor is able to generate momentarily drawbar pull by 10% higher compared to the maximum drawbar pull that will generate without implement in the same conditions.

Using local extremes method counting, waveform the traction effort P_X in the time interval from 10 to 55 s. was developed spectrum tractor load drive system during ploughing [8]. Different load values percentage are shown in Fig. 7. The traction effort lowest value of the ranging from 5 to 6 kN occurring when the tractor works with plough is 30%. The largest share of 50% determined according to counting local extrema loads method in the range of 6,5 to 7,5 kN. Slightly larger load values in the range of 8 to 8,5 kN a group of the share 18%. The highest traction effort values that occurred during ploughing are 9 kN, whose participation is not exceed 2%.

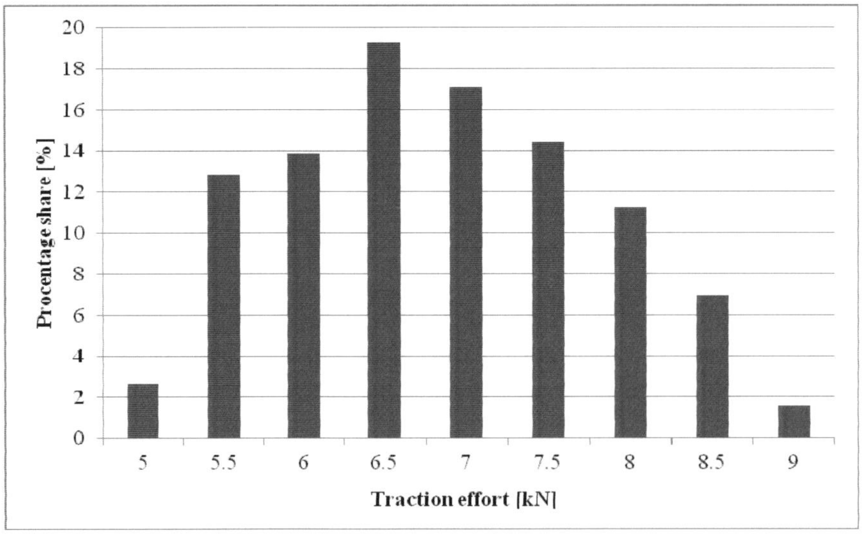

Fig. 7. Load spectrum driveline of the tractor during ploughing

Presented load spectrum tractor driveline during ploughing may be used as one of the elements to estimate the spectrum loads occurring in total tractor working live.

4. Summary

Load spectrum driveline development is an essential element in the research phase of newly constructed power transmission tractors. The method proposed determining in this article the load spectrum based on the most energy-intensive agricultural tillage which is ploughing, allows to determine the any tractor load spectrum performing this tillage. In order to accomplish this task, it is necessary to enter the geometry parameters and mass of the both: tractor and plough. The next work stage is to verify the results obtained from the model proposed simulations with tests field results performed by towing the tractor with the plough.

References

[1] Koniuszy A, Optimazation of the tractor work U 912. *MOTROL* 8/2006; p. 168-175 (in polish)

[2] Karparvarfard SH, Rahmanian-Koushkaki H. Development of a fuel consumption equation: Test case for a tractor chisel-ploughing in a clay loam soil. *Biosystems Engineering*; 2015. p. 23-33

[3] Kostecki P, Borowiak P, Nowowiejski R. Static and dynamic pressure exerted by silty soil onto the woking surface of a plouhshare. *Tribologia* 6/2012. p. 61-73 (in polish).

[4] Janulevicius A, Juostas A, Pupinis G. Tractor's engine performance and emission characteristics in the process of ploughing. *Energy Conversion and Management;* 2013. p. 498-508

[5] Simikic M, Dedovic N, Savin L, Tomic M, Ponjican O. Power delivery efficiency of a wheeled tractor at oblique drawbar force. *Soil & Tillage Research*; 2014. p. 32-43

[6] Kolator B. The method for determining the forces that act on the tractor when working with the mounted implement. *Inżynieria rolnicza.* 4/2008; p. 381-387 (in polish)

[7]Abo Al-kheer A, Eid M, Aoues Y, El-Hami A, Kharmanda MG, Mouazen AM. Theoretical analysis of the spatial variability in tillage forces for fatigue analysis of tillage machines. *Journal of Terramechanics*. 48/2011; p. 285-295

[8] Kocańda S, Szala J. *The base of fatigue calculation. PWN.* Warsaw; 1997 (in polish).

[9] Grzelak K, Śnieżek L, Tchórz T. The influence of tractor drive system modernization on tractor capacity. *Proceedings of 19th International Scientific Conference. Transport Means;* 2015. p. 271-276

[10] Łopatka M. J, Muszyński T, Rubiec A. Loads analysis of skid steer robot drive system. *18th International Conference on Methods and Models in Automation and Robotics MMAR, IEEE*, New York; 2013. p. 362-367

Forum on Innovative Technologies and Management for Sustainability (ITMS'2016)

ktu 1922 panevėžys faculty of technologies and business

The 11th International Conference Intelligent Technologies in Logistics and Mechatronics Systems (ITELMS'2016), 28-29 April 2016, Panevėžys, Lithuania

Investigation of the load of three axled semi-trailer

Tautkus A.[a], Micevičienė D.[a], Vaitiekūnas T.[a,b]*

[a]*Kaunas University of Technology, Panevėžys Faculty of Technologies and Business, 33 Nemuno str., 37164 Panevėžys, Lithuania*
[b]*UAB Premeta, J. Janonio str. 1, LT-35101 Panevėžys, Lithuania*

Abstract

Vehicles can be loaded up to the maximum allowable capacity only in the case if the total load center of gravity is in a certain area. However, even when the load is less than the maximum weight lifting cargo must be stowed also is in precise places. Otherwise, the tonnage of the axle might be too high or the low. If the cargo center of gravity lies within the zone - the so-called load distribution curve – it can be guaranteed, that the axles areloaded correctly. Using the load distribution curve and the cargo distribution scheme, the distance from the front wall might be determine as well as the weight of the load for certain the semi-trailer can be detected. In the article the load of semi-trailers is simulated using the doftware „Science Truck body builder" that allows to draw the load distribution curve for a three-axle semi-trailer's model.

© 2016 A. Tautkus, D. Micevičienė, T. Vaitiekūnas.
Peer-review under responsibility of the Kaunas University of Technology, Panevėžys Faculty of Technologies and Business

Keywords: Three-axled semi-trailer; laod; overload of semi-trailer.

1. Introduction

The heavy vehicles – trucks' – center of gravity is a very important aspect or point for increasing the driving performance, vehicle stability on the road as far as braking force distribution. In the case goods are loaded outside the specific guidelines, the consequences might be catastrophic. For example, the heavy weight of the cargo is loaded into trailers' end so the towing vehicle drive axle will not have good adhesion to the road and it may happen that the powered wheels will not even reach the road base or adhesion will be very poor. If excessive weight is placed on the front of the semi-trailer it will result in the overweight on the drive axle, which will affect the front axle for trailer. Avoidance of these mistakes might lead to safe operation of the vehicle [1]. The load distribution curve of the semi-trailer serves for the optimal determination of the cargo placement.

The European manufacturers of semi-trailers present the load distribution curves as a loaders guide (see Fig. 1 pav.) However, this scheme gives only the approximate orientation to loading a semi-trailer: it allows to find out just where the heaviest load must be placed. The graph shows that in the front of the semi-trailer it is recommended to load cargo up to 7 tons in weight, in the midle - of even up to 30, and by the end of the truck – at least 2-3 tons [2].

* Corresponding author. Tel.: *+370-686-75101*
E-mail address: arunas.tautkus@ktu.lt

© 2016 A. Tautkus, D. Micevičienė, T. Vaitiekūnas.
Peer-review under responsibility of the Kaunas University of Technology, Panevėžys Faculty of Technologies and Business

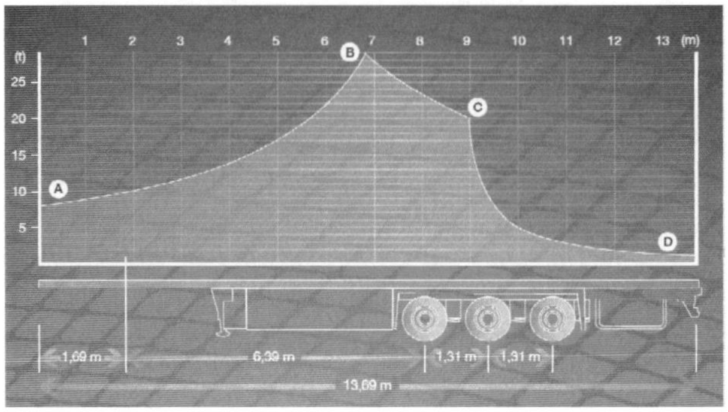

Fig.1. Example of load distribution diagram - semi-trailer (Schmitzh Cargobull) [2].

In the article the loading of different cargo is simulated as well as the load distribution curves for a three-axle semi-trailer model are drawn. The investigation is carried out according „Weight and dimensions of heavy commercial vehicles as established by Directive 96/53/EC" [3].

2. Simulation of gravity centre location for a truck and semi-trailer

The stability and security of the operation of a vehicle depends on the effects of inertia forces. Inertia focuses in the point of gravity. The position of the centre of gravity depends of the load size and positioning on a semi-trailer. The article presents the simulation of dynamic centre of gravity when, where cargo is pulled back from the front of the wall in a meter. The calculation schema of the center of gravity is given in Figure 2. Under considerations that firstly, the dimensions of the cargo are 2.4 m x 4.0 m, secondly, the centre of gravity of the load is dynamic, which is determined as the intersection of symmetry axis of the rectangular shape cargo; and finally the height of the load is always the same – 2 metres.

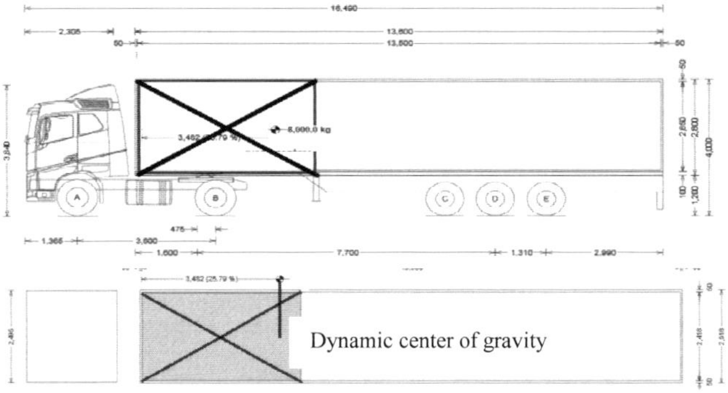

Fig. 2. Example for determining the gravity center for two axle vehicle and three axled semi-trailer loading: dimension of cargo1.5 x 4 m and weight 8 t.

Using the forementioned parametrs the simulation process was started. The weight og the load was contantely increased from 8 to 17 t, other parameters of the load were not changed. The total weight of the vehicle and the semi-trailer was m = 151, 704 t. At the time of the simulation of the load was moved from the front of the semi-trailer's wall till its end. The change in the centre of gravity curves is presented in Figures 3 and 4.

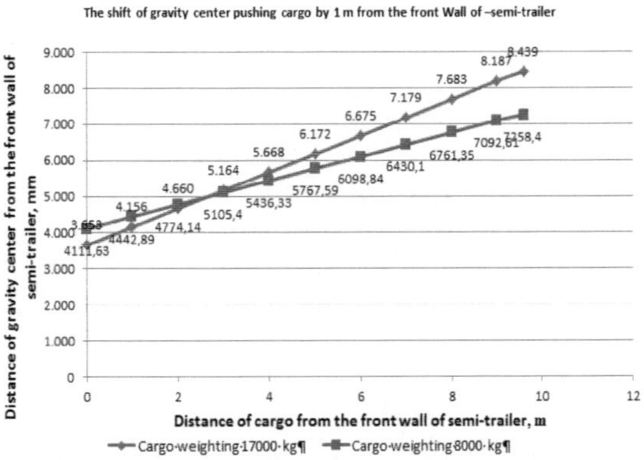

Fig. 3. The dependence of gravity center of two axle vehicle and three-axle semi-trailer's front axle (rear wheel drive) on cargo (of 8 and 17t in weight) location.

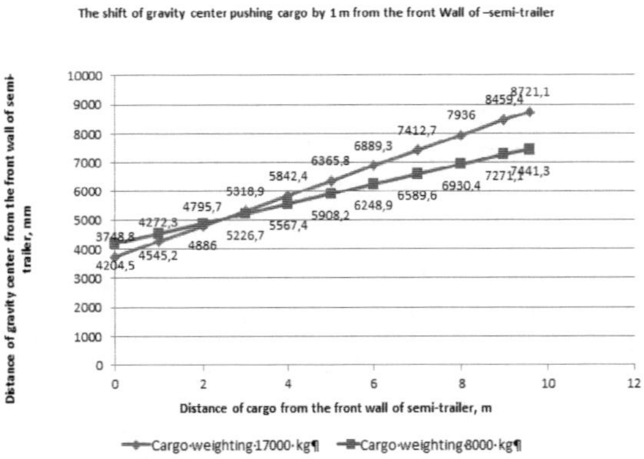

Fig.4. The dependence of gravity center of three axle vehicle and three-axle semi-trailer's front axle (rear wheel drive) on cargo (of 8 and 17t in weight) location.

From the charts the pattern, demonstrating the movement of the gravity centre, was determined for both simulation models: two axle vehicle and three-axle semi-tariler‘s as well as three axle vehicle and three-axle semi-tariler's. Figure 3. and Figure 4 show that the dynamic displacement of the centre of gravity evenly increases and moves towards the end (rear) of the semi-trailer. The model of two axle vehicle and three-axle semi-trriler has shown that after move of the cargo (both 8 and 17 in weight) from the front wall at 2.8, the dynamic centre of gravity in the trail matches. The dynamic centre of gravity in the trail consisting of three axle vehicle and three-axle semi-trailer matches after movement of the cargo at 2.6 m.

3. Simulation of axle load for a truck and semi-trailer

Vehicle‘s axle load depends directly on the location of the cargo on a semi-trailer. Modelling performed by the authors allowed to detect the variation patterns for load of both rear and wheel drive axles in dependence on the cargo loaction on semi-trailer. The simulation was carried out with two-and three-axle trucks and three axle semi-trailerhandling the same cargo characteristics.

Figure 5 and Figure 10 show the active load of the axles when the cargo (8 and 17 t) was pushed from the front wall of the semi-trailer to rear one. The diagrams allow to determine the position or location of the cargo at which the load of wheel drive axled exceeds the permissible limit – 10 t.

Figure 5 and Figure 6 also show that rear axis is not overloaded, that is it does not exceed the permissible load 10000 kg and evenly decreases. Figure 7 pav. demonstrates that the load of the wheel drive axle of two axled semi-trailer vehicle wll be overloaded – that is wll exceed the persmissible 11500 kg limit – only in the case if the cargo of 17 t wll be located pushed from the front semi-trailer‘s wall at 2.6 m. Modelling the same dituaction inderthe same conditions for three axled vehicle the results showed that the formentioned problem wll never arise - pusdhing the cargo from the front wall of the semi-trailer to its rear, the wheel drive axle wll never be overloaded: will never exceed permissible limit of 19000 kg.

The load of axles in two axle vehicle and semi-trailer (Figure 9) and the load of axles in three axle vehicle semi-trailer (Figure 10) areoverloaded under such conditions: when the loaction of the cargo exeeds is 7,5 – 9,6 m from the front wall, and the weight of the cargo – 17000 kg. The load increases evenly.

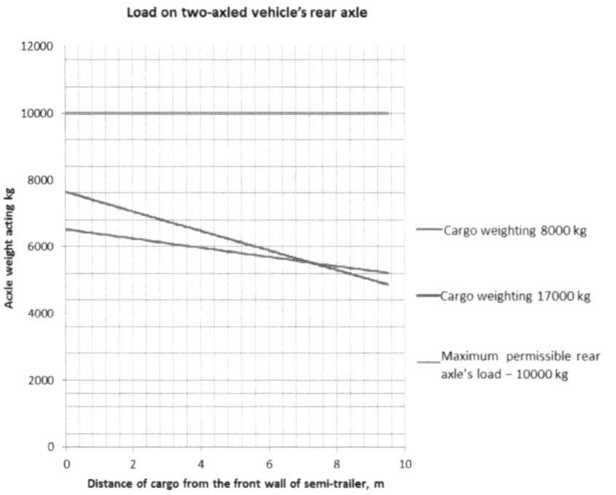

Fig.5. The dependence of two axle vehicle a front axle (rear wheel drive) load on cargo (of 8 and 17 t) location.

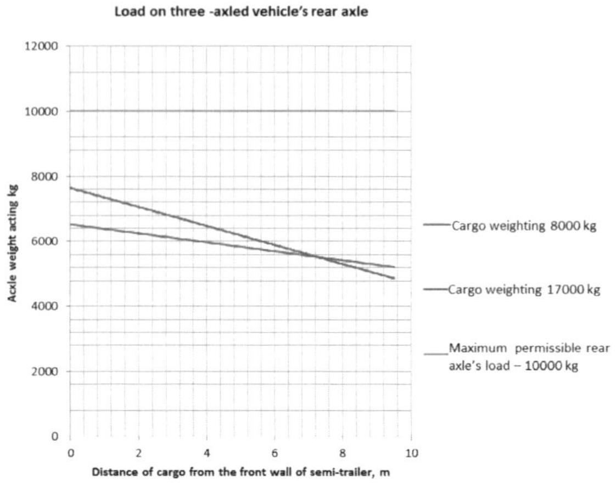

Fig.6. The dependence of three axle vehicle front axle (rear wheel drive) load on cargo (of 8 and 17 t) location.

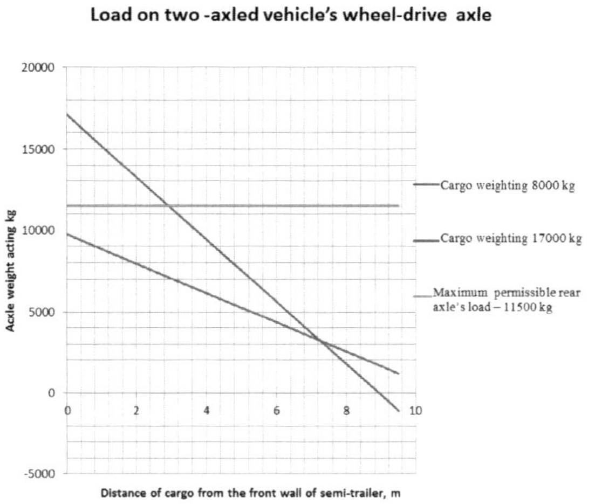

Fig.7. The dependence of two axle vehicle front axle (wheel drive) load on cargo (of 8 and 17 t) location.

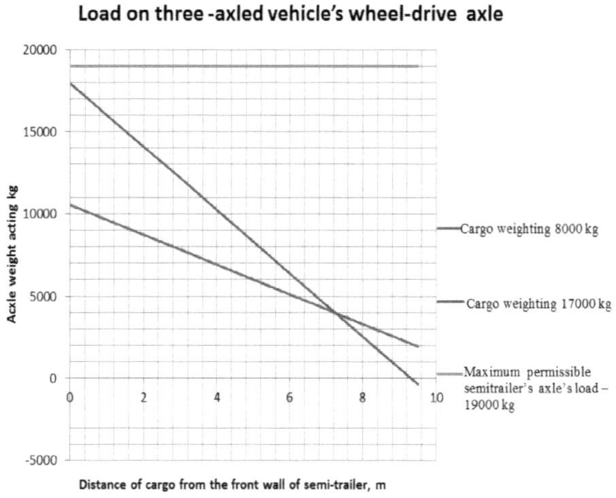

Fig.8. The dependence of three axle vehicle front axle (wheel drive) load on cargo (of 8 and 17 t) location.

Using software „Truckscience" [4] it was determined what the maximum weight [5], [6] of the cargo can be, if its width is 2, 4 m and length changes.

Pushing the load from the front towards the rear of the semi-trailer, it was found what the greatest weight of cargo in any place of the semi-trailer could be within the limits of the permissible axle loads. The situationed was simulated using as an example a cargo piece of 1.5 m wide and 1 m high with a length from 1 m up to 13, 6 m. The length of the load was increased every 1 m. Under such conditions the maximum permissible axle load limits were determined. The schemes of three axeled semi-trailer load were drawn that allow to detect the precise position of a cargo of different length.

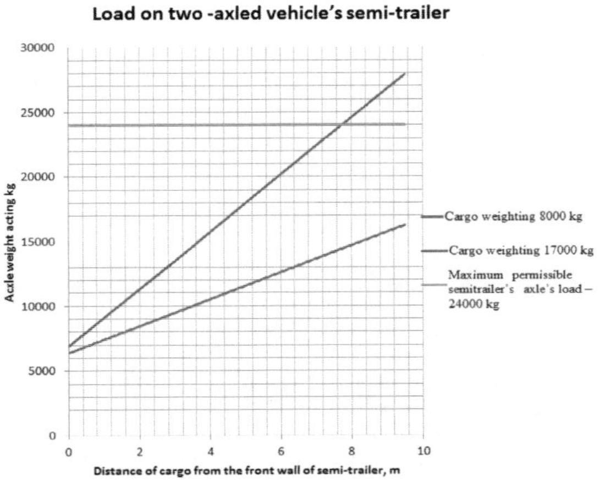

Fig.9 The load of axles in two axle vehicle semi-trailer (cargo of 8 and 17t).

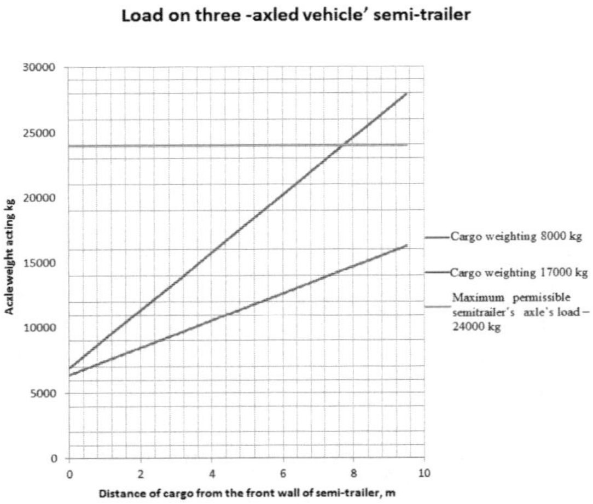

Fig.10. The load of axles in three axle vehicle semi-trailer (cargo of 8 and 17t).

Figure 11 demonstrates one possible example – a scheme of three axeled semi-trailer‘s load.

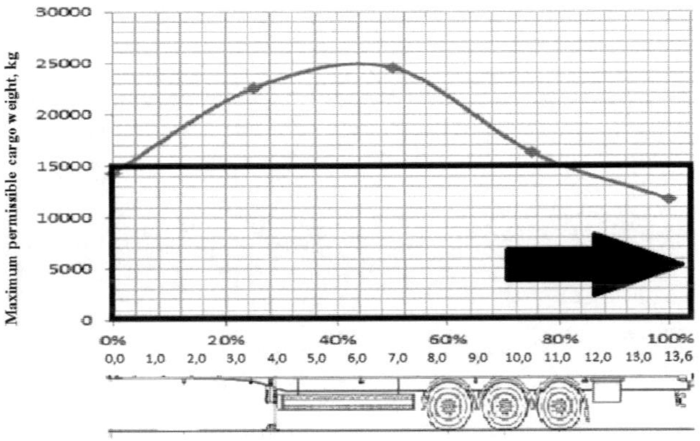

Fig. 11.Sheme of 13, 6 m long cargo load on three axle semi-trailer.

Figure 11 shows the patterns of axle load distribution. The maximum weight is achieved, when the Centre of gravity of the load is in the middle of a load, that is makes 50% of the length of the load.

Conclusions

1. After completion of the investigation, it was found that the center of gravity of the cargo pushing it from the front of the semi-trailer towards the rear scrolls smoothly toward the rear (end). That means that when the cargo is loaded at the rear of the semi-trailer, the center of gravity is the most distant from the front wall and the rigidity of the semi-trailer is less when the center of gravity is more distant from the driving axle.
2. Axle load of the semi-trailer change smoothly pushing the load from the front wall to the rear one: the load of rare and wheel drive axles constantly declines, and the load of the semi-trailer – constantly grows.
3. Pushing the 8000kg mass cargo loaded into the truck on a two-axle vehicle with three axle semi-trailer, the permissible axle load is not exceeded. However, pushing the cargo at the weight of 17000kg on the two-axle towing vehicle when its location reaches from 0 to 3 m from the front of the semi-trailer's wall, the wheel drive axle is loaded at 11500 kg and this exceeds the permitted range.
4. Semi-trailer axles are overloaded when cargo is moved from the front to the rear wall more than 8 m. When a cargo weighing 8000kg is loaded into the truck on three axed vehicle with three axle semi-trailer, the permissible axle load is not exceeded. However, if the cargo weights 17000kg and is moved towards the end of the semi-tariler, the permissible load is reached ath thedistance of 7.5 m. This means that it is

advised to transport the heavy freight by three-axled vehicle (because the axles are not overloaded), but it must be avoided to move the cargo on the end of the semi-trailer.

5. Partial and heavy freight must be transported by three-axed vehicles (it optimal) because the wheel driven axle can be overloaded .Complete cargo is advised to transport on two-axed vehicles avoiding the permissible limit of the total weight.

References

[1] International guidelines on safe load securing for road transport [online] [cit. 2016-02-02]. Available at: https://www.iru.org/sites/default/files/2016-03/en-safe-load-guidelines.pdf

[2] Neufahrzeuge/Ladungssicherung/SCS-LaSi-LT-PC.pdf [online] [cit. 2016-02-02]. Available at: http://www.cargobull.com/files/lt/filemanager_files

[3] Weight and dimensions of heavy commercial vehicles as established by Directive 96/53/EC and the [online] [cit. 2016-02-1]. Available at: http://ec.europa.eu/transport/modes/road/events/doc/2009_06_24/2009_gigaliners_workshop_acea.pdf

[4] Truckscience [online] [cit. 2016-02-02]. Available at: http://truckscience.com/axle-weight-calculator/

[5] Hansels TJ, Mullins WW. Heavy duty transmission and axle lubricants – mineral or synthetic for extended drain? Industrial Lubrication and Tribology; 2002. Vol. 54,6. http://dx.doi.org/10.1108/00368790210445641.

[6] Savaidis G, Malikoutsakis M, Davaidis A. FE simulation of vehicle leaf spring behavior under driving manoeuvres. International Journal of Structural Integrity;2013. Vol. 4. 1. http://dx.doi.org/10.1108/17579861311303609.

Forum on Innovative Technologies and Management for Sustainability (ITMS'2016)

panevėžys faculty of technologies and business

The 11th International Conference Intelligent Technologies in Logistics and Mechatronics Systems (ITELMS'2016), 28-29 April 2016, Panevėžys, Lithuania

Theoretical assessment of apartments energy performance characteristics depending on their territorial configuration

Tumavičiūtė G.[a]*, Kelpšienė L.[a]

[a]Kaunas University of Technology, Panevėžys Faculty of Technologies and Business, 33 Nemuno St., 37164 Panevėžys, Lithuania

Abstract

While analyzing the development of partitions in construction, three equal and closely interrelated components are distinguished: the ecological, economic and social aspects. Climate change and limited resources encourages to use thermal energy in residential accommodation more efficiently. In Lithuania almost 96 % of apartment building were built before the year 1993. The consumption of thermal energy in these buildings is about two times higher then in new construction apartments. Comparing updated (modernized) buildings also a huge difference in energy consumption is highlited. State, considering modernization of residential buildings a priority, encourages the renovation of apartment buildings and provides financial aid to investment projects. It is a fact that resources are not infinite, so it is important to continuously look for ways to conserve and use them as efficiently as possible. The consumption of thermal energy of a house can influenced by such factors: geographical location, climatic conditions and city‘s main fuel for the production of heat. Unfortunately it would be a very difficult task to change the geographical location of objects and you can not ‘‘book‘‘ desired weather conditions. In this case it is important to focus on other alternatives, which would be possible and realizable, encouraging environmental protection, conserving finances and improving people‘s level of comfort. Apartment renovation is one of the means.

© 2016 G. Tumavičiūtė, L. Kelpšienė.
Peer-review under responsibility of the Kaunas University of Technology, Panevėžys Faculty of Technologies and Business

Keywords: apartment renovation; update (modernization); energy performance.

1. Introduction

One of the issues of today is still the renovation of apartments. While analyzing renovation through the perspective of improvement of energy consumption, you can not separate such aspect as fostering the quality of resident‘s living environment.

The residential sector is the largest final energy consumer and is responsible for 30% of total carbon dioxide emissions in Hungary. In order to address the general poor condition of the building stock and resulting inefficiency in energy use, from 1990 onwards, the government and local authorities 111 initiated energy efficiency support programmes in the residential buildings sector [1].

Only about 4% of apartments in Lithuania are classified as consuming the least heat (~9 kWh/m^2), i.e. new construction or modernized (updated), or having individual heating control and accounting [2]. Thus, the remaining approximately 96% of apartments are built according to Construction Technical Standards that were valid until 1993 or with large windows, which barely satisfy modern requirements. Their residents have to pay a substantial amount in order to keep a comfortable environment in the apartments. Most of the apartments are provided with thermal energy through central heating systems. In order to ensure better living conditions in both comfort and financial aspects, these systems need to be updated (modernized).

The EES market potential in the residential sector indicates the presence of an additional annual market volume of about 190 M€ in the EU-27 for investments with payback times below 3 years (or of about 1640 M€

* Corresponding author. E-mail address: gerdusia@gmail.com.

© 2016 G. Tumavičiūtė, L. Kelpšienė.
Peer-review under responsibility of the Kaunas University of Technology, Panevėžys Faculty of Technologies and Business

for investments with payback times below 8 years). The largest part of this potential comes from EESs related to wall and roof insulation and installation of energy efficient boilers in multifamily buildings. Although specific barriers in the residential sector still hinder national EES markets development in important ways, positive market signals have been registered in Germany, Denmark, France, Flanders (BE), Hungary, Romania and UK [3].

The government approved the proposal to set up a Public Investment Development Agency in 28 November, 2012. The European Union and national resources will be used more efficiently and faster for implementation of renovation of multi-apartment buildings. At the same time it aims to develop the JESSICA financial model in other areas: urban development, environment, sports infrastructure, education, culture, tourism and transport [4].

The first stage massive apartment renovation list was published in 2013 in the Ministry of the Environment website, in which almost all of Lithuania's cities expressed the desire to participate in the program – 840 apartment buildings from different cities [5]. Following the successful results the major part of apartments were updated (modernized). So it is no wonder that during second stage the number of willing participants, who live in houses built in Soviet times, increased more than two times – 1680 homes from 58 municipalities of the country have submitted applications to participate in the program [6].

2. Heat demand dependence on apartment's geographical position

Energy efficiency improvement of the housing sector is usually associated with renovation. Renovation – is renewal and modernization of buildings and residential environment, development of new quality. Renovation is like a separate part of the reconstruction work, which includes construction and installation work complex, which aims to improve the existing buildings physical, mechanical, aesthetic and other performance properties, seeking greater energy efficiency [7].

Specific technical requirements of renovation projects are divided into two main groups of measures: 1) building's energy efficiency increasing and 2) other building's renewal (modernization) measures (Fig. 1) [8]. The latter group has two criteria, and the first – is characterized by an abundance of requirements, which is closely related to renovation of structures and building systems. However these measures should be applied with consideration to object's, in this case apartment building, present condition. Therefore the criteria table once more proves that apartment buildings mostly affected by complex building structures and heating systems renewal (modernization).

Comparing the following two cities, Palanga and Šiauliai, many respondents who were asked about the difference between these cities, responded that the climate and financial standard of living are different. In order to rationally evaluate, you can select heating of apartment buildings as the optimum option. Thus while assessing these two Lithuania's cities, it is understood that in financial terms the city of Palanga is more expensive, although smaller than Šiauliai. Examining average heating costs monetary difference is obvious, which shows that during winter in Palanga apartment resident will pay more than twice than the one living in the city of Šiauliai.

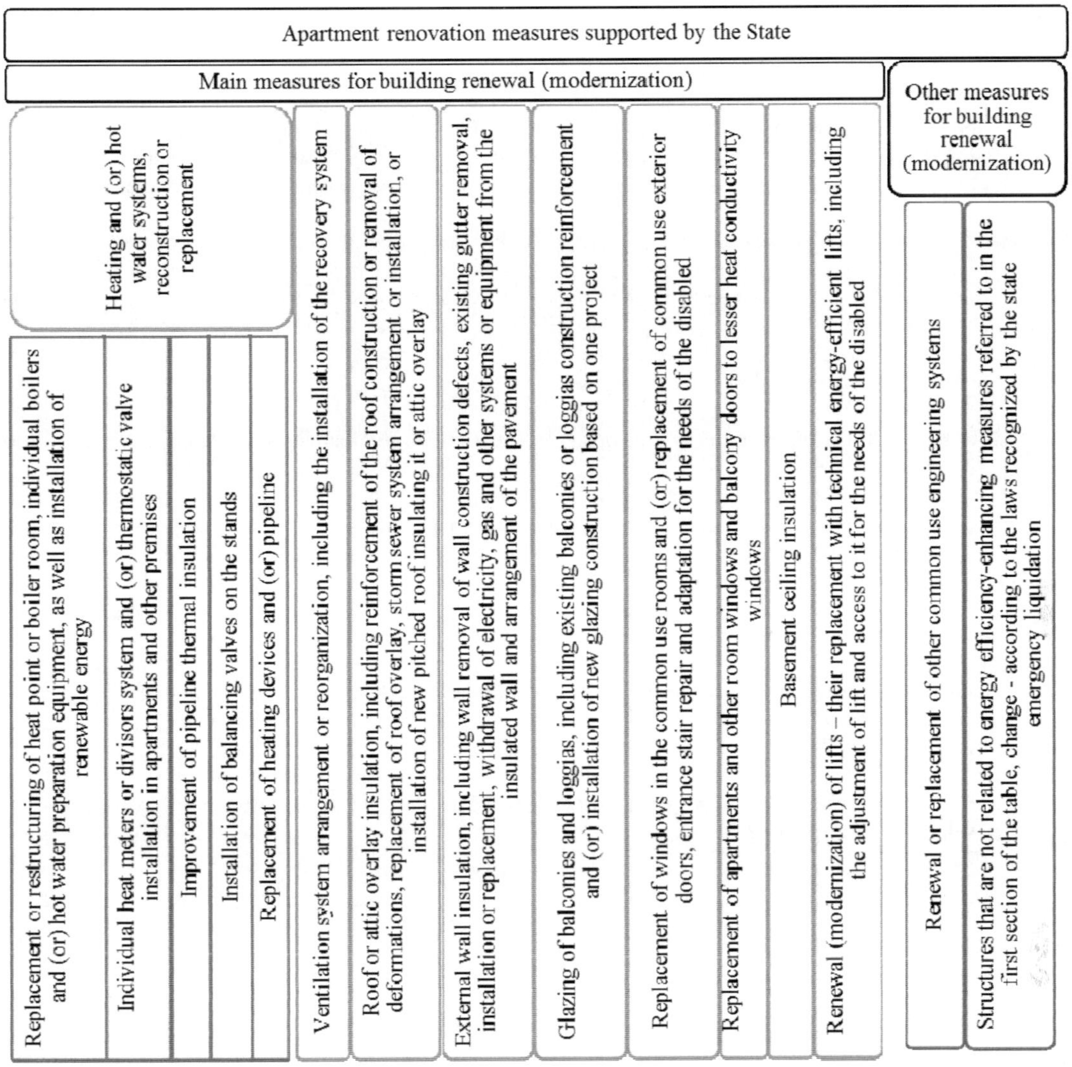

Fig. 1. Apartment building renewal (modernization) measures supported by the State. The figure is compiled on the basis of Republic of Lithuania Governments resolution "On the apartment building renewal (modernization) program approval"

One of the factors affecting the prices is main fuel – constantly rising prices of natural gas, whose consumption in both Palanga and Šiauliai is more than 50% of all used fuel types (Table 1) [9]. The heating is cheapest in those Lithuanian cities where the main fuel for heat production is local biofuels, which is more than twice cheaper than gas. In Šiauliai the percentage of biofuel used out of the total used fuel is higher, i.e. 10,17%, than in city of Palanga in which only more than 3% use this type of fuel. And the benefit of biofuel is obvious – it's cheaper than gas.

Table 1. Heating production sources of Palanga ir Šiauliai cities.

City		Palanga			Šiauliai	
Company name		UAB „Litesko“ branch „Palangos šiluma“	UAB „Lamberta“	Total	UAB „Šiaulių energija“	Total
Number of boilers (units)		10	3	13	8	8
Installed capacity (MW)		104,41	0,55	104,96	286,95	286,95
Installed capacity by fuel type (MW)	Biofuel	6,38	0	6,38 (3,54%)	50,19	50,19 (10,17%)
	Natural gas	98,03	0	98,03 (54,32%)	254,16	254,16 (51,52%)
	Other fuel	76,04	0,55	76,59 (42,14%)	189	189 (38,31%)

Another factor which may lead to higher heating prices of apartment buildings, are climatic conditions. The further from the coastal area, the hotter the summer, however in winter it is vice versa. This is because due to the heating effect of the Baltic Sea the temperature at the seaside is higher in comparison to other part of the country, in this case – Šiauliai. However a closer examination of the coastal zone, shows that during winter it snows a lot more and quite often. Moreover Palanga very often has windy days, as it is geographically located in the coastal plain area, sand plains and coastal pinewoods plains [10]. On average only 8-10 days out of the year are calm, which would be brave to say about Šiauliai. Winds are quite strong, their average speed in Palanga is 4,5-5,0 m/s, dominated by south and southeast winds. About 40-60 days out of the year wind blows at the speed higher than 15 m/s. Pattern change of wind parameters is mostly determined by the distance from the sea and coastal topography. Due to coastal topography effects the average wind speed blowing from the sea at the height of 100 meters 20 kilometers from the coast fell by 22% compared with the wind speed at the sea coast. Sometimes storms rage at the seaside with hurricane winds. And all of these weather conditions are given special attention while designing objects, because Technical Building Regulations show that it is very important to pay attention to the calculation of snow and wind load for these precise areas.

3. Analysis of energy for heating demean and price dependence on modernization and climatic conditions

It is a fact that majority of the population lives in old construction, non-insulated and a lot of heat consuming apartment building. Heating is the maintenance of positive, higher than outside, inside temperature during the cold season. From the room through the walls and natural or forced ventilation heat is transfered to the outside air. The compensation of these losses in heating system is called indoor heating. Apartments of such buildings always to be served 2-3 times more heat compared to newly constructed or renewed (modernized) residential buildings in order to ensure climate conditions regulated by hygienic norms.

The apartment buildings chosen for assessment, were those that were built according to Construction Technical Standards that were valid until 1993 in Palanga and Šiauliai. During the evaluation tables were created of not renewed and renewed (modernized) apartment buildings energy energy efficiency. Two apartment building were selected, with at least partially similar characteristics in order to rationally evaluate, i.e. three or four floor [12, 13] apartments in Palanga of year 1964 and 1965. Meanwhile in Šiauliai city most common five floor apartments were chosen, that were built in the year 1970 and 1971 [14]. The main facade of all the buildings was directed toward northeast.

Table 2. Palanga ir Šiauliai cities apartment heat consumption and payment for heating.

Renewed (modernized) and not renewed (not modernized) apartments					
		Palanga city		Šiauliai city	
Overall information	Renewed (modernized)	No	Yes	No	Yes
	Adrress	8 Oškinio st	5 Oškinio st	17 Gegužių st	19 Gegužių st
	Building year (y.)	1964	1965	1970	1971
	Position of the buildings main facade	Northeast (NE)			
	Number of floors (f.)	4	3	5	5
Period		January 2015			
Average monthly temperature (°C)		1,0		-0,8	
The average consumption of heating energy (kWh/m²)		26,238	11,981	10,790	6,805
Average heating cost (Eur/m²)		2,09	0,95	0,58	0,36
Heating price for residents including VAT (Eur/MWh)		79,55		53,41	

Comparing different cities, Palanga and Šiauliai, renovated apartments, it is obvious that the average heating energy consumption between different areas differs more than two times (see Table 2). More precisely, energy consumption in Palanga is almost 60% higher, possibly due to previously mentioned reason – climatic conditions, also characteristic features of buildings are a part of the high consumption. As a result both average cost of heating per square meter and the heating price for residents with VAT per megawatt-hour is higher at the seaside. All of this can be explained with the earlier mentioned reason, i.e. cities different used fuel types.

Fig. 2. 5 Oškinio st, Palanga. From left: not renovated building, apartment during renovation, renewed (modernized) building.

The exact same assumption can be made while evaluating renewed (modernized) apartments (Table 2). The average amount of consumed heating energy is more than 40% higher in Palanga 5 Oškinio st (Fig. 2) in comparison with apartment in Šiauliai 19 Gegužių st. However it is obvious that old apartments, that use a tremendous amount of heating energy to heat up the apartments need to be renovated, because it reduces even percentile difference between cities average consumed amount of heating energy and heating costs per square meter, i.e. respectively the difference in average heating costs between Palanga and Šiauliai not renovated apartments decreased by almost 30% in comparison to percentile difference of these cities renovated buildings.

However evaluating separately Šiauliai and Palanga cities apartments of both categories, shows that coastal city has achieved higher benefit while renovating apartments, since the average amount of consumed heating energy decreased almost 60% from similar not renovated building (Table 2). Moreover the average cost for heating is more than two times lower after renovating a building in Palanga, whereas residents of Šiauliai save about a third of their income to pay for the average heating cost.

It is obvious that many praise and welcome the municipality of Palanga, seeing that solely by 2014 January data, Palanga was one of the cities showing most initiative on the issue of building renewal (modernization), which can not be said about Šiauliai. Moreover during 2015 intensive work was put into coastal city‘s apartment renovation and all this successfully continues.

In the Republic of Lithuania Governmets resolution "On the apartment building renewal (modernization) program approval" it is mentioned that the aim of the program is to reduce heating energy consumption in apartment buildings, built according to Construction Technical Standards that were valid until 1993, up to the year 2020 no less than 20% which is more than 1000 GWh per year [1].

Conclusions

Renewal (modernization) – investment in fixed assets, which aims to extend it‘s usefulness and (or) to improve qualitative properties, applying and (or) implementing new technologies. Apartment renovation is a way to reduce power and heating consumption. The main purpose for buildings renewal (modernization) is to reduce energy consumption, but at the same time to restore and improve buildings and it‘s engineering systems physical properties. This implementation not only reduces heating costs, but ensures a better indoor microclimate and reaches higher comfort level. And all of this affects the issue of surrounding environment. However renovation contributes to the change of the landscape according to residential time requirements and taste.

It is obvious that the weather can influence apartment building heating costs, although the climate does not differ distinctly in the country, there can be felt a partial difference, which likely determines a part of the heating energy cost in the cities located in the coastal zone. This result was proven by comparison of heating energy consumption as well as average heating energy cost per square meter of apartment buildings located in the coastal zone and in further located cities.

It can be argued that it is useful to renovate apartments in city of Palanga, since more of the residents money is saved then of those who live in city of Šiauliai. However there must be a possibility to modernize (renew) old apartment buildings in all municipalities, what has been carried out so far.

Apartment residential fund modernization is necessary throughout Lithuania. This is especially true in coastal zone, where average whether temperatures during the cold period are higher, but the effect of the wind increases heating energy loss by two or more than two times.

References

[1] Czakó V. Evolution of Hungarian residential energy efficiency support programmes: road to and operation under the Green Investment Scheme. Energy Efficiency, 2012.

[2] Lietuvos šilumos tiekėjų asociacija. Available from Internet: <http://www.lsta.lt/lt/pages/apie-silumos-uki/silumos-suvartojimo-analize> [2015-03-07].
[3] Labanca N., Suerkemper F., Bertoldi P., Irrek W., Duplessis B. Energy efficiency services for residential buildings: market situation and existing potentials in the European Union. Journal of Cleaner Production. Available online 7 March 2015.
[4] Kelpšienė L., Pelikša M. Multi-criteria evaluation of housing modernization in time of economic recession and recovery, 2015. 15th International Multidisciplinary Scientific GeoConference SGEM2015.
[5] Statybos, būsto ir interjero gidas. Available from Internet:
http://lt.lt.allconstructions.com/portal/categories/368/1/0/1/article/15132/pirmasis-renovacijos-etapas-840-daugiabuciu [2015-03-07].
[6] Lietuvos Respublikos Aplinko ministerija. Available from Internet:
http://www.am.lt/VI/article.php3?article_id=13635 [2015-03-07].
[7] Klimantavičiūtė A. Apartment block energy efficiency evaluation after completing modernization project. Proceedings of the 16 th Conference for Junior Researchers „Science – Future of Lithuania“, 20 – 22 March 2013, Vilnius, Lithuania. ISSN 2029-7149 online /ISBN 978-609-457-536-5.
[8] LR Respublikos Vyriausybės nutarimas. Dėl daugiabučių namų atnaujinimo (modernizavimo) programos patvirtimo, 2004.
[9] Valstybinė kainų ir energetikos kontrolės komisija. Available from Internet:
<http://www.vkekk.lt/siluma/Puslapiai/silumos-zemelapis/silumos-zemelapis.aspx> [2015-03-07].
[10] Galvonaitė A., Kilpys J., Kitrienė Z., Valiukas D. Lietuvos kurortų klimatas. Lietuvos hidrometeorologijos tarnyba prie Aplinkos ministerijos. Vilnius, 2015. P. 106.
[11] Marčiukaitis M., Erlickytė-Marčiukaitienė R., Tumosa A. Vėjo greičio kitimo dėsningumų Lietuvos pajūrio regione tyrimas. Lietuvos mokslų akademija, 2009.
[12] UAB „Litesko“. Available from Internet: <http://www.litesko.lt/litesko_house_info> [2015-03-07].
[13] Valeika V. Daugiabučio namo B. Oškinio g. 5, Palanga atnaujinimo (modernizavimo) projektas, 2013.
[14] AB „Šiaulių energija“. Available from Internet: <http://mano.senergija.lt/NamoInfo/> [2015-03-07].

Forum on Innovative Technologies and Management for Sustainability (ITMS'2016)

panevėžys faculty of technologies and business

The 11th International Conference Intelligent Technologies in Logistics and Mechatronics Systems (ITELMS'2016), 28-29 April 2016, Panevėžys, Lithuania

Impact of the entrance vestibule on energy demand of a building

Zdanys P.[a], Susinskas S.[a]1, Stasiskis A.[a]

[a]*Kaunas University of Technology, Panevėžys faculty of Technologies and Business S. Daukanto str. 12, 35212, Panevezys, Lithuania*

Abstract

The article addresses the issues of energy efficiency of public buildings through application of dynamic simulation of the building energy demand. The main obje ct of the Research is the influence of the entrance vestibule of the public building on the building's energy demand. By means of the computer programs DesignBuilder and EnergyPlius, the energetic simulation has been done in order to determine the energy demand. For the public building, two different simulations of the building premises' layout and changes of their use purposes have been done. We have presented the Research results and conclusions.

© 2016 P. Zdanys, S. Susinskas, A. Stasiskis.
Peer-review under responsibility of the Kaunas University of Technology, Panevėžys Faculty of Technologies and Business

Keywords: entrance vestibule; atrium; simulation of energy demand; DesignBuider; EnergyPlius; calculated total energy consumption; dynamic methods of detailed simulation.

1. Introduction

A great part of initial energy consumption is used for creation of a proper indoor climate of the premises. In order to increase the efficiency of energy consumption and to motivate using the renewable energy sources, various models of energy supply to buildings and assessment of such energy efficient use are created. Various simulation means and methods of optimization are applied.

Although, in the recent years, the efficiency of energy consumption has greatly increased, however, the biggest cost-saving opportunities are in the sectors of residential and public buildings. Here, it is possible to save approximately 27% to 30% of the current consumption of final energy [1]. The energy consumption of a building is important because of this is the most expenditure throughout the entire life of the building. The factors of the highest influence upon the building energy consumption include climatic conditions, thermal comfort conditions, thermic, air tightness, architectural and structural peculiarities of the building, mode of the building usage, features of installed microclimate conditioning and energy supply systems to ensure the microclimate of the premises. This means that in order to determine an optimal level of the efficiency of the building energy consumption, it is necessary to consider a set of the energy end-use efficiency measures and the measures of integration of renewable energy.

The building is a complex system with many physical phenomena taking place in it. It is the complex system closely interconnecting outdoor weather conditions, structural elements, engineering systems of the building and people. The dynamic methods of detailed modeling must be used for assessment of all these links. The energy performance of the building is determined to a large extent at an early stage of the design, therefore the design decisions taken at this stage have a decisive impact on the construction costs and energy consumption of the building in the phase of its usage.

* Corresponding author. Tel.: +370-45-434-247; fax: +370-45-465-738.
E-mail address: ptvf@ktu.lt

© 2016 P. Zdanys, S. Susinskas, A. Stasiskis.
Peer-review under responsibility of the Kaunas University of Technology, Panevėžys Faculty of Technologies and Business

In 2013, Lithuania adopted a Technical Construction Regulation STR 2.05.01:2013 'Building Energy Performance Designing' [2]. The regulation is prepared in accordance with the Directive 2010/31/EU of the European Parliament and of the Council of 19 May 2010 on the Energy Performance of Buildings (OJ 2010 L 153, p. 13). The regulation conforms to the requirements of the Directive 2010/31/EU of 16 January 2012 delegated by the Commission, which supplements the Directive 2010/31/EU of the European Parliament and of the Council on the Energy Performance of Buildings by setting a comparative methodology framework for calculating the optimum level in terms of consumption and minimum energy performance requirements for buildings and building elements.

Pursuant to Technical Construction Regulation STR 2.02.02:2004 'Public Buildings' [3], energy saving and heat conservation may be regulated by the following five means: 1. determination of requirements for construction products applied to protections (their thermal resistance, resistance of water vapours circulation, emission of a layer reflecting infrared rays); 2. determination of requirements for protections and Construction engineering systems' equipment characteristics (e.g. heat transmission of walls, roofs, doors and windows, impermeability of doors and windows, boilers, ventilators, cooling equipment); 3. determination of requirements for specifications of Construction and Construction engineering systems' use (e.g. specific thermal loss, Construction air transmission, air circulation, general efficiency of heating and cooling systems); 4. determination of requirements for generated capacity of Construction engineering systems (e.g. annual quantity of heating energy necessary for heating and (or) cooling of building premises up to defined temperature, taking into consideration indoor heat generation and transmitted quantity of solar energy); 5. determination of requirements for energy capacity supplied to Construction energy systems, in order to define characteristics and efficiency of such systems' use (e.g. consumption of heating and (or) cooling energy), taking into consideration the peculiarities and price of energy source.

In order to control the energy consumption by the means provided for in Technical Construction Regulation STR 2.02.02:2004 'Public Buildings', the requirements for Construction and its parts are defined. Such requirements shall be related to characteristics of construction products, taking into consideration the users' needs: 1. demands for heated rooms and other premises, regarding all necessary comfort requirements under winter and summer conditions; 2. determination of demands for indoor air quality; 3. calculation of predicted consumption of hot water for domestic needs.

The entrance vestibule is an unheated room next to the outer door of the building functioning as antechamber between the street and heated interior of the building. The entrance vestibule is both a buffer zone before entering the heated premises of the building, and architectural solution. Atrium is a space with a ceiling illumination that can be designed as two or several stories high. Atrium is architectural and performs the function of the building's entrance vestibule.

The obligatory construction and equipment of the entrance vestibules in the buildings is provided for only by Technical Construction Regulation STR 2.02.01:2004 'Residential Buildings' [4]. The construction of entrance vestibules in the building of other purposes is not regulated.

2. Subject, Aim and Objectives of the Research

The subject of the Research is the influence of the entrance vestibule of the public building on the building's energy demand.

The aim of the Research is to perform a complex analysis of the influence of the building entrance vestibule on the energy demand of indoor microclimate systems.

Objectives of the Research:

1. To create two different models of public buildings and to calculate total energy consumption.
2. According to the created model of the building, to carry out simulations for different design decisions of building entrance.
3. To process the simulation results and perform the analysis of the influence of the building entrance vestibule on total energy demand of the building.

3. Methodology of the Research

The building is a complex system with many physical phenomena taking place in it. Structural elements, engineering systems and people present in the building are exposed to outdoor weather conditions. In turn, the quantity of heat (cold) accumulated in the building envelopes is dependent on the amount of internal heat inflows that occur from the energy supplied to the building by people and engineering systems [5] (Figure 1.).

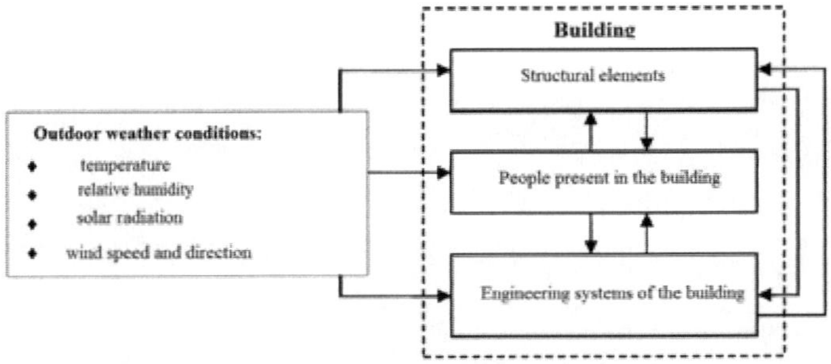

Fig.1. Building like a system.

Quoting Motuziene V. [5] the figure 2 shows a simplified view of the main processes influencing on the parameters indoor air.

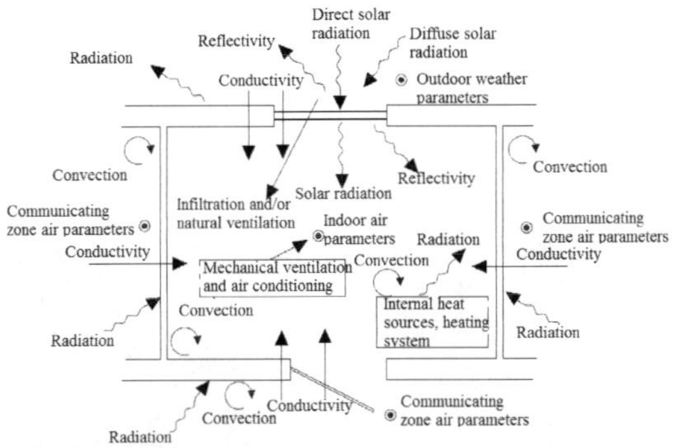

Fig.2. Indoor energy flows.

So, we have the complex system closely interconnecting outdoor weather conditions, structural elements, engineering systems of the building and people. The dynamic methods of detailed simulation must be used for assessment of all these links (Figure 2).

4. Creation of calculation models

Theoretical models of two buildings were created basing on typical buildings' layouts and using a program DesignBuilder – for the administrative building with designed entrance atrium and for the administrative building with designed entrance vestibule (Figure 3).

The administrative building with atrium, the total area of the building is 900,59 m^2. The building has two storeys and designed separate office blocks, each having own utility rooms (kitchens, toilet areas). Atrium is a space with ceiling illumination that is designed over two storeys and performs a function of the building's entrance vestibule.

The other model is the administrative building with entrance vestibule, the total area of which is 399,46 m^2. The building has one storey and designed office blocks that are connected with a common corridor; the workers of all offices use a common kitchen and toilet areas. The vestibule is designed in the entrance of the building.

When creating the computer models the following zones of people activities (occupation) have been distinguished: office blocks, common areas, kitchens and toilet areas; the number of people in zones has been chosen in accordance with design standards. For simulated buildings, the same protective constructions with coefficients of protective heat transmission were chosen in accordance with provisions of Technical Construction Regulation STR 2.05.01:2013 'Building Energy Performance Designing' [2] applied to the buildings with class A of energy performance (Table 1).

Table 1. Data of buildings' theoretical model.

Building data	Description of building data
Climatic data	The climatic data of Kaunas typical meteorological year have been used for the energy simulation of the building and entered into the data base of the international network of meteorology data IWEC (International Weather for Energy Calculations) [6].[2]
Protective constructions of the building	The protective construction thermal transmittance coefficients U, W/(m^{2}K) are used for energy performance class A of the building: exterior walls - 0,15; roofs, dividing walls – 0,11; floors on the ground – 0,16; windows – 1,3; doors – 1,4. Standard values of air exchange $n_{50.N}$ (1/h) at the pressure difference of 50 Pa – 1.
Indoor characteristics of the premises	It is accepted that in winter the designed indoor air temperature of 20^{0}C is maintained in the premises where people are working, and the air temperature of 12^{0}C is maintained outside the working hours. In summer a ventilation system maintains the air temperature of 24^{0}C, and it is turned off outside the working hours. The illumination level of 500 lux is maintained in the premises.
Parameters of indoor climate systems	A water radiator heating systems is equipped in the building, the heat source – gas boiler having the efficiency of 85 per cent. We use a mechanical ventilation system with 0,65 per cent efficiency of heated air return, and the consumption of energy used for recuperative heat exchanger's fan does not exceed 0,75 Wh/m^3. Electric energy is used for premises cooling, work ratio of a cooling system – 1,8. A system of natural ventilation is functioning in utility rooms.

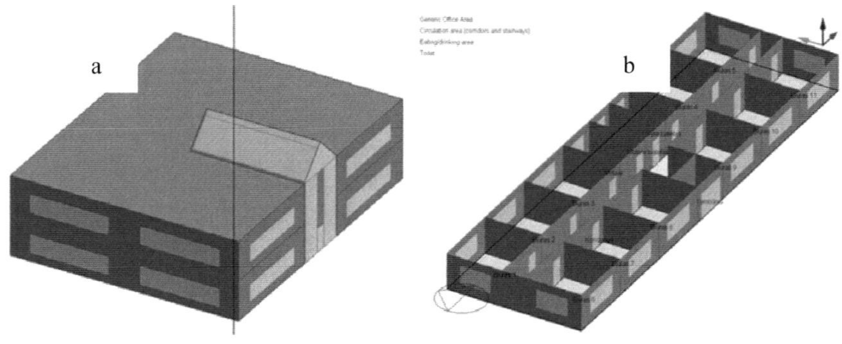

Fig. 3. Geometrical models of the buildings in the program DesignBuilder: a) Model of administrative building with atrium; b) Model of administrative building's storey with entrance vestibule.

5. Research

A computer program EnergyPlus performing integrated simulation has been chosen for the Research. This means that elements of the building, the building and its systems have been worked out simultaneously. Thereby, the influence of change of one parameter on another parameter has been assessed in every time step in the conditioned space. For physically real simulation, all the elements must be interconnected [5]. EnergyPlius program has a number of functional elements which are interrelated with contours.

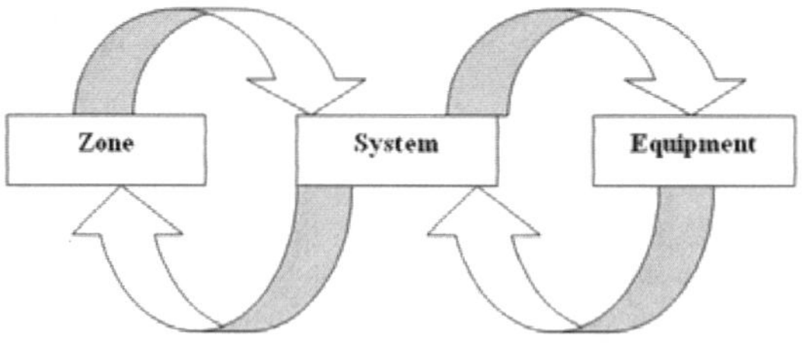

Fig. 4. Scheme of simultaneous solution EnergyPlus Engineering Referece.

---

[2]IWEC database used by the program EnergyPlius, Lithuanian territory are only Kaunas meteorological data. Since the Lithuanian territory is not high, in the region of large climatic fluctuations are not modeling assumes that the weather data will be suitable for the whole territory of Lithuania.

The counters are divided into use and supply parts, and the solution scheme depends on proper combination of use and supply parts. When the computer program EnergyPlus simulates the solution by using the method of covering with thermal capacity of zone, the information of previous time steps is used to predict the system response and update the temperature of the zone at the present time step.

Upon creation of theoretical model of the selected building with the program Design Builder [7] and simulation with the program EnergyPlius [8], we obtained the calculated the annual energy consumption of the whole building and per square meter of the useful space of the building. The program presents also the energy demand for separate engineering systems – heating, ventilation, cooling, water heating, and other figures and graphic data necessary for the analysis (Figure 5).

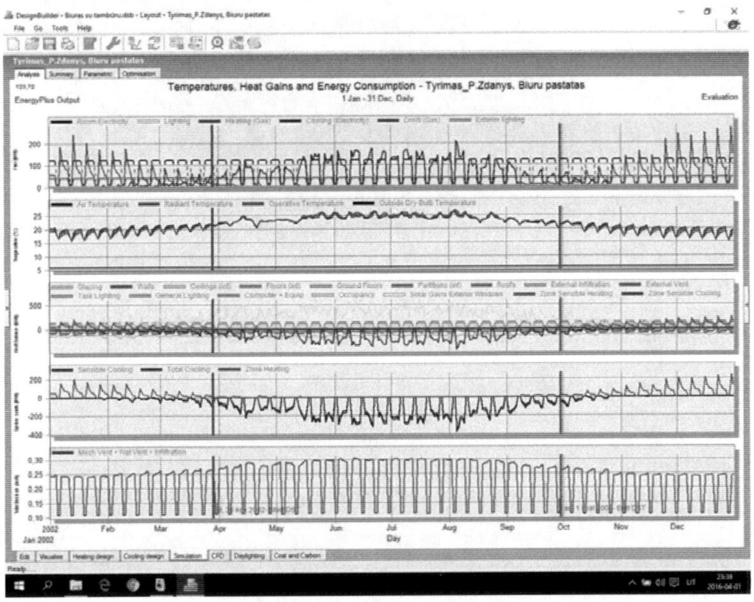

Fig. 5. Window of EnergyPlus simulation results.

In the course of the Research we have simulated both buildings and determined the annual energy demand of these buildings (Version 1). In order to explore the impact of the interior vestibules designed in the buildings on the calculated total energy demand, we have deleted the entrance vestibules in the theoretical models of the buildings and the repeated simulation of energy consumption was conducted by the means of the computer program EnergyPlus (Version 2). The received results are presented in the part of analysis of simulation results.

6. Analysis of simulation results

The program EnergyPlus performed the simulation of the building energy performance at selected time step for the period of a year, and the output data were provided in the program DesignBuilder. Processing of the data was done with the program Microsoft Excel. The scheme of the calculation and simulation of the selected building is showed in the figure 6.

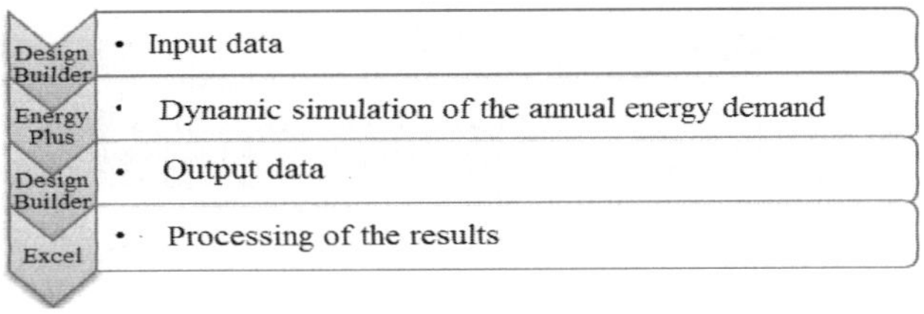

Fig. 6. Scheme of programs used for analysis.

Version 1. The administrative building with atrium (Figure 7), the total area is 900,59 m^2. Upon the simulation with the program EnergyPlius, we have established that the calculated total energy consumption per one square meter of the building's useful area is 117,69 kWh/m^2 per year, total energy demand – 105994,47 kW per year.

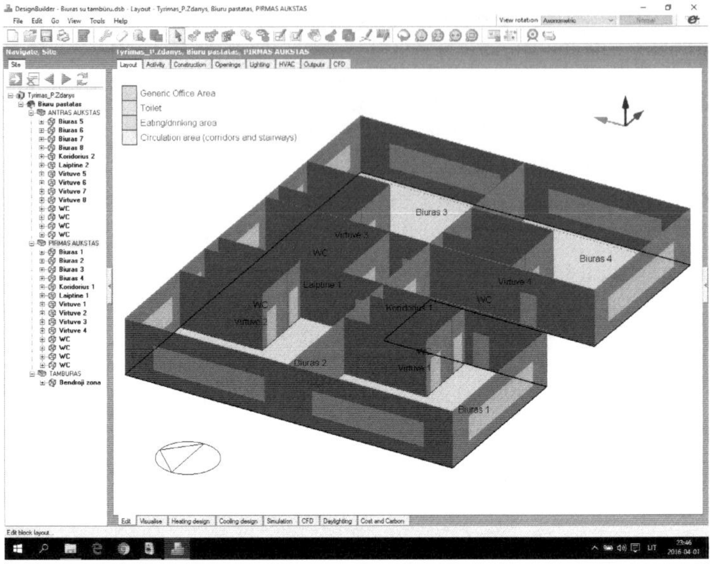

Fig. 7. Layout of the first storey of the administrative building with atrium in the program DesignBuilder.

The administrative building with entrance vestibule (Figure 8), the total area is 399,46 m^2. Upon the simulation with the program EnergyPlius, we have established that the calculated total energy consumption per one square meter of the building's useful area is 96,69 kWh/m^{2}per year, total energy demand – 37939,25 kW per year.

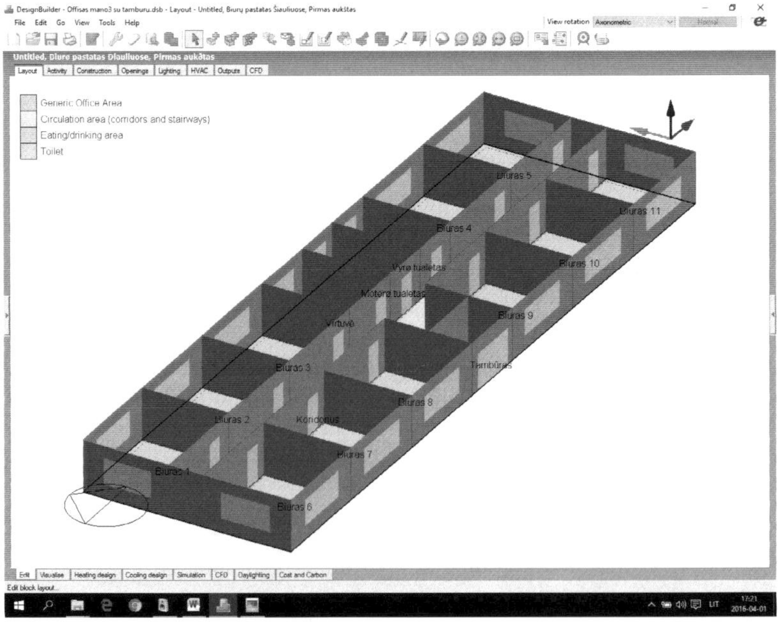

Fig. 8. Layout of the first storey of the administrative building with entrance vestibule in the program DesignBuilder.

The detailed calculated total energy consumption of the buildings is presented in Table 2.

Table 2. Input data and results of the analysis of the Version 1.

Building	Total area of the building, m²	Area of the entrance vestibule, m²	Annual building energy demand, kWh	Annual energy consumption, kWh/m² per year	Allocation of energy consumption	Energy consumption, kWh
Administrative building with atrium	900,59	49,71	105994,47	117,69	Heating	8914,69
					Ventilation	26707,92
					Preparation of hot water	9.340,55
					Electrical equipment	33.559,19
					Illumination	27.472,11
Administrative building	399,46	9,994	37939,25	96,69	Heating	14033,54
					Ventilation	30,18
					Preparation of hot water	1106,20
					Electrical equipment	13117,07
					Illumination	9652,27

The obtained simulation data is presented in the example 9.

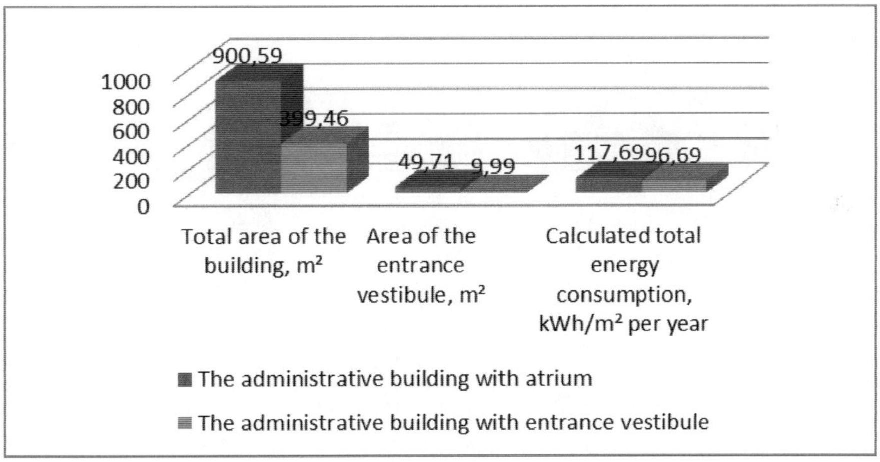

Fig. 9. Simulation data of the buildings with entrance vestibules.

Version 2. The administrative building (without entrance vestibule), the total area is 836,57 m². Upon the simulation with the program EnergyPlius, we have established that the calculated total energy consumption per one square meter of the building's useful area is 116,13 kWh/m² per year, total energy demand – 97147,40 kW per year.

The administrative building (without entrance vestibule), the total area is 400,22 m². Upon the simulation with the program EnergyPlius, we have established that the calculated total energy consumption per one square meter of the building's useful area is 96,09 kWh/m² per year, total energy demand – 38455,84 kW per year.

The detailed buildings' energy consumption is presented in Table 3.

Table 3. Input data and results of the analysis of the Version 2.

Building	Total area of the building, m^2	Area of the entrance vestibule, m^2	Annual building energy demand, kWh	Annual energy consumption, kWh/m^2 per year	Allocation of energy consumption	Energy consumption, kWh
Administrative building (without atrium)	836,57	-	97147,40	116,13	Heating	9912,44
					Ventilation	20618,78
					Preparation of hot water	9.322,17
					Electrical equipment	33.207,30
					Illumination	24.086,71
Administrative building (without entrance vestibule)	400,22	-	38455,84	96,09	Heating	14533,96
					Ventilation	34,44
					Preparation of hot water	1106,20
					Electrical equipment	13117,07
					Illumination	9664,17

The obtained simulation data is presented in the example 10.

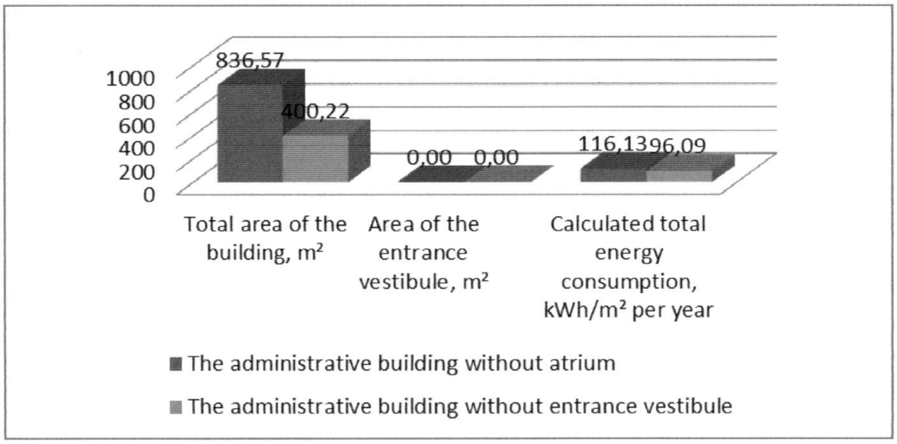

Fig. 10. Simulation data of the buildings without entrance vestibules.

The analysis of the results of both Researches shows that the entrance vestibule in the buildings has minor impact on the calculated total energy consumption of the buildings.

When simulating the buildings without entrance vestibules (atrium and vestibule), the calculated total energy consumption of the administrative building with atrium decreases in 1,33 per cent, and the calculated total energy consumption of the administrative building with entrance vestibule decreases in 0,62 per cent (Figure 11).

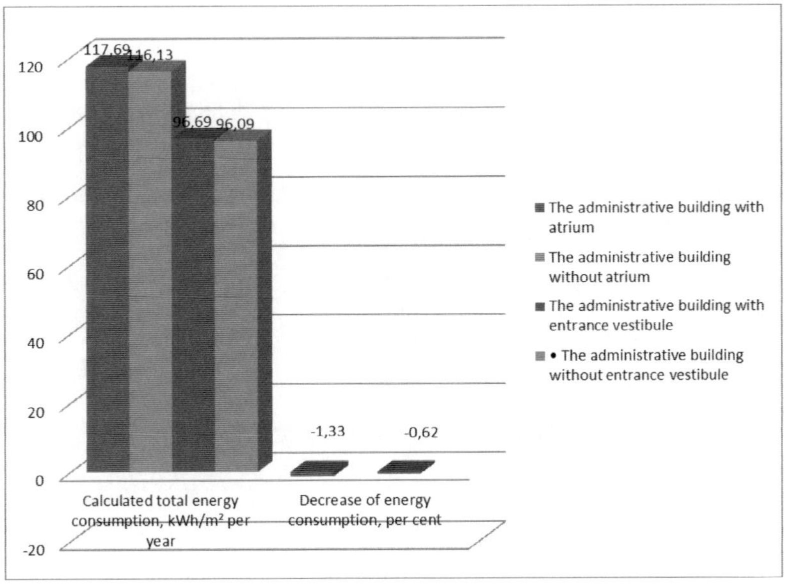

Fig. 11. Decrease of calculated total energy consumption in the buildings wthout entrance vestibule.

Upon comparison of detailed energy consumption in accordance with parameter defined for energy consumption of engineering systems of the building, we can see that the annual demand of energy consumption vary for separate systems (Fig. 12).

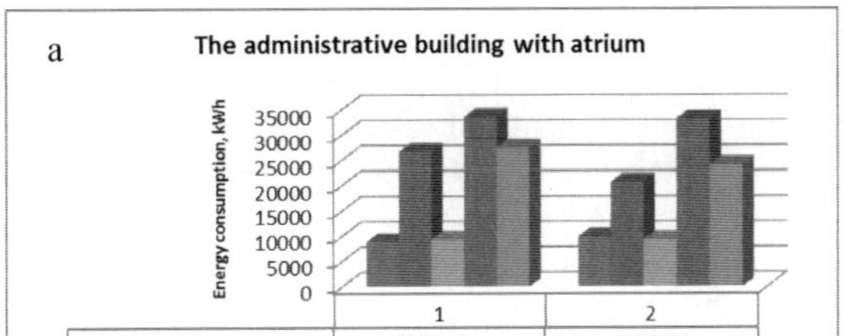

	1	2
Heating	8914,69	9912,44
Ventilation	26707,92	20618,78
Preparation of hot water	9340,55	9322,17
Electrical equipment	33559,19	33207,3
Illumination	27472,11	24086,71

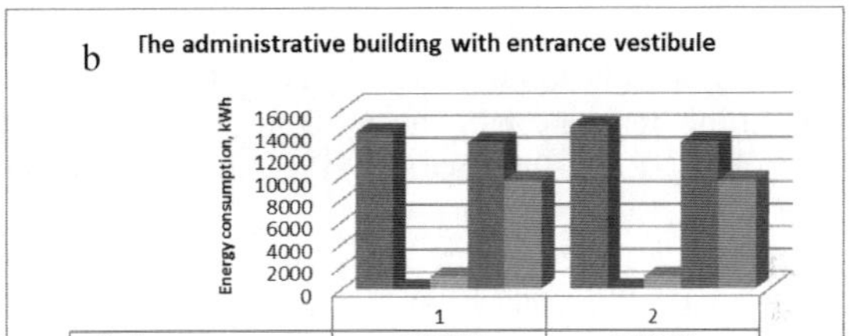

	1	2
Heating	14033,54	14533,96
Ventilation	30,18	34,44
Preparation of hot water	1106,2	1106,2
Electrical equipment	13117,07	13117,07
Illumination	9652,27	9664,17

Fig. 12. Changes of calculated total energy consumption of the building for engineering systems, kWh: a) The administrative building with atrium; b) The administrative building with entrance vestibule.

The presented examples (Figure 13) show the percentage reallocation of energy demand for separate engineering systems ensuring the indoor climate of the building.

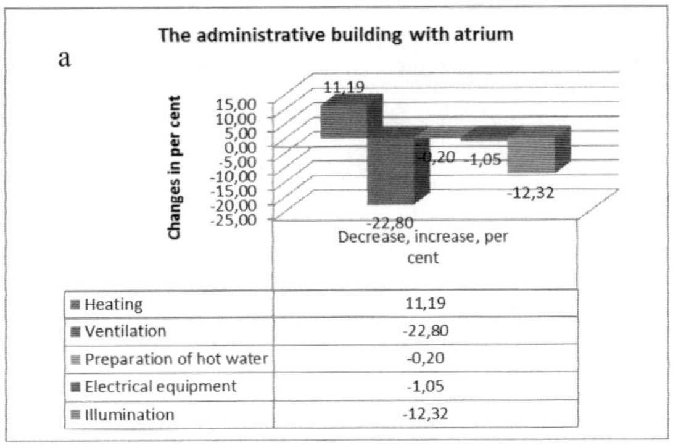

Heating	11,19
Ventilation	-22,80
Preparation of hot water	-0,20
Electrical equipment	-1,05
Illumination	-12,32

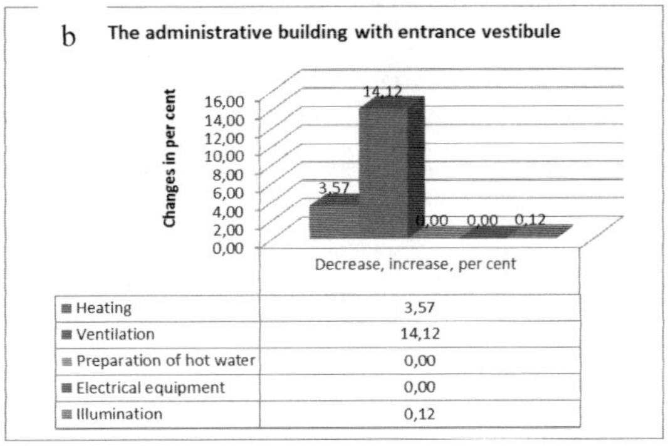

Heating	3,57
Ventilation	14,12
Preparation of hot water	0,00
Electrical equipment	0,00
Illumination	0,12

Fig. 13. Changes of calculated total energy consumption of the building in per cent: a) The administrative building with atrium b) The administrative building with entrance vestibule.

Upon generalization of the results of both Researches, we can see that the designing of entrance vestibule does not have any significant impact on the building calculated total energy consumption, and the changes of energy consumption for engineering systems are determined directly by changes of purposes of premises use and demands of separate indoor climate engineering systems installed in such premises and ensuring a proper indoor climate of such premises (Figure 14).

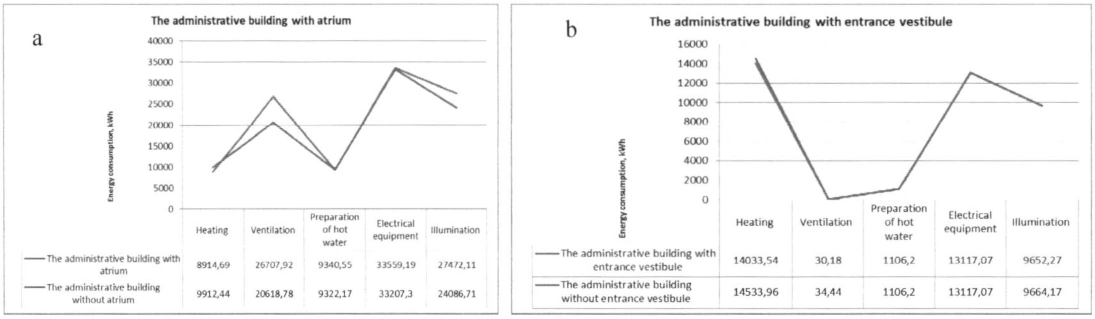

Fig. 14. Changes of calculated total energy consumption for engineering systems:
a) The administrative building with atrium; b) The administrative building with entrance vestibule.

7. Conclusions

1. The building is a complex system with many physical phenomena taking place in it. It is the complex system closely interconnecting outdoor weather conditions, structural elements, engineering systems of the building and people. The dynamic methods of detailed modeling must be used for assessment of all these links.

2. The energy performance of the building is determined to a large extent at an early stage of the design, therefore the design decisions taken at this stage have a decisive impact on the construction costs and energy consumption of the building in the phase of its usage.

3. In the analyzed buildings without entrance vestibule, the annual calculated total energy consumption is slightly less (-1,33; -0,62 per cent) if calculating the energy consumption per square meter of the useful space of the building.

4. After deleting the entrance vestibules in the analyzed buildings, the energy demand for heating increased averagely in 7,38 per cent, for ventilation – decreased averagely in 5,22 per cent.

5. The changes of energy consumption for engineering systems depend directly on changes of purposes of premises use and demands of separate indoor climate engineering systems installed in such premises and ensuring a proper indoor climate of such premises.

References

[1] *Commission of the European Communities 2011*. Communication of commission: Energy Efficiency Plan 2011[online] [accessed on 17 February 2016]. Internet access: http://europa.eu/rapid/press-release_MEMO-11-149_en.htm?locale=en.

[2] *Technical Construction Regulation STR 2.05.01:2013* 'Building Energy Performance Designing' [online] [accessed on 17 February 2016]. Internet access: http://vtpsi.lt/node/1153 .

[3] *Technical Construction Regulation STR 2.02.02:2004* 'Public Buildings' [online] [accessed on 17 February 2016]. Internet access: http://www.vtpsi.lt/node/1063>.

[4] *Technical Construction Regulation STR 2.02.01:2004* 'Residential Buildings' [online] [accessed on 17 February 2016]. Internet access: http://www.vtpsi.lt/node/1062 .

[5] Motuziene V. *Complex Analysis of the Influence of Glazing on Energy Demand of Public Buildings*: Doctoral Dissertation. Vilnius Gediminal Technical University. Vilnius: Technika;2010. p. 29 – 59

[6] *International Weather for Energy Calculations*. [online] [accessed on 1 March 2016]. Internet access: https://www.ashrae.org/resources--publications/bookstore/international-weather-for-energy-calculations.

[7] *DesignBuilder - Simulation Made Easy*. [online] [accessed on 1 March 2016]. Internet access: http://www.designbuilder.co.uk/.

[8] *U.S. Departament of Energy 2015*. Engineering Reference. The Reference to EnergyPlus Calculations [interaktyvus] [accessed on 7 March 2016]. Internet access: https://energyplus.net/sites/default/files/pdfs_v8.3.0/EngineeringReference.pdf.

Author Index

© Medimond

Kaunas University of Technology
K. Donelaičio St. 73
44249 Kaunas, Lithuania

ISBN 978-1-7138-4112-8

9 781713 841128

CURRAN ASSOCIATES INC.
proceedings.com